LAKE BAIKAL

A Mirror in Time and Space for Understanding Global Change Processes

With presentations by

Genki Inoue, Kenji Kashiwaya, Takayoshi Kawai, Kimiyasu Kawamuro, Masayuki Kunugi, Kazuo Mashiko, Yoshiki Masuda, Koji Minoura (editor), Hiroshi Morino, Takejiro Takamatsu, Yasunori Watanabe, Takahito Yoshioka and Norio Yoshida

LAKE BAIKAL

A Mirror in Time and Space for Understanding Global Change Processes

Edited by

Koji Minoura

The 1998 BBD Baikal Symposium of
the Japanese Association for Baikal International Research Program
(JABIRP), Yokohama, November 5th - 8th, 1998

2000

ELSEVIER

Amsterdam - London - New York - Oxford - Paris - Shannon - Tokyo

ELSEVIER SCIENCE B.V.
Sara Burgerhartstraat 25
P.O. Box 211, 1000 AE Amsterdam, The Netherlands

First edition 2000

Library of Congress Cataloging in Publication Data
A catalog record from the Library of Congress has been applied for.

ISBN: 0 444 50434 6

∞ The paper used in this publication meets the requirements of ANSI/NISO Z39.48-1992 (Permanence of Paper).
Printed in The Netherlands.

Preface

Lake Baikal is the largest and oldest lake on Earth. Its water volume is about 23,000 km^3, compared with the 18,000 km^3 of Lake Tanganyika and the 22,000 km^3 of all five Great Lakes in North America combined. Lake Baikal originated about 3.5 million years ago, and it never froze completely during the glacial ages. Thus, its organic evolution has progressed tremendously, both qualitatively and quantitatively ways, and it has been the initial object of investigation by many scientists over the past several centuries.

In 1988, the Soviet Academy of Sciences (now the Russian Academy of Sciences) decided to establish the Baikal International Centre for Ecological Research (BICER) at its Siberian Branch, and the first official meeting of international board members was held in Irkutsk in December 1990. After several discussions the Japanese researchers decided to join and support the BICER, and in March 1991 they established the Japanese Association for the Baikal International Research Project (JABIRP). The Baikal Drilling Project (BDP) was proposed in 1991, and Japanese researchers joined the project a year later.

The International Programme for Biodiversity Science (DIVERSITAS) was established in 1991 under the International Union of Biological Science (IUBS), the Scientific Committee of Problems in the Environment (SCOPE) of the International Committee of Scientific Unions (ICSU), and the United Nations Educational, Scientific and Cultural Organization (UNESCO), a year before the United Nations Conference for Environment and Development (UNCED), usually referred to as the "Earth Summit", was held in Rio de Janeiro, where two international conventions were signed. An International Network for DIVERSITAS in the Western Pacific and Asia (DIWPA) was proposed in 1993 and established in 1994. Lake Baikal and environs is one of the main sites in the DIWPA region.

Thus, since 1991 many Japanese scientists have traveled to the lake to conduct research with Russian and scientists from other countries.

Needless to say, scientists belonging to the academy, universities, museums, etc., around Lake Baikal have long concentrated their efforts on many studies in and around the lake. I personally was attracted to the lake and its biological communities by reading a book entitled, “Biology of Lake Baikal”, in the series “Binnengewässer”, as an undergraduate. A book by

Professor Kozhov later made a very strong impression on me, and I remember wanting to learn Russian mainly to be able to read the book in the original Russian. It should also be remembered that many people are living in the region, and thus our joint research needed to be conducted primarily by scientists in the region and be related to the future comfort of the lives of the residents around the lake. On the other hand, Lake Baikal and environs is of enormous value to the globe itself, and thus our research should also be for true international by that I mean inter-regional, or global interests. The BICER should not only be the site of bilateral research but the site of real interregionally based research.

In the year of the 10th anniversary of the BICER, the international joint symposium of the BICER, BDP, and DIWPA, 'Lake Baikal: A Mirror in Time and Space for Understanding the Processes of Global Change', was held in Yokohama from November 4 to 8, 1998. This volume is based on material presented at the symposium, but most articles have been considerably revised based on discussions during and after the symposium.

I would like to thank all of the participants in the symposium for reading their papers and for their cooperative and positive discussions on all of the issues. Special thanks are due to the Russian scientists who have long been conducting research on the Lake, especially to Professor Mikhail Grachev, the first director of the BICER and the former director of the Institute of Limnology, who, unfortunately, was unable to attend the symposium because of an accident several months before. Thanks also to Professor Koji Minoura, the editor of the book and the secretary-general of the symposium, to Dr. Takayoshi Kawai, the secretary-general of the JABIRP, and to many others for their help in holding the symposium.

16 August 1999

President of the JABIRP and the Chairperson of the DIWPA
Hiroya KAWANABE
Lake Biwa Museum

Introduction

In November 1998, the BICER (Baikal International Center for Ecological Research), BDP (Baikal Drilling Project), and DIWPA (Diversitas Western Pacific and Asia) Joint International Symposium on Lake Baikal convened in Yokohama, Japan, on the tenth anniversary of the establishment of the BICER. More than 180 scientists attended the symposium, and 64 of them were from abroad.

A lecture meeting was held at the Museum of Natural History in Toyohashi, Central Japan, prior the Symposium, where public lectures on scientific topics afforded participants a good opportunity to become familiar with Lake Baikal and its great potential for wonderful discoveries in science. Following the Symposium, a special meeting on zoology was organized under the title: Animal Community, Environment and Phylogeny in Lake Baikal, and provided an outstanding occasion for researches and students to review the latest developments in the biological field.

It is more important now than ever-before for scientists from different disciplines who are studying Lake Baikal to come together for discussions. The three international scientific associations, the BICER, the BDP, and the DIWPA, decided to hold a symposium in Japan in late autumn 1998 to allow networking by scientists from a wide variety of fields.

Outline of the symposium

Lake Baikal lies in the middle of Siberian taiga, which consists of boreal conifers and forms the northern end of the east-Asian green belt that extends to the tropical rain forest of Southeast Asia. Throughout the long history of basin development the lake has been a theatre of evolution and speciation, and currently sustains more species than any of the world's other freshwater lakes. Because of its distinctive character, Lake Baikal is recognized as the best field for elucidation of biological problems awaiting solution. Theoretical and experimental studies on the extant biotic community will shed strong light on the contemporary subjects of species diversity and ecological complexity.

The limnological conditions of Lake Baikal have been under the control of continental climates because of its location in the interior of the continent, far removed from the influence of oceans. The lake sediment is therefore expected to provide a means of documenting the long history of changes in the Asian climate. Understanding paleoclimatic changes has become increasingly important because of the links between atmospheric circulation and terrestrial vegetation. Proxy paleoclimatic data from the geological record will allow verification of the global effect on the evolution of taiga. In addition, the geological information obtained from drilled cores is expected to yield indispensable to elucidating the origin of Lake Baikal and its environs.

Lake Baikal has recently been affected by human activities both on a global and a local scale. The importance of Lake Baikal as a large freshwater resource makes it urgent to study and understand the biological, physical, and chemical mechanisms determining its dynamics in time and space, and to assess the role of anthropogenic changes occurring in the system.

Theme and scientific topics of the symposium

The theme for the symposium, "Lake Baikal: A mirror in time and space for understanding global change processes," reflects the present challenge facing the scientific communities studying Lake Baikal to clarify the mechanisms of the global system and the evolution of life. Each of the associations established specific topics for sessions in keeping with this theme.

The symposium program was composed of three scientific sessions: an Earth Science Session, a Biology Session, and a Limnology Session. Topics concerning neutrino physics were an interesting focus of the Limnology program. It was suggested that new topics would address the frontiers of scientific study of Lake Baikal in the 21th century.

Proceedings of the symposium

The discussions of the scientific topics related to Lake Baikal were interdisciplinary, bringing together evidence from geology, paleontology, chemistry, biology, limnology, and physics, and thus it was felt that the symposium on Lake Baikal should make fostering of this interdisciplinary debate its main aim. Every scientist is required to respond to demands for relevant knowledge and solutions by a public that is ever more concerned with scientific information. In view of this situation, the Organizing Committee decided to publish selected scientific papers in the proceedings of the symposium as debates of the symposium.

The volume of the scientific proceedings consists of three parts, Paleoenvironment and Rift Basin History (Part 1), Physicochemical

Limnology (Part 2), and Evolution and Biodiversity (Part 3).

The limnological conditions of Lake Baikal are under the influence of the continental climate because of its location in the interior of the continent, far removed from the influence of oceans. Lake sediments from such a setting are therefore expected to represent an opportunity to document the long-term history of the East Asian climate. Considerable progress in reconstructing the past glacial-interglacial climate has been made during the last 20 years, including the establishment of detailed chronologies and stratigraphic correlations of paleoclimatic events. However, the Quaternary climatic changes have been investigated mostly in the marine realm, and thus the climatological response of continents is not yet fully understood. The drainage area of Lake Baikal is so large that lake sediments are expected to provide one of the best records of paleoclimate fluctuations in the eastern portion of the Asian continent. In this context, bottom sediments of Lake Baikal have been examined to evaluate the effect of climate on productivity, circulation, and terrestrial vegetation. Part 1 consists of 13 papers comprising analyses of lithology, sedimentology, mineralogy, paleontology, and geochemistry. Geological and geochemical aspects of cored and dredged deposits from the lake bottom are expected to elucidate paleoenvironmental and paleoecological processes in East Asia that have been under the influence of global climatic oscillations during the late Cenozoic.

Recently, Lake Baikal has been suffering from anthropogenic impacts responsible for rapid environmental changes both on a global and local scale, making elucidation of its biological, physical, and chemical mechanisms, which determine the lake's dynamic processes both in time and space, an urgent task. Furthermore, the chemical and biological samples from the lake will provide indispensable information for assessing pollution levels in the modern lake and its environment. The papers in Part 2 describe important findings for evaluating the causal effect of hydrochemical impacts in response to human activities and recent global changes. The huge volume of clean freshwater stored in the lake is a great potential resource for potable water, and limnological understanding of the lake will contribute to the lacustrine integrity of Baikal.

Lake Baikal is the oldest lake and largest freshwater reservoir in the world. As a result of its exceptionally long geological history, the lake has been a theatre of evolution and speciation of organisms, and it currently harbors most more species than any other lake in the world. Based on its unique nature, Lake Baikal was recently designated a World Heritage site and is regarded as a hotspot for evolution, speciation, and biodiversity. With its tremendously peculiar biota, Lake Baikal is now awaiting modern analytical approaches to the profound problems of speciation and evolution. These approaches, combined with theoretical and experimental analyses on the extant biotic community, will shed strong light on the contemporary subjects of species diversity and ecological complexity. The papers in Part 3 present new results and interpretations in answer to these problems.

Acknowledgements

Several acknowledgements should be made in connection with preparations for the BICER, BDP, and DIWPA Joint International Symposium: first, the members of the Japanese Association for the Baikal International Research Program (JABIRP), and the Science and Technology Agency of Japan, who gave their enthusiastic support and secured financial assistance, and second, the President of the National Institute for Environmental Studies, Professor Gen Ohi, who made available the facilities for the successful Workshop held in conjunction with the symposiums, and the Proceedings would never have been published without the enthusiasm and support of Dr. Osamu Nishikawa of Tohoku University. The contributions of all these persons are gratefully acknowledged. Last, but by no means least, I wish to thank Miss Yuko Watanabe of the National Institute for Environmental Studies for patiently transforming the various manuscripts into the camera-ready form that follows and for a level of editorial assistance that essentially rendered the editor redundant.

19 August 1999
Koji Minoura
Sendai, Japan

Table of Contents

Lake Baikal
K. Minoura (editor)
2000 Elsevier Science B.V.

Baikal drilling project

Kuzumin, M. I.[1]*, Williams, D. F.[2], and Kawai, T.[3]
[1] Vinogradov Institute of Geochemistry, SB RAS, Irkutsk, Russia,
fax: (3952) 46 40 50,
E-mail: mikuzmin@igc.irk.ru
[2] University of South Carolina, Columbia, SC, 29208, USA
[3] National Institute for Environmental Studies, Tsukuba, Japan
* - correspondence

Abstract

A brief history of the "Baikal drilling project" is presented here. The aim of this project is to study the paleoclimate in Central Asia through a comprehensive study of Lake Baikal sediment. A drilling rig that operates in an environmentally friendly manner has been specially manufactured for this project. The rig is capable of drilling a sediment core of up to 1000 m at a depth of 900 m below the lake surface. Four boreholes have been drilled to date.

The sedimentation pattern of the samples is dependent on the climate and topographical features of the area. Dense terrigenous clays formed during cold glacial periods, while sediments containing large amounts of diatom fossils were deposited during the interglacial periods. This alternating sediment pattern is typical of underwater uplifts (e.g. the Academician Ridge) that are isolated from the lakeshore by deep basins. A significant amount of sedimentation in the deep basins arises from turbidite flows, which also bring a large amount of fossilized vegetation. Gas hydrates ($CH_4 \cdot 6H_2O$), which were collected in 1997 for the first time in fresh water, also form in the deep basins. A continuous 5 Ma paleoclimatic record has been obtained from the Academician Ridge. This record correlates well with the oceanic oxygen curve. The paleoclimate of Central Asia has been reconstructed using the distribution of diatoms and biogenic silica content. The Lake Baikal paleoclimatic record is continuous and well dated and can be regarded as an excellent source of information on the paleoclimatology of continental interiors.

Introduction

The international program entitled "Global changes in the environment and climate of Central Asia based on comprehensive studies of Lake Baikal sediments" was initiated in 1989. The short title of the project is the "Baikal Drilling Project." The present article will describe the main results of the project as well as provide a brief history.

Lake Baikal is an ancient rift lake that started forming nearly 40 Ma ago. The Asian continent was broken into a series of small plates when the Indian and Eurasian plates collided. The Baikal rift system formed on the boundary of the small Amur plate and the Eurasian plate as a result of the relative movement of these plates. Lake Baikal is located in the center of this rift system.

Lake Baikal is located at a high latitude on the Asian continent (Fig. 1). The lake consists of three deep basins, separated by underwater uplifts. The Northern basin (maximum depth = 900 m) is separated from the Central basin (maximum depth = 1,634 m) by the underwater Academician Ridge. The Central basin is separated from the Southern basin (maximum depth = 1,400 m) by the Selenga-Buguldeika saddle, which was mainly formed by the deposition of sediments from the Selenga River (the largest river flowing into the lake).

Lake Baikal is a promising site for the study of paleoclimatology because continuous sedimentation has been occurring on its floor for millions of years. The lake's geographical position also makes it sensitive to changes in solar radiation. When the inclination of the Earth's orbit and its

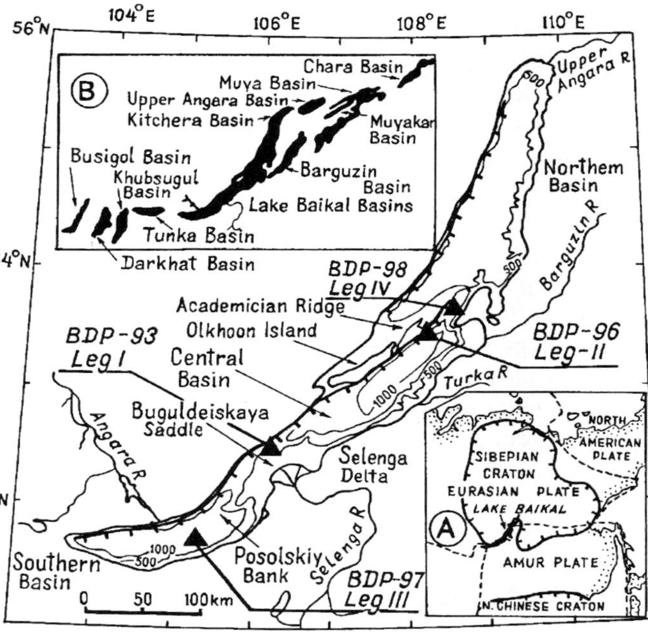

Fig. 1. Map of Lake Baikal. The black triangles indicate the locations of drill sites. V – sedimentation rate (in cm/thousand years).

precession are taken into account, the largest changes in solar radiation are experienced in the area of Lake Baikal. Changes in these parameters should therefore be clearly shown in the "climatic records" of Lake Baikal. The location of Lake Baikal, an area with a distinctly continental climate in the center of the Asian continent, also makes it an ideal site to study seasonal climatic changes. Lake Baikal remained free from glaciation, resulting in the continuous deposition of sediments on its floor over the last 30-35 Ma. Lake Baikal is the only lake with a history of several million years in the northern hemisphere.

Since Lake Baikal is the only site where a continuous climatic continental record for the northern hemisphere can be obtained, it has attracted the attention of the scientific community around the world.

BRIEF HISTORY OF THE BAIKAL DRILLING PROJECT

The Baikal rift zone has been studied by Russian scientists for many years. As a result of these efforts, the main geophysical characteristics, geological history, and sedimentation features of Lake Baikal are known. In 1988, however, a new stage in the study of Lake Baikal was initiated. Prof. Gratchev, Director of the Limnological Institute, invited both Russian and foreign scientists to collaborate in a comprehensive study of Lake Baikal. As a result of an initiative by Prof. Zonenshain, Russian scientists prepared a proposal entitled, "Deep ecology, paleoecology and geodynamics of Lake Baikal." The proposal involved a comprehensive geological and geophysical investigation of Lake Baikal's history and sedimentation. The program included studies requiring the use of "Pisces" submersibles. As a result of this program, scientists from the South Branch of the Oceanology Institute and the Limnological Institute obtained multi-channel seismic profiles of Lake Baikal in 1989. The profiles indicate the presence of an extremely thick (up to 8 km) sequence of sedimentation.

After the XXVIII Session of the International Geological Congress in 1989, Prof. D. Williams (University of South Carolina) contacted a group of Russian scientists and proposed that a drilling project in Lake Baikal be undertaken on a collaborative basis. Prof. S. Horie (Japan), head of the first drill project to be performed in Biwa Lake (Japan), participated at that meeting.

The technical part of the program was developed by the "Nedra" Drilling Enterprise. This Enterprise was in charge of deep continental drilling in Russia and had executed the drilling of a superdeep borehole on the Kola Peninsula. In 1992, a large group of Japanese scientists (JABIRP) represented by Dr. T. Kawai joined the project. German scientists participated in the project from 1995 to 1997 as associated members. Many institutions are involved in the project on the Russian side, but the majority of Russian scientists are from the Irkutsk Scientific Center (Institute of Geochemistry, Limnological Institute, and Institute of the Earth's Crust). The program has also been supported by the academics N. Koptyug and N. Dobretsov as well as the Russian Ministry of Science and Technology. Before the commencement of the drilling operation on Lake Baikal, through geophysical and geological investigations were performed, and the "Baikal" rig was designed and constructed.

Geophysical investigations, conducted by Russian and American scientists in 1989 and 1992, have identified the structure of the Baikal sedimentary sequence. Three horizons were revealed. The lower horizon, in the South and Central basins, has a thickness of up to 4-5 km and is seismically transparent. Two upper horizons exhibit good layering and can be successfully used for paleoclimatic investigations (Zonenshain L.P., et al., 1992; Hutchinson D.P., et al., 1993; Moore P.C., et al., 1997).

The underwater geological investigations focussed mainly on the underwater uplifts: the Posolskaya bank and, particularly, the Academician ridge. A basal horizon, containing beach pebbles in clay, was found at the very bottom of the sediment layers on the Academician Ridge. The age of this horizon was determined using sporepollen analysis and identified as the Late Miocene (5-10 Ma). Studies have suggested that a land barrier, which became the Academician ridge, was destroyed at that time, and the Northern basin, which is significantly younger than the Central and Southern basins, started to develop. Before the Academician Ridge subsided, the Barguzin River, which presently flows into the Central basin, probably had a different riverbed that crossed the Academician Ridge. The river delta sediments (up to 7.5 km) form a thick sedimentary sequence on the southern margin of the ridge. Well-stratified sediments occur on the top of the ridge. These sediment layers were deposited in non-turbid conditions and consist of material from the lake's water column.

Studies on the composition of the sediments and the sedimentation rate have mainly been conducted within the framework of BICER. The sedimentation rate in different parts of the lake varies from 0.12-0.2 to 0.03-0.04 mm/year. The lowest sedimentation rate was found on the Academician Ridge and the highest rates were observed on the Selenga-Buguldeika saddle and in the deep lake basins. A pattern consisting of two characteristic layers was found in the uppermost portion of the Baikal sediments (Bezrukova et al., 1991). The first layer of this pattern is composed primarily of biogenic silts that contain an abundance of diatoms. Beneath this is a layer of terrigenous sediments, mainly clays, that contain only a small number of diatoms. The diatom silts were likely formed during warm interglacial periods, while the terrigenous ones were probably generated

Fig.2. Drilling rig on Lake Baikal.

during cold glacial periods. These findings indicate that the Baikal sediments clearly reflect climatic changes.

Parallel to the above studies, specialists at the "Nedra" Drilling Enterprise were designing and constructing the "Baikal" drillrig. The drill rig is environmentally friendly, which is an important requirement of all ventures in Lake Baikal. The last version of the complex (Fig. 2) was assembled in 1997 on a 1000-ton barge. The drill is capable of drilling a borehole that is up to 1000 m deep at a depth of 900-1000 m below thelake's surface. The core recovery rate can be as high as 95-98 %. In 1998, oceanic and continental drilling specialists from the USA and Germany evaluated the drilling operations at Lake Baikal very highly.

Four boreholes have been drilled since 1993. The drill sites were as follows: 1993 - Buguldeika-Selenga saddle (water depth = 351 km, 100 m core), 1996 - Academician Ridge (water depth = 320 m, 300 m core), 1997 - South Basin (water depth = 1,427 m, 200 m core), and 1998 - Academician Ridge (670 m core).

Features of sedimentation in different topographical features structures of Lake Baikal and Baikal paleoclimatic record from the Academician Ridge

Detailed descriptions of all the cores have been published in a number of articles appearing in Russian and international journals (BDP Members, 1995; BDP Members, 1998; Kuzmin *et al.*, 1997; Kuzmin *et al.*, 1998). Therefore, this article will only describe the main results of the Baikal Drilling Project. In particular, the sedimentation conditions for the different topographical features of Lake Baikal will be outlined.

The first borehole was drilled on the Buguldeika saddle at a location 7 km southeast of the Buguldeika river's mouth. Geological investigations have shown that during the Manzurka erosion-tectonic stage, the Buguldeika River flowed out of Lake Baikal and into the Lena River. This ancient river has been given the name "Pramanzurka" (BDP Members, 1995). Later, in the Neobaikalian stage, large tectonic movements and the growth of the Primorsky Ridge (located on the western flank of Lake Baikal) resulted in the reconstruction of the river network, and the riverbed of the Buguldeika, which flows into the lake, was formed (BDP Members, 1995). These findings have been confirmed by seismic profiles, which show that the sedimentary sequences near the drill site are divided into two seismic-stratigraphic complexes separated by nonconformities. These nonconformities are found beneath the 100 m mark, which means that only the upper seismic-stratigraphic horizon was penetrated.

Fig. 3 shows a cross section of the Buguldeika core. The sedimentary sequence at the drill site is composed of dense, fine-grained silty clays containing terrigenous and biogenic material. A repeating pattern of sedimentary layers is present. Each pattern unit has a layer that is enriched with diatoms and a layer that contains mainly clay-like terrigenous material. This pattern of two layers is continued for up to 100 meters.

Clay and diatom silts are different in terms of biogenic silica content, because diatoms contain biogenic silica. Since silica is a magnetic mineral that influences the magnetic susceptibility of the layer, this information can be used to compare the Baikal cross sections with the oceanic oxygen curve (Colman *et al.*, 1995)

The main characteristic of the Buguldeika cross section is an increase in the amount of coarse-grained material at the bottom. This feature can be explained by an intensive reworking of the Buguldeika riverbed during its early stage of development (BDP Members, 1995). The Buguldeika cross section is also characterized by the presence of so-called turbidite layers (see Fig. 3), which is 1-2 cm thick. These flows are probably connected with seasonal floods of the Buguldeika River.

The sedimentary cross section obtained from the Academician Ridge was quite different from that of the Buguldieka saddle. The Academician Ridge is separated from the shore by deep basins. The ridge rises over these basins by 400-600 m. These basins hinder the supply of coarse-

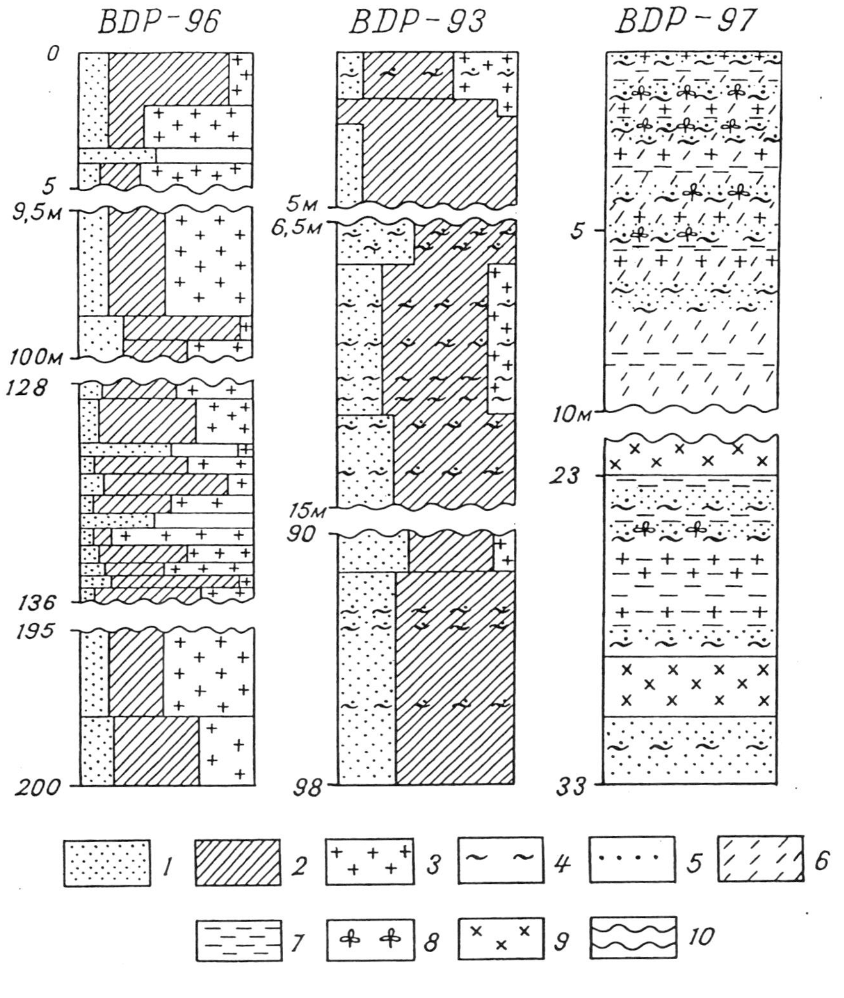

Fig.3. Cross sections of the cores from Academician Ridge (BDP-96), Selenga-Buguldeika saddle (BDP-93) and the South basin (BDP-97).
1. Fine sand and silt, 2. Clay sized material, 3. Diatoms, 4. Lower boundary of turbidite interlayers, 5. Coarse-grained material in lower part of turbidite sediments, 6. Clay mud from deep basins, 7. Clay interlayers in sediments of deep basins, 8. Fossilized vegetation, 9. Waste sediments, 10. Gaps in cross-sections

grained material from the shore, so the sediment in this region is mainly composed of material from the water column itself. The sediment on the top of the ridge near the 1996 and 1998 drill sites is about 1000 m. Moreover, geophysical investigations have shown a well-stratified sedimentary sequence separated by two nonconformities. Some investigators consider the lower nonconformity, located at a depth of 400 m, to have an age of 1.5-2 Ma (Kazmin *et al.*, 1995). However, the data from the drilling indicates that this nonconformity is even more ancient (BDP Members, 1998) and suggests that the deep Baikal basins have been in existence for at least 5 Ma. In fact, the 200-m core from the Academician Ridge has been dated at 5 Ma.

The cores from the Academician Ridge and the Buguldeika site both contain silty-clay-biogenic sediments. However, the nonconformities and breaks seen in other samples have not been found in the BDP96 Hole 1 core. Turbidite layers are also absent. When coarse-grained material is present, it occurs as separate lenses and likely results from the deposition of sandy material in ice once the ice has melted. The cross section has a repeating pattern, consisting of alternating diatom silt deposits and terrigenous clay sediment layers. This pattern continues to a depth of 200 m (see Fig. 3). The lithological features in the core from the Academician Ridge indicate constant sedimentation conditions during the deposition of the entire sedimentary sequence. In other words, the material in the sediment did not originate from the shore and instead was supplied by material in the water column. Cross sections like those obtained from Academician Ridge are the most suitable for paleoclimatic investigations.

The sedimentary sequences obtained from the deep basins exhibited yet another type of characteristic cross section. A sample from the central part of the South basin was obtained in 1997 (see Fig. 1). In addition to deep lacustrine sediments, which contain diatom silts or terrigenous clays, turbidite interlayers containing gravel and sandy material were abundant. The lower boundary of the interlayers is sharp, uneven, and washed. Turbidites gradually transit to deep lacustrine sediments further up the core. The interlayers are marked by clear gradation layering, progressing from a coarse-grained material in the lower region to a fine-grained material in the upper region of the interlayer. This gradation indicates the deposition of material from temporal water flows, which transfer material from the shore into the deep lake basins. Similar turbidite flows are found on ocean margins, resulting in L.P. Lisitsyn's so-called "avalanche sedimentation" (Lisitsyn, 1991). The transfer of large amounts of sedimentary material by temporal water flows can be described as an underwater avalanche. The turbidite flows also transfer a large amount of plant debris, leaves, and grass into the lake. This organic material is buried by the sediment and can become a source of organic hydrocarbons. The high pressures at the bottom of the basins then cause the hydrocarbons to turn into gas hydrates. This process was theoretically suggested by Dr. Golubev and was predicted by geophysicists using seismic data. Drilling in 1997 confirmed these suspicions. Gas hydrate samples were collected at depths of 121 m and 161 m. They were then analyzed in a number of Institute laboratories in Novosibirsk (Kuzmin *et al.*, 1998). The composition of the gas hydrate was determined to be $CH_4 \cdot 6H_2O$, the carbon isotope of which is methane. Thus, the Baikal Drilling Project has confirmed the formation of gas hydrates in fresh water for the first time, although the formation of gas hydrates in oceans and marginal seas is quite common.

As these results show, the sedimentation pattern is significantly different in the various topographical regions of Lake Baikal, which is a typical rift lake. Deep-seated lacustrine sedimentation is found in uplifts such as Academician Ridge, which is separated from the shore by deep basins. The sedimentary cross-sections obtained from this region are the most informative in terms of investigating the paleoclimatic record because the sedimentation pattern in this area only depends on the environment.

As mentioned above, deep depressions in the rift lake exhibit an avalanche sedimentation patter. This observation can be compared to those of passive oceanic margins, where such sedimentation patterns have been previously reported (Lisitsyn, 1991). The inclination angles of these oceanic margins are between 4° and 8°, and turbidite flows are found for thousands of kilometers. The slopes of the deep Baikal basins have an inclination that varies between 15° and 30°, so the turbidite flows at the bottom of the Baikal basins completely overlap. Such cross sections are of great importance for determining the dynamics of how rift basins are formed, investigating the formation of hydrocarbons, and studying the features of sedimentary continental basin formation. The cross sections obtained from the Buguldeika saddle region exhibit a pattern that is intermediate to those of the Academician Ridge and the deep lake basins.

Interesting paleoclimatic data has been obtained from the Buguldeika saddle (BDP Members, 1995) and the Academician Ridge. This article will only discuss the BDP96 core, which can be considered to be a model for continental paleoclimates. As a continuous record, the BDP96 core surpasses the information obtained from marine cores (BDP Members, 1998).

A precise method for determining the age of sediments is required to interpret paleoclimatic signals. Unfortunately, available carbon dating techniques can only provide ages up to 30-50 Ka. Other methods of absolute age determination are only useful for limited intervals. However, the age of sediments can be reliably determined using measurements of paleomagnetism. Several epochs of reverse magnetization are known to have existed on Earth. When the magnetization of the Earth is reversed, the South and North Poles change positions. Paleomagnetic measurements of the BDP96 core were performed by three groups: a Russian-American team, a Japanese team, and a German team. All of the results correlated well, indicating the high quality of the core. Four paleomagnetic epochs were identified in the core: Bruhnes, Matuyama, Gauss, and Gilbert. Consequently, the age of the core was determined to be 5 Ma. The quality of the core was further confirmed by Japanese investigators who used the cryogenic magnetometer for BDP96 achieve core with an archive core. In addition to the geomagnetic epochs, the scientists also distinguished several excursions, i.e. short-term deviations from the average paleomagnetic direction. These excursions include previously documented and new events (preliminarily called BDP96-15, BDP96-17, etc.) that were identified using a cryogenic magnetometer (Kravchinsky *et al.*, 1998). The U-Th method for determining the age of core samples (developed at the Institute of Geochemistry) allows ages of up to 1 Ma to be determined. This method will allow the ages of many excursions to be more precisely defined. The overall findings indicate that the cross section from the Academician Ridge can be regarded as a model for the Cenozoic paleomagnetic scale (Kravchinsky *et al.*, 1998).

As described earlier, the 200 m borehole from the Academician Ridge has a cross section that has been dated at 5 Ma. Within this time interval, a constant sedimentation rate has been maintained (4 cm per 1000 years). The sedimentation conditions of the distant past are similar to more recent ones, and no significant variations have been found.

The results of the Baikal core investigations indicate significant variations in the amount of diatoms, biogenic silica and a number of other sediment characteristics that are related to climatic change. This conclusion was verified by the correlation between the Baikal records and the marine oxygen isotope curves, which reflect climatic variations resulting from changes in solar insolation that arise when the Earth's orbital parameters change (according to the Milankovitch theory; for comparison of records, see BDP Members, 1995; Kuzmin *et al.*, 1997). The curve for diatom variation in Baikal sediments (BDP96 Hole 1 core) and the marine oxygen isotope curve (ODP 667) are compared in Fig. 4. The similar tendencies of Baikal and marine climatic parameters are obvious. The marine and Baikal records can also be compared using a special spectral-comparative analysis of both curves (Williams *et al.*, 1997). Climatic cycles of 100, 44, 23, and 19 Ka that are associated with the location of the Earth in its solar orbit can be distinguished for both curves. Thus, continental and oceanic climatic changes have been connected to astronomical factors for the last 5 Ma.

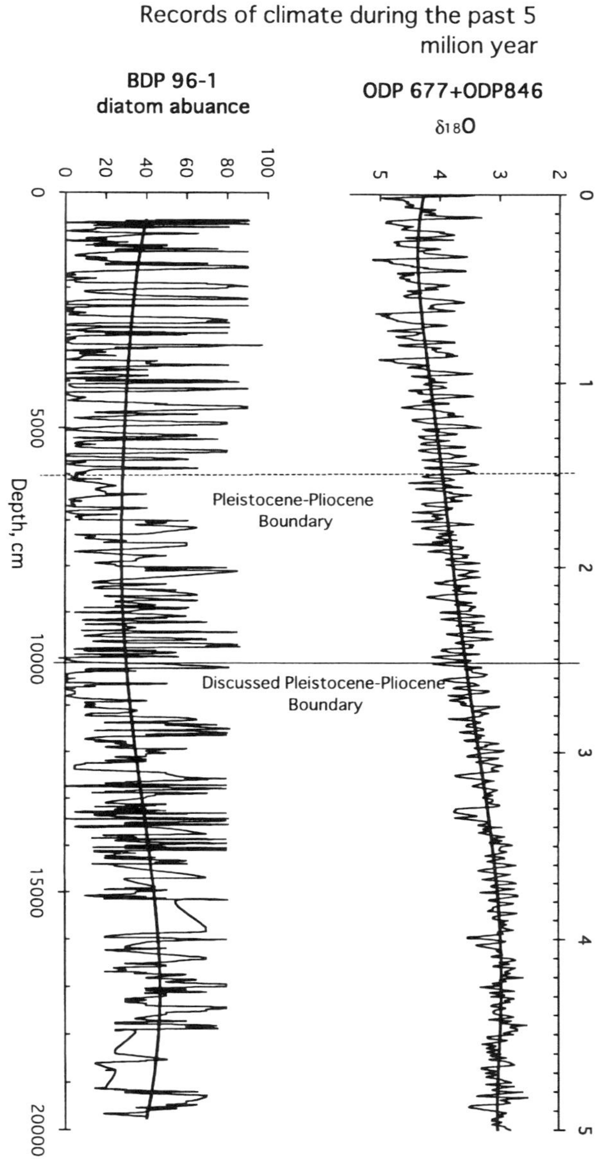

Fig.4. Plot of averaged diatom content in Baikal sediments and the marine oxygen isotope curve (ODP 677 +846 $\delta^{18}O$)
The solid line represents the average values for diatom content and $\delta^{18}O$.

Two significant cold climatic minima are found in the Baikal record against a background of a general tendency towards a decrease in heat (BDP Members, 1998; Karabanov *et al.*, 1999, in press; Bezrukova 1999). The more ancient of the two cooling periods is dated at 2.8-2.5 Ma and lasted for 300,000 years. The second cooling period is dated at 1.75 to 1.45 Ma and also lasted for 300,000 years. Judging from the productivity of Lake Baikal diatoms, which is closely associated with climatic characteristics, a period of warming appears to have followed the first cooling period. This period of warming corresponds to the Pliocene era. After the second period of cooling, the climate became much colder overall (Karabanov *et al.*, 1999, in press).

The episodes of cooling agree well with data collected from paleontological analyses (Bezrukova *et al.*, 1999) and an analysis of diatom species in the composition of the samples (Khursevich *et al.*, 1998). The ancient period of cooling is characterized by a considerable decrease in the number of arboreal species; for example, broad-leaved species completely disappeared. Grass vegetation was abundant, suggesting that the climate became significantly colder and drier (Bezrukova *et al.*, 1999). Paleologists (-?-Palinologists) believe that a change from forest assemblages to dispersed forests on mountain slopes occurred. Cold steppes and moss bogs also formed on the shores of Lake Baikal.

The diatom species during this cold period are characterized by the disappearance of the Stephanopsic genus and the appearance of a new genus of algae called Tertiarius. The climate warming that occurred after 2.4. Ma in turn led to the complete disappearance of the Tertiarius genus, and a new species called Cyclotella praetempetei appeared (Khursevich *et al.*, 1998).

During the second (1.75-1.45 Ma) period of cooling arboreal vegetation decreased, grass vegetation increased, and broad-leaved trees completely disappeared. The landscape consisted of a forest-tundra, suggesting the development of mountain glaciers in the Baikal region (Bezrukova *et al.*, 1999, in press). The diatoms (Cyclotella praetempetei) that were typical of the previous warm period were replaced by diatoms of the same genus (Cyclotella comtaeformicu)(Khursevich *et al.*, 1998).

Thus, paleoclimatic analyses based on quantitative estimations of diatom or biogenic silica content as well as data on the presence of various diatom species and paleontological analyses indicate the presence of two cooling episodes during the Late Cenozoic. The cooling episodes have been dated at 2.8-2.5 Ma and 1.75-1.45 Ma. These episodes have also been found in marine records and have been observed in Alaska, Iceland, Europe, Western Siberia and a number of other places (Karabanov *et al.*, 1999, in press).

As a result of investigations on the continuous Baikal record and its accurate dating, the age of several global events influencing Central Asia, the Eurasian continent, and the northern hemisphere have been more precisely dated.

The Baikal Drilling Project was supported by the Russian Foundation for Basic Research, Ministry of Science and Technology of the Russian Federation, US National Science Foundation, Science and Technology Agency of the Japanese government. The authors are grateful to all listed organizations as well as to all participants of the projects for their help in performing the drilling program.

References

BDP Members, 1995, Results of drilling the first bore hole on the Buguldeika saddle. Geology and Geophysics, 36(2) , 3-32.

BDP Members, 1998, Continuous climatic record in sediments of Lake Baikal for the last 5 Ma. Geology and Geophysics, 39(2), 139-156.

Bezrukova E.V., Yu.A. Bogdaov, D.F. Williams *et al.*, 1991, Deep changes of ecosystem of the Northern Baikal in the Holocene. Dolkady AN SSSR, 321(5), 1032-1037.

Bezrukova E.B., H.B. Kulagina, P.P. Letunova *et al.*, 1999, Direction of change of flora, vegetation and climate in the Baikal region for the last 5 Ma using the data of palynological investigations of 200-m core. Geology and Geophysics, 5, 739-749.

Colman S.M., J.A Peck, E.B. Karabanov *et al.*, 1995, Continental climate response to orbital forcing from biogenic silica record in Lake Baikal. Nature, 378, December 21/28, 769-771.

Hutchinson D.P., A.J. Golmshtok, L.P. Zonenshain *et al.*, 1993, Features of Lake Baikal sedimentary sequence using multi-channel seismic profiling [1989]. Geology and Geophysics, 34(10/11), 25-36.

Kazmin V.G., A.Ya Golmshtok, K. Klitgord *et al.*, 1995, Structure and development of Academician Ridge using seismic profiling data. Geology and Geophysics, 36(10), 164-176.

Karabanov E.B., M.I. Kuzmin, A.A. Prokopenko *et al.*, 1999, Global cooling of climate in the Asia in the Late Cenozoic in accordance with the sedimentary record from Lake Baikal.-Doklady RAN, (in press).

Khursevich G.K., E.B. Karabanov, D.F. Williams *et al.*, 1998, Pliocene-Pleistocene geochronology and biostratigraphy of bottom sediments of lake Baikal: new data of deep drilling. In "Paleoclimates and evolution of paleogeographic settings in geological history of the Earth." Petrozavodsk, pp.87-88.

Kravchinsky V.A., J. A. Peck, J. King, S. Nomura, A. Tanaka, M.I. Kuzmin, D. Williams and T. Kawai, 1998, The Late Cenozoic magnetic-stratiographic scale of the Central Asia using the data of deep drilling on

Lake Baikal, In: Global changes of the environment, eds. Dobretsov N.L., V.I. Kovalenko, Novosibirsk, SB RAS, pp. 73-77.
Kuzmin M.I., M.A. Gratchev, D.F. Williams *et al.*, 1997, Continuous record of paleoclimates for the last 4.5 Ma from Lake Baikal (first information). Geology and Geophysics, 38(5), 1021-1023.
Kuzmin M.I., G.V. Kalmychkov and V.F. Geletyi, 1998, Discovery of gas hydrates in Lake Baikal sedimentary sequence. Dolkady RAN, 362(4), 541-543.
Lisitsyn A.P., 1991, Processes of terrigenous sedimentation in seas and oceans. M. Nauka, 271 pp.
Moore P.C., K.D. Klitgord, A.J. Golmshtok and E. Weber, 1997, Sedimentation and subsidence patterns in the Central and North basins of Lake Baikal from seismic stratigraphy. Geological of America Bulletin, 109(6), 746-766.
Williams D.F., J. Peck, E.B. Karabanov *et al.*, 1997, Lake Baikal record of Continental Climate response to orbital insolation during the past 5 million years. Science, 278, 1114-1117.
Zonenshain L.P., A.J. Golmshtok and D.P. Hutchinso, 1992, Structure of Baikal rift. Geotectonics, 5, 63-77.

Lake Baikal
K. Minoura (editor)
2000 Elsevier Science B.V.

Changes in Lake Baikal water levels and runoff direction in the Quaternary period

Mats, V. D.[1]*, Fujii, S.[2], Mashiko, K.[3], Osipov, E. Yu.[1], Yefrimova, I. M.[1], and Klimansky, A. V.[1]
[1]Limnological Institute, SD RAS,Russia, 664033, Irkutsk, Ulan-Batorskaya, 3, POB 4199, fax: 7-3952-466933, e-mail: info@lin.irkutksk.ru
[2]Fujii Laboratory for Environmental Geology, Toyama, Japan, fax:86-764-22-2974, e-mail: fujisan2@shift.ne.jp
[3]Department of Biology, Teikyo University, Hachioji, Japan, fax: 86-426-78-3430, e-mail: kmashiko@main.-teikyo-u.ac.jp
(*corresponding author)

Abstract

Interaction between the Lake Baikal water level and tectonism in the surrounding area, specially the Prebaikalye area, is discussed. More specifically, the changing drainage process of Lake Baikal from the Lena River system to the Angara River system are discussed.

There is reliable evidence of water level lowering, such as topography showing many fjord-like features, but they cannot be assessed accurately. Rises in water level are assessed 120-150 m above the present level in the middle Pleistocene, about 200 ka.

Uplifts in the western side of the Baikal depression began in the late Pliocene ca. 3 Ma and caused restructuring of the river network of Western Prebaikalye. Development of stream captures and young formation in late Pleistocene, extremely rugged relief of slope zones, developed against the background of relicts of an ancient smooth relief.

The level of Lake Baikal rose as a result of a tectonically determined rupture of the Lena runoff in the direction along the ancient RiverManzurka valley. A high terrace reaching 200 m above the present level formed, and this was followed by increasing water in Lake Baikal in the middle Pleistocene ca. 200 ka. The Kultuk-Irkut runoff channel also began to flow into the Yenisey River system in the middle Pleistocene-early late Pleistocene.

The position of the modern and ancient Kultuk-Irkut runoff sills is such that lowering of its level by more than 2 m would have made Lake Baikal a drainless reservoir. Based on geologic-geomorphologic data the modern Angara effluent is presumed to have formed ca. 50-60 ka. This is supported by molecular biology studies of gammarid populations.

Introduction

Water level changes play a major role in reservoir development, and under and above water terraces are the main indicators of such changes. Terraces have been the subject of studies by many researchers who have investigated Lake Baikal geologically and geomorphologically (Chersky, 1886; Tetyaev, 1915; Dumitrashko, 1952; 1968; Ladokhin, 1959; Palshin,1959; Eskin et al., 1959; Mats, 1974; Kononov & Mats, 1986; Kononov, 1993; Mats, 1993; Mats et al., 1998).

There are three problems: 1) a number of terraces have been identified, reaching heights of 283 m (Chersky) or even 600-700 m (Tetyaev, Dumitrashko); 2) only 4 principal terraces exist throughout Lake Baikal, with sharply varying heights due to young tectonic movements (Lamakin, 1968); 3)10-12 terraces are distinct and they have a maximum height of 200 m above water level (Kononv & Mats, 1986)). Terrace formation is attributable to tectonic and hydrological factors. Colman (1998) contends that climate-determined, changes in the level of Lake Baikal never exceeded 2 m and that "a major rise in the lake level is as unlikely as a major fall". He explains the presence of a staircase of terraces solely on the basis of tectonic deformations and believes that the terraces linked to definite climatic phases were identified based on radiocarbon dating whose data are unconvincing.

Accordingly, the discussion concerning the Lake Baikal terraces can be reduced to the following issues:

1. How many terraces are there in Lake Baikal? How high and old are they?
2. Were the terraces caused by changing water levels or by tectonic movements?
3. If the water level did change, was it because of climatic factors, tectonic factors, or both?

The existence of terraces higher than Lake Baikal terrace IV has been established (Mats, 1990). The terrace-like platforms described by N.V.Dumitrashko as high ancient terraces of Lake Baikal have been found to be of tectonic origin (Pavlovsky, 1937; Ufimtsev, 1992). Lower terraces have been fairly well studied (Mats, 1974; Imetkhenov, 1987), but the problem of higher terraces has not yet been fully resolved.

This paper deals with a number of specific issues related to the terraces. Although a number of recent papers have been devoted to these topics (Kononov & Mats, 1986; Kononov, 1993; Colman, 1998), they have not exhaust the issues. The data presented below may help to advance the search for answers to the above questions. We intend to discuss the following: 1. water level changes in Lake Baikal; 2. restructuring of the river network, with the Buguldeyka River as an example; and 3. the evolution of the discharge of Lake Baikal waters.

A large number dating data have been collected, such as Tandetron AMS of Carbon-fourteen dating data, rock magnetism dating data for drilling core and molecular biological chronology data after acting of BICER (Baikal International Center for Ecological Research). Some numerical treatment are done about geological tectonics et al., when these data are used for the former data.

Water level changes in Lake Baikal

To establish the role of hydrological and tectonic factors in terrace formation, it is necessary to first ensure that there is convincing proof of past water level changes. In the case of Lake Baikal, we can be sure that water levels have both risen and dropped.

High water level

Information on past rises is based on a variety of data. Palaeontologically dated (Mats et al.,1982) early Pleistocene sediments in Nyurga Bay (Olkhon island), including ancient Lake Baikal sediments with valves of endemic diatoms (Chernyaeva, 1990), are located near the lake water edge. A distinct abrasion platform spreads out over the Nyurga section at a height of about 80 m. Its surface is scattered with small flat pebbles, including unweathered granites, and the age of the pebble composition shows that they are young (middle Pleistocene).

On Cape Tiya (city of Severobaikalsk), palaeontologically dated (Bazarov et al.,1982) middle Pleistocene sediments, including lacustrine with Lake Baikal diatoms, are found in the section of an 80 m terrace. Around Frolikha Bay near the mouth of the Biraya River and at other spots, these sediments form socles of 40-50 m and lower Lake Baikal terraces. Such hypsometric ratios of ancient and younger sediments are evidence of a rising water level in the lake in the middle Pleistocene. This is also evidenced by signs of accumulation of middle Pleistocene sands in the Selenga River delta and in other areas of the eastern coast when the water level had risen (Imetkhenov, 1987).

This evidence of rising water levels in the lake is consistent with data on Lake Baikal outflow via the ancient Kultuk-Irkut valley at that time (Kononov & Mats, 1986). The surface of the valley now reaches an absolute height of as much as 700 m, and it is entrenched by a canyon formed by the latest tectonic uplift along the fault delimiting the Southern Lake Baikal trough. According to a large-scale topographic map the canyon is 110 m deep and this figure provides an estimate of the latest tectonic uplift. Thus, the water level rise in Lake Baikal in the middle

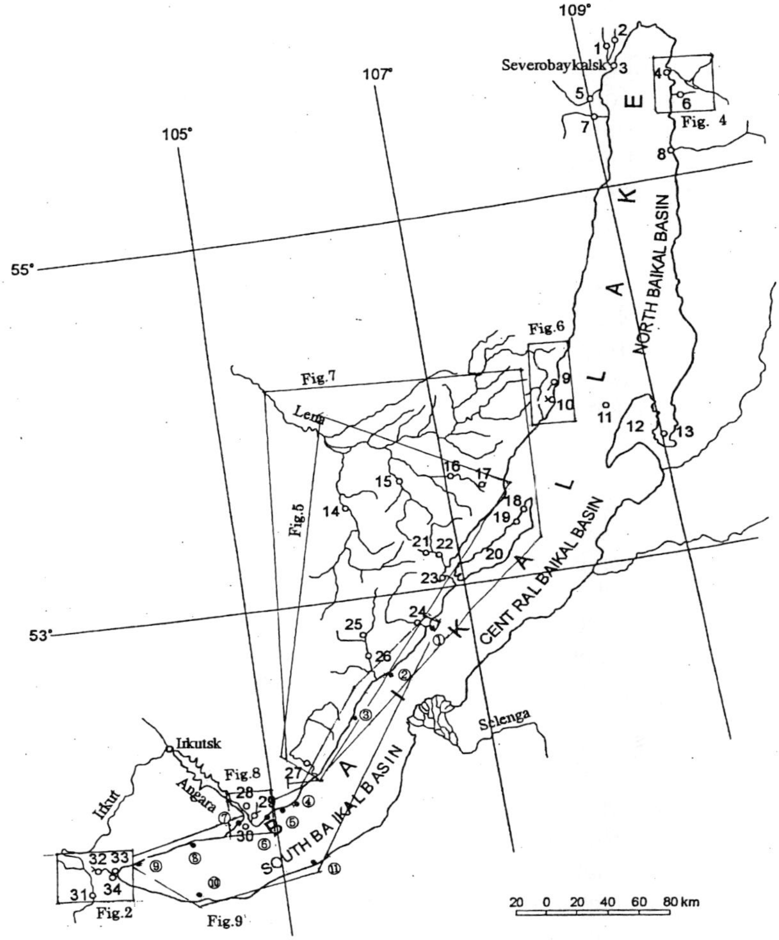

Fig.1. General overview map
1: Tiya R. 2: Nyurundukuan R. 3: C. Kurla. and C. Tya. 4: Frolikha B.
5: Sludinka R. 6: Biraya R. 7: Rel R. 8: Tompuda B. 9: Solntse-pad R.
10: Shartly R. 11: Big Ushkany I. 12: Svyatoy Nos Pen.
13: Tshivyrkuty B. 14: Manzurka R. 15: Lev. Ilikta R. 16: Pr. Ilikta R.
17: Kurga R. 18: C. Sasa 19: Peschanka B.(Nyurga B.) 20:Olkhon I.
21: Uspan R. 22: Sama R. 23: Mukhor B. 24: Anga R. 25: Alguy Vil.
26: Buguldeyka R. 27: Goloustnaya R. 28: Nikola Vil. 29: Krestovaya R.
30: Listviyanski B. 31: Big Bystrayana R. 32: Ilcha R. 33: Kultuk Vil.
34: Kultutchnaya R.
R.: River B: Bay C.: Cape I: Island Vil: village

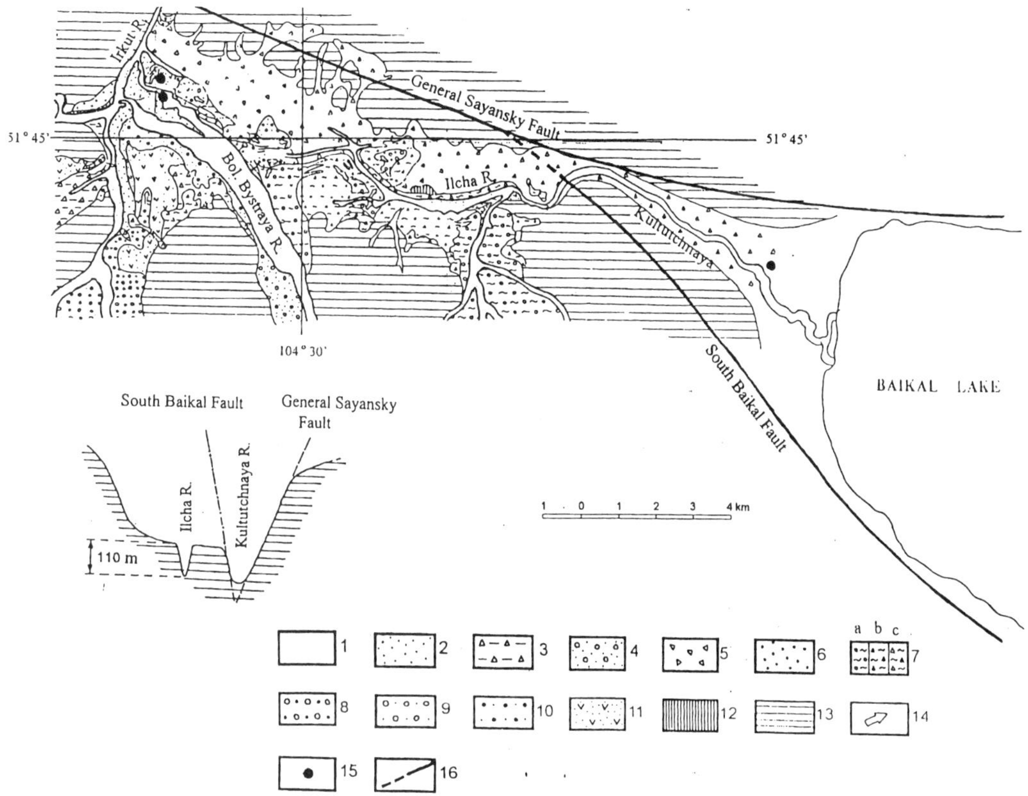

Fig.2. Schematic map of Cenozoic deposits in the area of Lake Baikal and the Bystraya Depression junction (Kononv & Mats, 1986, with changes)

1: alluvium of river beds and low flood lands, boulders, pebbles, sands
2: alluvium of high flood lands, sands, loam
3:delluvial-proluvial deposits of pre-valley sediments, proluvium of fans, trains of feet, bricks, loams (Q4)
4: alluvium of the Ist (6-8 m) terrace
5: Zyrkuzun unit, deluvial-proluvial deposits, arena, bricks, loams
6: alluvim of the 2nd (12-16 m)terrace, pebbles, sands
7: Bystraya unit
(a): alluvial-proluvial boulder pebbles deposits, deluvial-proluvial bricks, arena deposits
(b): deluvial deposits of slopes
(c): arena, loams
8: alluvium of the 3rd (30 m) terrace, boulders, pebbles
9: Illcha unit, upper part of cross section, alluvial boulders-pebbles deposits
11:lower?-middle Quaternary deposits, Ochre alluvial sands
12: Ochre conglomerates, bricks, arena deposits
13: Pre-Cambrian units
14: Directions of Pra-Irkut River flows
15: holes
16: fault

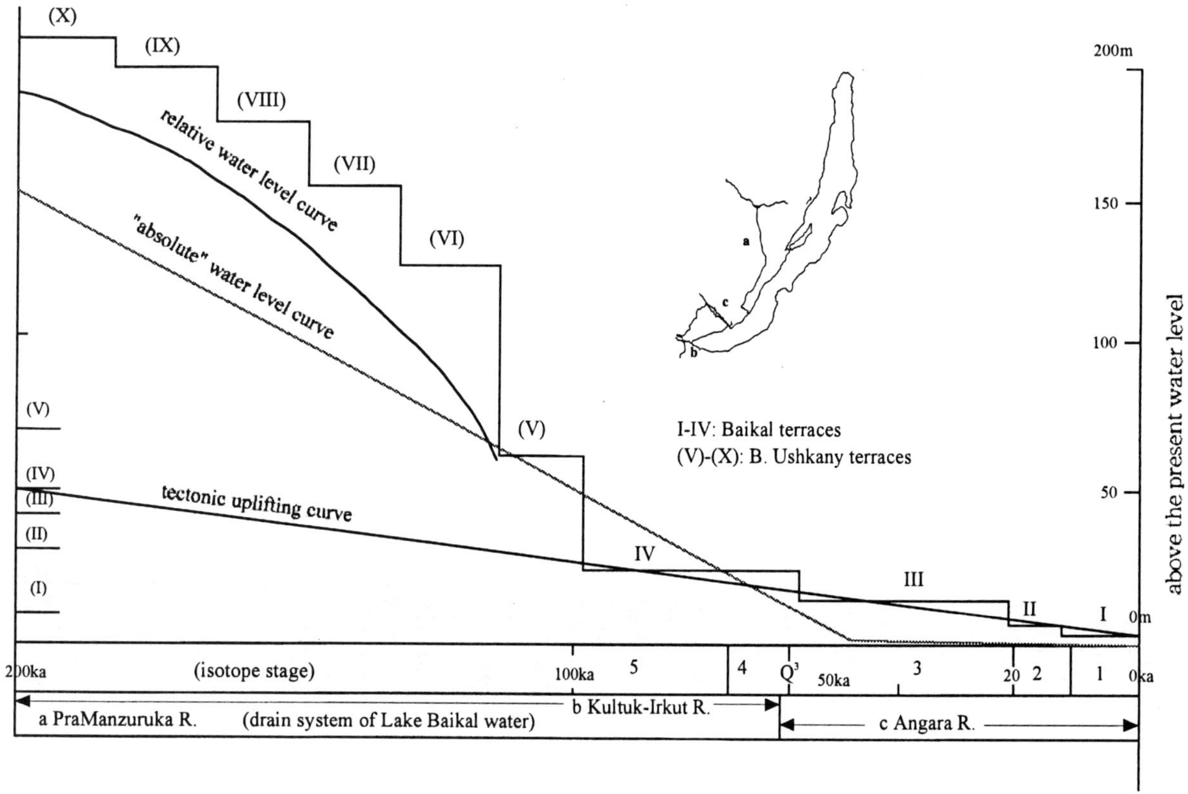

Fig.3. Schematic of water level changes in Lake Baikal in the late Quaternary

Pleistocene can be estimated to have been about 120-130 m (Kononov & Mats, 1986).

This value approaches the estimate based on terrace studies on Big Ushkany Island, where the highest terrace, retaining the cover of large pebbles in some places, is 670 m high. The first (Holocene) terrace on the island is 5 m high (Kononov, 1993), the typical value being 1.5-3 m (Mats, 1974; Fujii et al., 1994), meaning that the Big Ushkany terrace rose about 2.5 m over 10 ka. In the middle Pleistocene (ca. 200 ka) this rise would have been 50 m. Accordingly, the estimated water level increases in Lake Baikal in the middle Pleistocene would be approximately 150 m, and this is in sufficient concordance with the above estimate by the Kultuk-Irkut out-flow channel, and also in concordance with the presence of high terraces (up to 150-200 m).

They have been found on the west slopes of the Svyatoy Nos peninsula (Eskin et al., 1959; Ladokhin, 1959; Palshin, 1959; Mats et al., 1998) in several areas of the eastern coast of Northern Lake Baikal and on Olkhon Island.

Thus, a rise in water level to 120-150 m above the present level is considered realistic, and the formation of high terraces should be regarded as a result of both hydrologic factors (rise in level) and tectonic factors.

Low water level

Information on past water level falls is rather contradictory. The fact that the water level of Lake Baikal has dropped relative to the present day water level appears beyond doubt, but estimates of the magnitude, duration, and chronology of these drops are problematic. In a number of cases, subaerial sediments descend directly to the water's edge and below. Their relationships have been established in the area of Cape Kurla, where cover loams dated late Pleistocene are remnants of palaeolithic material that reach under water, and close relationships have been described by G.A.Vorobyova (1994) for the Buguldeyka River area, Mukhor Bay, and Chivyrkuy Bay. These findings support a lower Lake Baikal level at the end of late Pleistocene to early Holocene.

The peculiarities observed in the delta sediment structure can be explained by tectonic settling of the delta prism. The scant information on underwater "terraces" (Bukharov & Fialkov,1996) does not allow reliable assessment of the possible drops in level, because data on the platforms described as terraces are insufficient to regard them. Information on moraines located as deep as 300-400 m (Galkin, 1961; Lut, 1964) or even 500 m (Bukharov et al., 1996) has been reported in Frolikha Bay.

The fjord-like structure of some east coast bays in northern Lake Baikal is also cited as proof of a significant drop in the lake's level along with data showing maximum glaciation moraine and fluvioglacial deposits on the west coast of northern Lake Baikal (Rel-Slyudyanka area) below the level of the lake.

Numerous fjords exist around boreal seas. They used to be regarded as flooded inland valleys, but the multitude of fjords found at various depths cannot be explained by sea level variations. Fjords may have been formed by the action of glaciers on the sea bottom (Flint, 1971; Charlsworth, 1975).

This fact does not show lowering of the water level but the actiong of glaciers under the water. Florikha glacier extended below the water level and cut the bottom of the lake to a depth of 500 m as well as opinions of Flint and Charlesworth.

In summarizing the data on drops in the level of Lake Baikal, we must give a positive answer to the question of whether they ever took place.

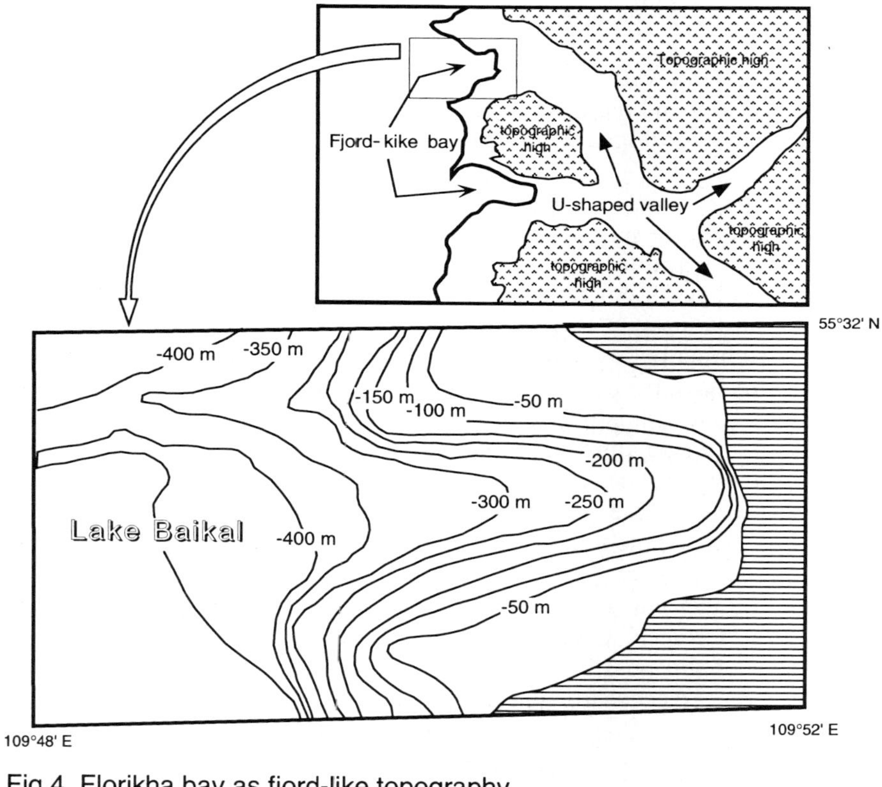

Fig.4. Florikha bay as fjord-like topography
(after USSR Navy, 1991-1992, No.62063 et al.)

Nevertheless, opinions on their magnitude and duration conflict with the sediment layer structure, which lacks any evidence of Lake Baikal ever being a drainless reservoir (Colman, 1998). Calculations made by L.Z. Granina at our request confirm Colman's findings and enable us to answer the question of how long it would take for evidence of a drainless Lake Baikal to appear in sediments. If Lake Baikal had been drainless for a certain period, then it should have begun to accumulate salts, giving rise to the emergence of new mineral phases and their precipitation. Chemical precipitation of calcium carbonate, in particular, should have occurred, followed by its accumulation in sediments. The calculations show that chemical precipitation should have occurred at ca. 10 ka, after the lake runoff stopped, i.e., when Lake Baikal water would be 1000s of times supersaturated with $CaCO_3$. This suggests that a drainless Lake Baikal could only exist for a short period geologically (10 ka or less).

Available data on the structure of runoff channels makes it possible to estimate decreases in its level, but they should not have been great enough to make Lake Baikal a drainless reservoir. After disruption of the Lena runoff channel and cessation of outflow along the ancient Manzurka Valley (Logachev, 1974; Kononov & Mats,1986), the Lake Baikal level rose as a result of increasing water.

Hydrological calculations performed by M.N.Shimarayev based on the Lake Baikal water balance show that its level would rise very rapidly after runoff were stopped (about 1 m a year). In any event, this value is several orders of magnitude more rapid than the admissible rate of tectonical uplifting. Thus, runoff disruption via the ancient Manzurka Valley was not geologically momentary. As the tectonic rise on the west side proceeded, and the corresponding runoff threshold was reached, the Lake Baikal water level rose too rapidly and once again discharged via the ancient Manzurka River. This process continued until its level reached the new discharge threshold via the Kultuk-Irkut valley, causing the discharge via the ancient R. Manzurka to gradually dry up.

The Kultuk-Irkut valley is filled with loose sediments, which drilling has revealed to reach 70 m deep (635 m above sea level) and still not touch the basement. In view of the latest tectonic uplift (110 m), the valley bed must have been located below 525 m. It is unlikely that the level of Lake Baikal dropped below this mark for any considerable time in the middle Pleistocene.

The sill at the Angara River effluent is no more than 2 m below the level of modern Lake Baikal, and in low-water periods people forded the Angara River at the site observed by V.A. Fialkov. Thus, it is impossible to believe that there was a significant drop below the modern level (more than 2 m) that lasted any considerable time (over 10 ka).

At the same time, some areas lie on over-deepened mouth areas of Lake Baikal's affluent valleys. Systematic data on over-deepened tributaries of Lake Baikal would favor a short drop in the level of the lake that most likely occurred in the late Pleistocene.

Restructurings of the Lena River and Lake Baikal systems' river networks

Restructurings of the river networks are linked to the development of the Lake Baikal rift relief, and they have been relatively well studied in west Prebaikalye (Pavlovsky & Frolova, 1941; Logachev, 1974; Anosov, 1964; Zamaraev et al., 1976; Kononov & Mats, 1986; Mats, 1993).

In the late Pliocene, re-orientation of the river network began from submeridional to sublatitudinal and diagonal due to growth of the Lake Baikal arched uplift. Best known is the late Pliocene-early Pleistocene system of the ancient Manzurka River, which constituted the left upper reaches of the Lena River system. Its development is related to the reduction of lake and the lake bog Palaeogene Neogene depressions in the Prebaikalye, whose southern parts were drained via the ancient Manzurka River system. This system also included the modern Buguldeyka River valley, which is a good example of the transformations the river network underwent in the Pleistocene. The modern upper and middle flow of the Buguldeyka River were a part of the ancient Manzurka River, which flowed from Lake Baikal near the mouth of the Goloustnaya River and emptied into the Lena River a short distance below Kachug village in the late Pliocene to early Pleistocene, (Logachev, 1974; Kononov & Mats, 1986).

In the early middle Pleistocene this outflow was disrupted, and the ancient Manzurka River valley split into a number of shallower valleys: the valley in its upper portion was used by the Goloustnaya River (affluent of Lake Baikal), while the lower portions of the valley along the former flow retained the direction of the Lena.

This portion also included the middle Pleistocene Buguldeyka River valley, which was the upper flow of the Manzurka River at the time. Due to increased uplifting of the slope portion of the Lake Baikal depression, in the late Pleistocene, an intense deep erosion entrenchment of Lake Baikal affluent began that, thanks to backward erosion, advanced actively deeper into the land massif of the western raised shoulder of the rift. One of the larger tributaries, whose valley used the submeridional fault zone, reached the wide, weakly entrenched middle Pleistocene Buguldeyka River valley, intercepted it, and redirected its waters into Lake Baikal. The site of interception is located a short distance below the village of Alaguy, 25 km from

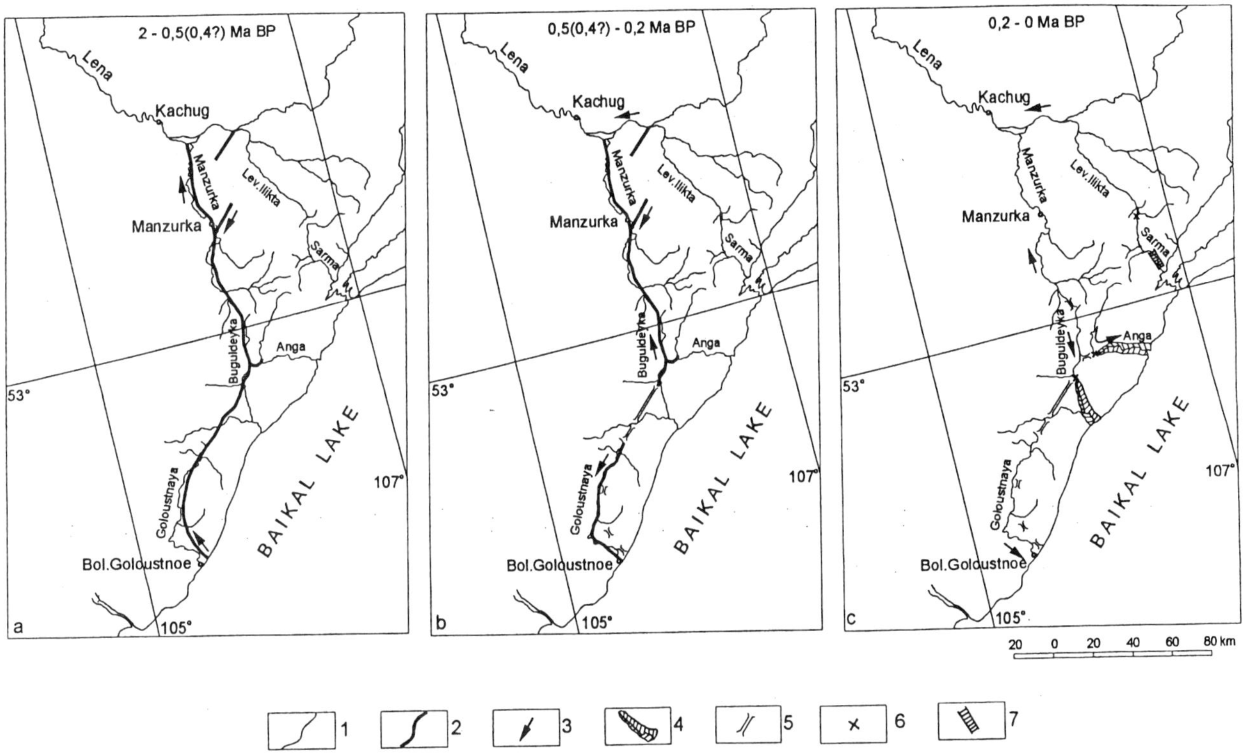

Fig.5. Evolution of the riverine net in the upper portion of the Lena River

a: Pre-Manzurka outflow, late Pliocene (2 Ma) early middle Pleistocene (0.5-0.4 Ma), b: Subdivision of the Paleo-Manzurka River into the Manzurka and Buguldeyka Rivers and overlapping of its upper portion by a Lake Baikal tributary (0.5-0.4?---0.2Ma), c: Present riverine net (0.2--0 Ma)

1: present hydrographic objects, 2: reconstructed hydrographic objects of the Pliocene-Quaternary, 3: direction of stream, 4: site of deep erosion incision, 5: abandoned valley, 6: overlap point realized, 7: former site of the valley the mouth of the Buguldeyka River. The breakpoint of the lengthwise profile of the Buguldeyka River is expressed with extreme contrast in the relief.

A deeply entrenched, steep V-shaped valley with tempestuous flow stretches below the point, a result of young (late Pleistocene) erosion entrenchment. Above it lies a wide, shallow, gently sloping valley with slowly flowing water that was excavated mainly in the middle Pleistocene by the water flow of the Lena River system.

Increasing water reached 200 m above the present water level including tectonic movement at Big Ushkany Island, because drainage of the Buguldeyka River into Baikal Lake ceased in the middle Pleistocene ca. 200 ka (200 ka was calculated from BDP93 drilling core data). The core changes character sharply 40 m below its top. Its age at this point is 200 ka according to the figures of Kashiwaya et al. (1997).

The restructuring of the river network described and related intercepts of the upstream Angara-Lena system occurred in an area on the western side of the southern and middle portion of Lake Baikal, up to the Sarma River, inclusive. These northeastward processes are only predicted and have never occurred. One of the most typical examples of a beginning intercept is linked to the head of the Lena River. It will be intercepted in the very near future, the Lena will be beheaded, and its uppermost tributaries will turn to Lake Baikal.

All this is linked to the young uplifts of the western frame of the depression. They began in the late Pliocene, ca. 3 Ma (the Olkhon phase of tectogenesis, Mats, 1990; Mats, 1993). Initially, the ancient Manzurka River, which then flowed out of Lake Baikal waters into the Lena River system, overcame the growing uplifts and cut a deep valley in it.

Between the end of the Pliocene and the early Quaternary, the uplifts stopped and the entrenchment that had formed earlier was filled with alluvial deposits of the Manzurka suite. At the end of the early Pleistocene, the uplifts increased again (the Primorsky phase of tectogenesis), and the Lena runoff channel was disrupted.

The thalweg of the ancient Manzurka River was deformed (Logachev, 1974), and its valley was divided into the Goloustnaya Valley, an affluent of Lake Baikal, and the valley of the Manzurka River, a Lena River tributary. The upper stretch of the latter was intercepted by a Lake Baikal affluent and is now a relatively ancient upper stretch of the valley of the Buguldeyka River, a Lake Baikal affluent.

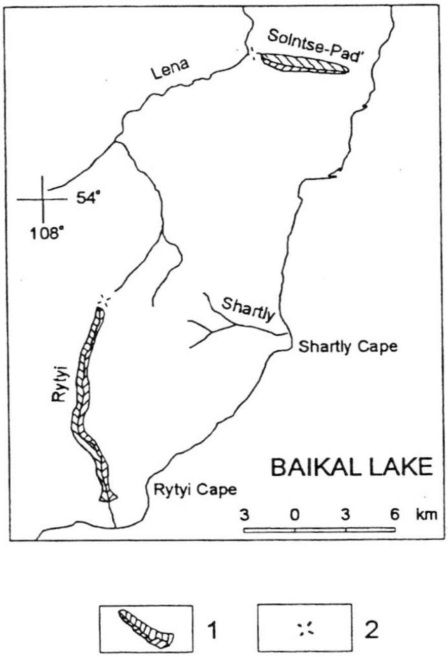

Fig.6. Future changes in the upper portion of the Lena River
1: site of deep erosion incision (canyon) 2: future overlap point

Quaternary evolution of the Lake Baikal runoff into the Yenisey River system

As noted above, the ancient Manzurka River runoff channel persisted until the Lake Baikal level rose due to tectonic uplift and reached the height of a new runoff sill in the area of the southern (Kultuk) end of the Lake Baikal depression.

Here, an ancient valley is clearly traced from the Lake Baikal depression into the Ilcha River valley and the Irkut River valley In the area where the ancient valley enters the Ilcha River valley, its bottom is deeply entrenched by the Ilcha River canyon. The northern slope of the ancient valley is formed by a steep escarp of the Main Sayansky fault, while in the south its bottom is fenced-in by a low accumulative range.

Deposit studies have shown that the Irkut River flowed into Lake Baikal along this valley at certain periods, and that in other periods Lake Baikal water flowed into the Irkut River. The Kultuk-Irkut runoff ceased to exist in the late Pleistocene because of the new lower discharge sill in Listvyansky Bay and the formation of the modern Angara outflow.

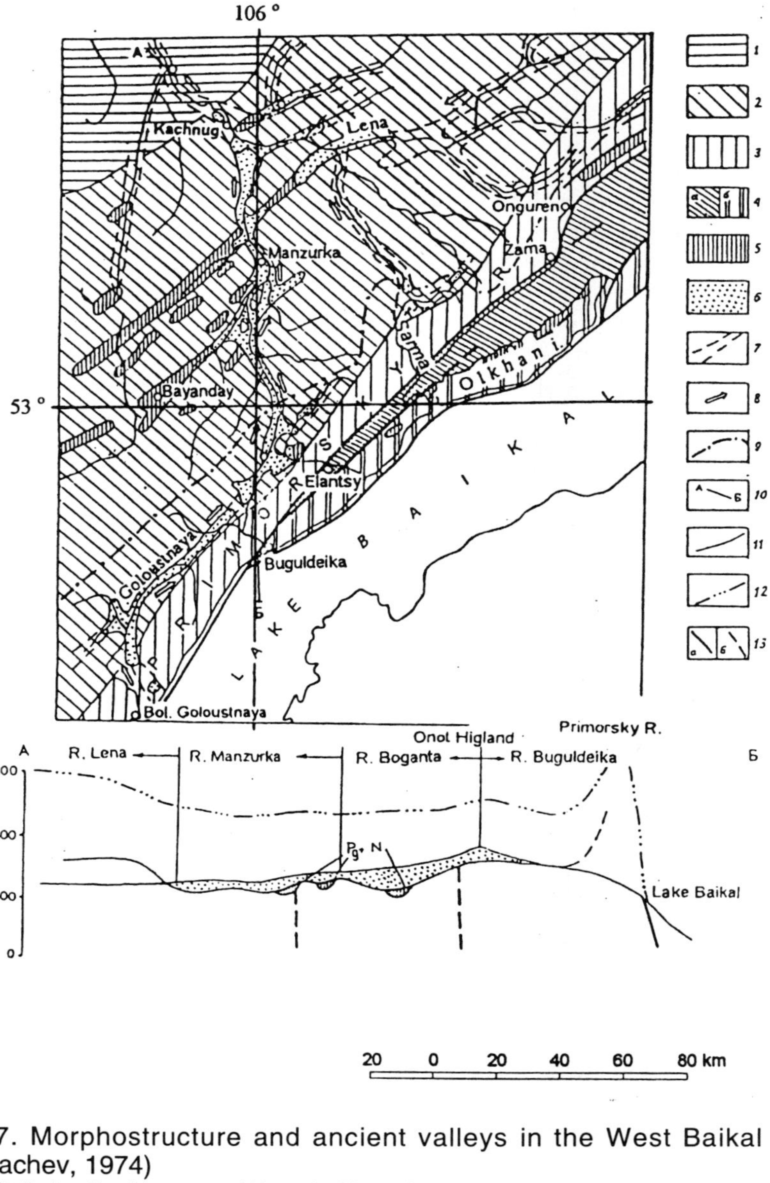

Fig.7. Morphostructure and ancient valleys in the West Baikal area (Logachev, 1974)

A longitudinal profile of the ancestral Manzurka River valley is shown at the bottom.

1: Upper Lena shield uplift, 2: Pre-Baikalsky depression, 3: side of Baikal dome, 4: (a) Maloe Mole Basin (b) raised Baikal and Olkhon blocks, 5: Paleogene and Neogene deposits, 6: Upper Pliocene Eopliocene Manzurka alluvium, 7: ancient river valleys, 8: direction of drainage along ancestral Manzurka; the profiles shown are contemporaneous rivers that use ancestral Manzuruka valley; arrows indicate the direction of their stream

The modern runoff of the Lake Baikal waters via the Angara River emerged very recently. Its formation was a consequence of tectonic settling of the basement block of Listvyansky Bay (Lut,1964) which began between the end of middle and the early late Pleistocene (ca. 150-120 ka) (Mats, 1990; Mats, 1993). The settling of the Listvyansky block partly uncovered the entrance into the submeridional, neotectonic Angara graben. The boundary faults of the graben are clearly visible on both slopes of the modern Angara Valley at the outlet and have been traced by geophysical studies in the basement of the Siberian platform on the territory of the city of Irkutsk.

The age of the Angara outflow formed has been estimated on the basis of geomorphological and molecular-biological data. Remnants of Lake Baikal terrace III (12-14 m) have been found near the northern entrance into Listvyansky Bay, but they are not visible on the bay shore: only the low terrace I stretches along the bedrock slope. Thus, the bay is younger than the terrace III. Additional evidence of this is river terrace II on the

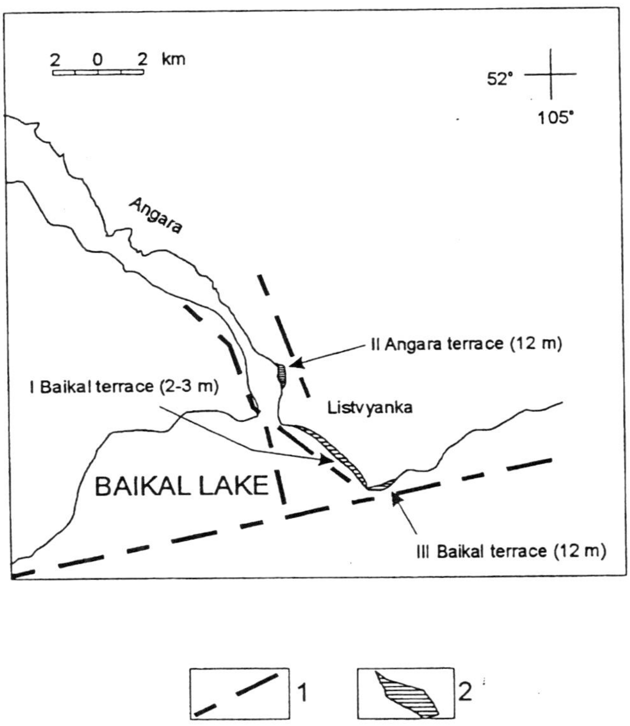

Fig.8. Schematic of the structure of Angara River outflow
1: fault 2: terrace

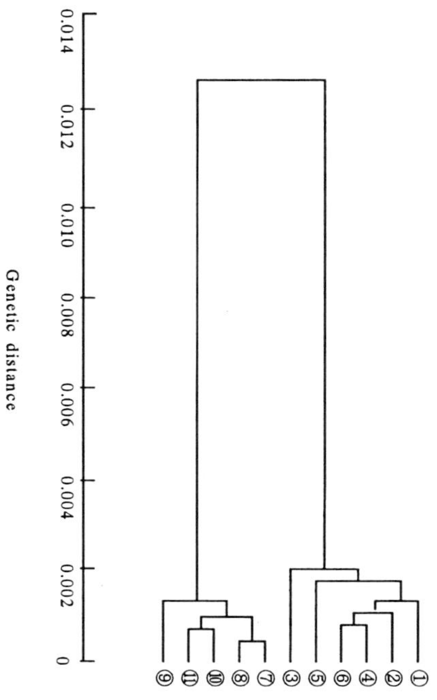

Fig.9. Results of a study of the gammarid population on the western coast of south Baikal
1–11: localities of samples collected in the south Baikal area shown in Fig.1.

right bank of the Krestovaya River near its mouth, which corresponds to the Lake Baikal terrace III in height and indicates that the mouth of the Krestovaya River extended somewhat deeper into Lake Baikal when Lake Baikal terrace III formed (Kononov & Mats, 1986).

Only two Angara terraces can be traced at the outlet of the Angara River. The upper one (12 m) corresponds to Lake Baikal terrace III. A sloping platform of the valley pediment and a steep escarp of the eastern fault of the Angara graben go higher. This implies that the Angara water flow has been in existence since the time when Lake Baikal terrace III began to form.

The lower layers of its section formed in the Karga time (isotpe stage 3), and the upper layers and their cover deposits formed in the Sartan time (stage 2, Mats, 1974; Imetkenov, 1987). Lake Baikal terrace II (6-8 m) is also dates from the Sartan based on radiocarbon and very explicit traces of intense cryogenesis. Cryogenic deformations have also been recorded in the upper portion of a section of Lake Baikal terrace III but have not been found on the surface of Lake Baikal terrace II. Accordingly, the age of the Angara outflow is estimated to be 50 ka, based on geologic-geomorphologic data.

Molecular biology studies of gammarid Eulimnogammarus cyaneus populations near the shores opposite the source of the Angara have shown that they separated ca. 60 ka (Mashiko et al., 1997). They appear to have been genetically dissociated since the Angara River arose at the present site, the rapid water flow acting as a barrier that prevented the migration of individuals across the river in this primarily lacustrine species of low mobility. Thus, the 50-60 ka value can be estimated as the time of formation of the Angara outflow from Lake Baikal.

Conclusion

Lake Baikal shores have unequivocally been established to expose a staircase of lake terraces. The formation of the lower terraces was governed by lowering of the water level in Lake Baikal in the Late Pleistocene and Holocene. The lowering was linked to the formation of the Angara outflow, but even the lower terraces were tectonically deformed in several places.

The presence of middle (up to 80 m) and higher (up to 200 m) terraces has also been reliably established. Their formation was governed by a combination of hydrological and tectonic factors, and their age has only been weakly substantiated. High terraces formed on Big Ushkany Island in the middle Pleistocene (ca. 200 ka) because the water level rose to 200 m above its level when the lake became a drainless reservoir as a result uplifting of Pre-Baikalye area.

Sufficiently convincing evidence exists of changes in the water level of Lake Baikal. Its tectonically determined rise reached ca. 50 m and most likely occurred during 200 ka. This rise in water level completely draw Big Ushkany Island, as well as caused formation of the higher Lake Baikal terraces, abrasion of the middle Pleistocene moraines (up to 150 m high), formation of an abrasive platform on Olkhon Island, and formation of high level facies of alluvial lake sands in the Selenga River delta. The Kultuk-Irkut runoff channel began to flow from Lake Baikal into the Yenisey River system.

Clear traces of Lake Baikal water level lowering also exist, but no reliable estimates of its magnitude have ever been obtained. The accepted limit of water level lowering in Lake Baikal appears clear from the lack of any evidence that Lake Baikal was ever drainless (Colman, 1998) for any prolonged period (>10 ka according to L.Z. Granina). The runoff sill at the Angara outlet was formed by a rock base that blocks the entrance into the Angara River bed. The surface of the base is about 2 m deep, i.e., about 454 m above the water level. Drilling in the ancient Kultuk-Irkut valley did not reveal the bedrock, which lies somewhere below the absolute mark of 635 m. Although it does not fully reveal the basement hypsometry across the entire section of the valley, the location of the deepest areas is hardly likely to differ greatly. Given the evidence of a possible younger tectonic uplift (110 m), the absolute height of the bedrock is about 500 m. This value appears to be the limit of possible Lake Baikal water level drops for any prolonged time in the middle early Pleistocene. At the same time, patchy data exist of a considerable drop in the level (ca. 100 m or more) of Lake Baikal in the late Pleistocene, but only for a short time (10 ka or less), based on information on over-deepened valleys of Lake Baikal tributaries.

The Lena River system is disrupted by uplifting of the Prebaikal area. Thus, an increase in water level to 200 m above the present level in the middle Pleistocene ca. 200 ka occurred as a result of cessation of drainage of Lake Baikal. The Kultuk-Irkut runoff then began to flow into the Angara River system. The age of formation of the modern Angara River is approximately 50-60 ka based on geomorphological data and molecular biology studies of the gammarid population around the mouth of the Angara River.

Acknowledgements
The authors are grateful to Dr.S. Colman (USGS) for his helpful discussion of the manuscript, to Dr. L. Z.Granina for her estimates of the period during which Lake Baikal could have been a drainless reservoir, and to Prof.M.N.Shimaraev for his estimates of the possible rate of rise in the lebel of Lake Baikal. The authors are also indebted to Prof. S. Osadchy of Irkutsk State University for his useful discussion in the field.

References

Anosov, V.S.(1964) Some data on ancient riverain net in South-Western and Central Prebaikalye. New data on geology, oil and gas availability and fossils of Irkutsk region. Moscow: 247-251.

Bazarov, D.B., Budaev, R.Ts. & Kalmykov, I.P.(1982) On the age of Pleistocene terraces of north-western coast of Lake Baikal. Logachev, N.A.(ed.). Late Pleistocene and Holocene Periods of the South of Eastern Siberia. Nauka, Novosibirsk, 155-158.

Bukharov,A.A. & Fialkov, V.A.(1996) Geological structure of Lake Baikal bottom. Observation from "Pisces". Nauka, Novosibirsk, 112.

Charlsworth,J.K.(1975)The Quaternary Era. Edward Arnold, London, 2 vols, 1700pp.*

Chernyaeva, G.P. (1990) The lake history from data on diatom flora. Kvasov,D.D. (ed.). History of Lakes of the USSR, Ladozhskoye, Onezhskoye, Pskovsko-Chudskoye, Baikal, Khanka. Nauka, Leningrad, 213-217.

Chersky, I.D. (1886) The report on geological survey on the Lake Baika shoreline. Izd. Vost. Sib. Otd. IRGO, Irkutsk, 405.

Colman M.S.(1998) Waterlevel changes in Lake Baikal, Siberia: Tectonism versus climate. Geology, 26, 531-534.*

Dumitrashko, N.V. (1952) Geomorphology and paleogeography of Baikal mountain province. Izd. AN SSSR, Moscow, 189.

Eskin, A.S.,Palshin, G.B.,Grechishev, E.K. & Galazy,G.I.(1959) Geology and some Problems on neotectonics of Ushkany Islands on LakeBaikal.

N.A.Logachev (ed) Materials on Geology of Eatern Siberia. Proceedings of Eastern Siberia Geol. Inst. Irkutsk 2:129-152
Flint, R.F.(1971) Glacial and Quaternary Geology. John Wiley and Sons, New York, 892 pp.*
Fujii,S., Nakamura,T.,and Mats,V.(1994) Holocene terrace around Baikal Lake. Summaries of Researchers Using AMS at Nagoya University, 6, 161-166
Galkin, B.I. (1961) On the problem of the glaciation charcater on the coast of Lake Baikal. Logochev,N.A.(ed).Materials on Geology of Meso-Cenozoic Deposits of Easthern Siberia. Irkutsk, 3, 50-59.
Imetkhenov, A.B.(1987) Late Cenozoic deposits of the Lake Baikal shores. Nauka, Novosibirsk, 150.
Kashiwaya,K., Nakamura,T., Takamatsu,N., Sakai,H., Nakamura,M. (1997) Orbital signals found in physical and chemical properties of bottom sediments from Lake Baikal. Jour.Paleolimnology,997, 18, 239-297.*
Kononov, E.E. (1993) High terraces of Lake Baikal. Geol. & Geophys., 34, 201-209.
Kononov, E.E. & Mats, V.D. (1986) The history of the Baikal drainage. Izv. Vuzov. Geol. & Razved., 6, 91-98.
Kulchitsky,A.A.(1985) Pleistocene glaciations in the mountains of North-Western Prebaikalye in the BAM zone (on the example of the river Kunerma basin). Geology. & Geophysics., 2, 3-10.
Ladokhin, N.P.(1959)On the problem of ancient glaciation of Prebaikalye. Logachev, N. A.(ed.). Proceedings of Eastern-Siberian Geol. Inst., Irkutsk, 2, 153-173.
Lamakin,V.V.(1968) Neotectonics of the Baikal depression. Nauka, Moscow, 247.
Logachev, N.A.(1974) The Sayan-Baikal Stanovoy Upland. N.A.Florensov (ed.). Uplands of Prebaikalye and Transbaikalye. Nauka, Moscow, 72-162
Lut, B.F.(1964) Geomorphology of the Baikal bottom. N.A. Florensov (ed.). Geomorphology of the Baikal Bottom and Its Shores. Nauka,Moscow, 5, 123.
Mashiko,K., Kamaltynov,R.M. & Sherako,D.Y.(1997) Genetic separation of a gammarid (Eulimnogammarus cyaneus) population by localized topographic changes in ancient Lake Baikal. Arch. Hydrobiol., 139, 379-387.*
Mats, V.D.(1974) Baikal terraces of lower complex. Votintsev,K.K. (ed.). Nature of Baikal, Leningrad, 31-56. otd. RGO, 1, 243-244.
Mats, V.D., Pokatilov, A.G.,Popova,S.M.(1982) Central Baikal in the Pliocene and Pleistocene. Nauka, Novosibirisk:192
Mats, V.D.(1990) The original and evolution of the Baikal basin. Kvasov, D.D.(ed.). History of Lakes of the USSR: Ladozhskoye, Onezhskoye, Pskovsko, Chudskoye, Baikal, Khanka. Nauka, Leningrad, 167-191.

Mats,V.D.(1993) The structure and development of the Baikal rift depression. Earth Science Review, 34, 81-118.*

Mats,V.D., Khlystov,O.M., De Batist,M., Smoliansky, E.N. (1998) Structure and development of international dam northern central Baikal basin on the base of comparative studies of its on land fragments and underwater one. BICER, BDP and DIWPA Joint International Symposium on Lake Baikal at Yokohama.

Palshin, G.B.(1959) On the problem of distribution of terraces on Lake Baikal. Tkachuk,V.G. & Grechishev,E.K.(eds.). Proceedings of Eastern Siberian Department AN SSSR, Series Geol., 10, 3-21.

Pavlovsky,E.V.(1937) Lake Baikal Depression. Izv. AN SSSR, Series Geol., 2, 351-375.

Pavlovsky, E.V. & Frolova, N.V.(1941) Ancient valleys of Angara-Lena watershed. MOIP Bull., Series Geol., 1-2.

Tetyaev, M.M.(1915) Lake Baikal in its nearest past. Geolog. Vestnik, 1, 76-79.

Ufimtsev, G.F.(1992) Morphotectonics of the Baikal rift zone. Nauka, Novosibirsk, 216.

Vorobyova G.A. (1994) Some data on the Lake Baikal water level in the Late Pleistocene and Holocene. Ufimtsev, G.F.(ed.). Baikal and Mountains around It (Cenozoic Geology, Geomorphology, Neotectonics and Geological Monuments of the Nature). Abstracts of Irkutsk Geomorphological Workshop, Institute of the Earth Crust Irkutsk, 92-94

Zamaraev, S.M., Adamenko, O.M., Ryazonov, G.V., Kulchitsky, A.A., Adamenko,R.S. & Vikentjeva, N.M. (1976) The structure and history of Prebaikalian piedmont depression. Nauka, Moscow, 134.

*: written in English;   no mark: written in Russian

Lake Baikal
K. Minoura (editor)
2000 Elsevier Science B.V.

Paleomagnetic and Rock-magnetic studies on Lake Baikal sediments -BDP96 borehole at Academician Ridge-

Sakai, H.[1]*, Nomura, S.[1], Horii, M.[2], Kashiwaya, K.[2], Tanaka, A.[3], Kawai, T.[3], Kravchinsky, V.[4], Peck, J.[4], and King, J.[5]
[1]Department of Earth Sciences, Faculty of Science, Toyama University Gofuku 3190, Toyama 930-8555
[2]Department of Earth Sciences, Faculty of Science, Kanazawa University, Kakuma, Kanazawa 920-1192
[3]National Institute of Environmental Studies, Onogawa 16-2, Tsukuba, Ibaragi 305-0053
[4]Institute of Geochemistry, Favorskogo str., Irktsuk 664033, Russia
[5]Graduate School of Oceanography, University of Rhode Island, South Ferry Rd. Narragansett, RI 02882-1197, USA
(*corresponding author)

Abstract

Paleomagnetic and rock-magnetic studies were conducted on two sedimentary cores, BDP96-1 (length: 200 m) and BDP96-2 (100 m), drilled at the Academician Ridge of Lake Baikal.

Comparison of the paleomagnetic inclination records with the geomagnetic polarity time scale showed that the sedimentary sequence covers the age of the past 5 million years. The study was conducted on discrete samples and on quarter-core samples. Path-through measurement of the quarter core samples revealed detailed geomagnetic variation, such as the double polarity transitions around the B/M boundary.

The average sedimentation rate was estimated from the depth-age relation to be 3.8 cm/kyr, with a correlation coefficient of 0.997-0.999. This high correlation suggests that the sedimentation at Academician Ridge during the past 5 million years has been continuous in a quiet environment.

Magnetic susceptibility is closely related to changes in the content of biogenic silica and shows a clear correlation with glacial-interglacial change. Susceptibility measurement is relatively quick and nondestructive, making it a valuable means of paleoclimatic study of Lake Baikal sediment. Changes in magnetic minerals (species, size) should also be taken into consideration in these studies.

Introduction

Lake Baikal is located in eastern Siberia (104-110°E, 51-56°N) and is one of the deepest (1643 m), most voluminous (23,000 km^3), and oldest freshwater lakes in the world. It is an important and unique site for paleoclimatic studies because of its high-latitude, continent-interior setting, and its long, continuous stratigraphic record. Grosswald (1980) suggested that Lake Baikal was never completely glaciated during the glacial periods, so that a continuous sedimentary record can be obtained even during the glacial periods. The sedimentary sequence of Lake Baikal is more than 5,000 m thick and believed to cover the age since the middle Miocene. Paleoclimatic records from continental regions are much fewer in number than records from marine regions. This makes Lake Baikal sediment particularly valuable, and it may provide a source of continental climate information over a long period.

The Baikal Drilling Project (BDP), in progress since 1993, is an international investigation of the paleoclimatic history and tectonic evolution of the sedimentary basin. In this paper, we describe a paleomagnetic study of the BDP96 cores drilled at Academician Ridge in the central part of Lake Baikal (Fig. 1). The Angara River, situated in the southern basin, is the only river draining Lake Baikal. The Selenga River, in the southern central portion of Lake Baikal is the largest river draining into the lake and carries

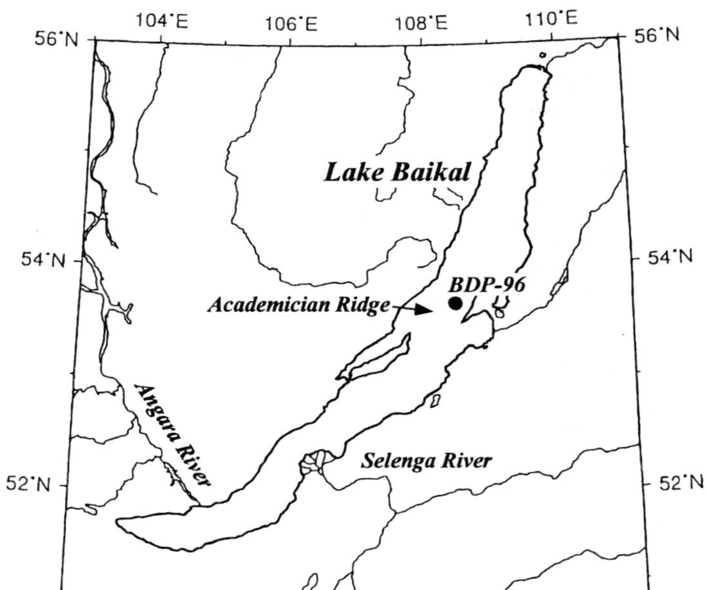

Fig. 1: BDP96 at Academician Ridge in the central part of Lake Baikal. The drilling site (53°41'38"N, 108°21'06"E) is at the depth of 382 m.

a large amount of sediment into it. Academician Ridge is away from these rivers and is a structual and bathmetric high that is isolated from direct fluvial and downslope sedimentation.

This study had two purposes. One was to examine the magnetostratigraphy and determine the age-scale of the sedimentary sequence, and the other was to study the history of paleoclimate based on the magnetic properties of the sediment.

Samples of BDP96 cores from Academician Ridge

BDP96 consists of two cores, BDP96-1 (length: 200 m) and BDP96-2 (100 m). The drilling was conducted by piston coring in the upper portion (depth <60 m), by percussion coring in the middle portion (60-120 m), and by rotary coring in the lower portion (120-200 m). Core recovery was 95% for BDP96-2, 90% for the upper 119 m of BDP96-1, and 70% for the rotary coring portion.

The BDP96 cores were divided into sub-cores 2 m in length, and each sub-core was cut in half lengthwise (split). The samples for the paleomagnetic study were discrete samples (DS) and quarter-core samples (QC) extracted from the half lengthwise core. The DS samples were divided into three groups. In this paper, we present the data obtained from the Japanese DS samples and the QC samples. Figure 2 is a schematic diagram of the paleomagnetic samples. The DS samples were collected from the core in 10 cm^3 plastic cube cases at 20-cm intervals. The QC samples were collected from the piston core portion, where the hardly disturbed cores at the drilling were not used.

Several short cores (~10m) were drilled around Academician ridge before BDP96. The rockmagnetic data of the St. 18 short core obtained near the BDP96 site (Sakai et al., 1997) is referred to in this paper.

Experimental methods and AF demagnetization

Remanent magnetization was measured by using a path-through-type cryogenic magnetometer (2G-760R), and magnetic cleaning was achieved by the AF demagnetization method. Magnetic susceptibility was assessed with a Bartington MS-2 meter, and the anisotropy of susceptibility (AMS) was measured with a Sapphire SI2B unit.

All of the discrete samples were AF demagnetized stepwise to 40 mT in 5 mT steps. Secondary magnetization was eliminated from most of the samples by demagnetization to 20 mT. The samples collected from the lower portion of the core had the weakest remanent magnetization and showed instability at high-field AF demagnetization.

The AF demagnetization experiment was conducted on the QC samples up to 40 mT in 10-mT steps. Stepwise AF demagnetization and measurement was performed at 1 cm intervals with a 2G-760R automatic demagnetization system. The results show that the main secondary component in the QC samples was also eliminated by demagnetization to 20 mT. The paleomagnetic data after 20 mT AF demagnetization are used below.

Paleomagnetic inclination

Variation of inclination with depth and magnetostratigraphy

Figure 3 shows the change in inclination with depth in the discrete samples. Since sedimentological study suggests that the BDP96-1 core lacks the surface sequence to a depth of 6.3 m, the inclination data have been

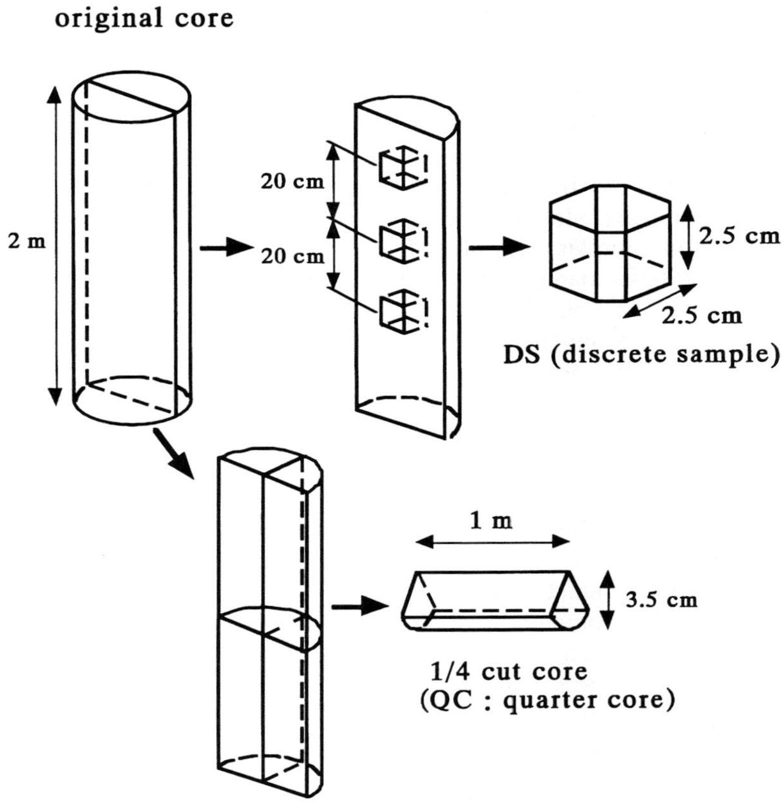

Fig. 2: Discrete samples and quarter-core samples were collected from the half lengthwise core for the paleomagnetic study.

plotted by taking its absence into account. The inclination changes with depth show a clear polarity reversal pattern, which is concordant with the preliminary data obtained by Baikal Drilling project II Members (1997).

Fig. 4 shows the inclination changes in the QC samples. The changes in inclination are almost perfectly consistent with the data from the DC samples. There are several abrupt inclination changes in Fig. 4. They may have been caused by regions of disturbance and/or core-breaks which we were unable to examine when making the measurements.

The DC samples during Matuyama reversed polarity in Fig. 3 show more scattered inclination than the other core portions, especially in the BDP96-1 core. Referring to the AF demagnetization results shows that these samples have larger unstable magnetization than other samples. One of the major reasons for the scattered inclination is a problem in the sampling, that is, the samples were collected from disturbed and/or mis-oriented areas. The thermal demagnetization experiment in chapter 5-4 suggests the presence of two different kinds of magnetic minerals in the sediments. Another possible reason for the scattered inclination is the existence of hard secondary magnetization against the AF demagnetization. This may be important, and further study is necessary.

In Figs. 3 and 4, the inclination changes are compared with the geomagnetic polarity time scale of Cande and Kent (1995). BDP96-1 includes the geomagnetic polarity epochs of Brunhes normal polarity, Matuyama reversed polarity, Gauss normal polarity, and Gilbert reversed polarity. BDP96-2 includes the Brunhes and Matuyama polarity epochs. Most of the geomagnetic events during the above polarity epochs have been identified. The comparison shows that the BDP96 covers the age of the past 5 million years. Table 1 shows the geomagnetic polarity epochs, events, and the corresponding depth of the BDP96 cores.

Sedimentation rate

The graphs for BDP96-1 and BDP96-2 in Fig. 5 show the correlation between depth and age for the assigned geomagnetic polarity boundary in Table 1. Straight lines are produced by the least squares method on the plots in the diagram.

The average sedimentation rate at BDP96 was estimated from the linear relation in Fig. 5 to be 3.8 cm/kyr. The correlation coefficient of linearity is 0.997 for BDP96-1 and 0.999 for BDP96-2. The high correlation coefficient in the depth-age graph suggests that the sedimentation at Academician Ridge has not suffered any major disturbances, such as produced by crustal movement, during the past 5 million years, and this may be important in assessing the tectonic history of Lake Baikal.

The AMS (anisotropy of magnetic susceptibility) of the samples was

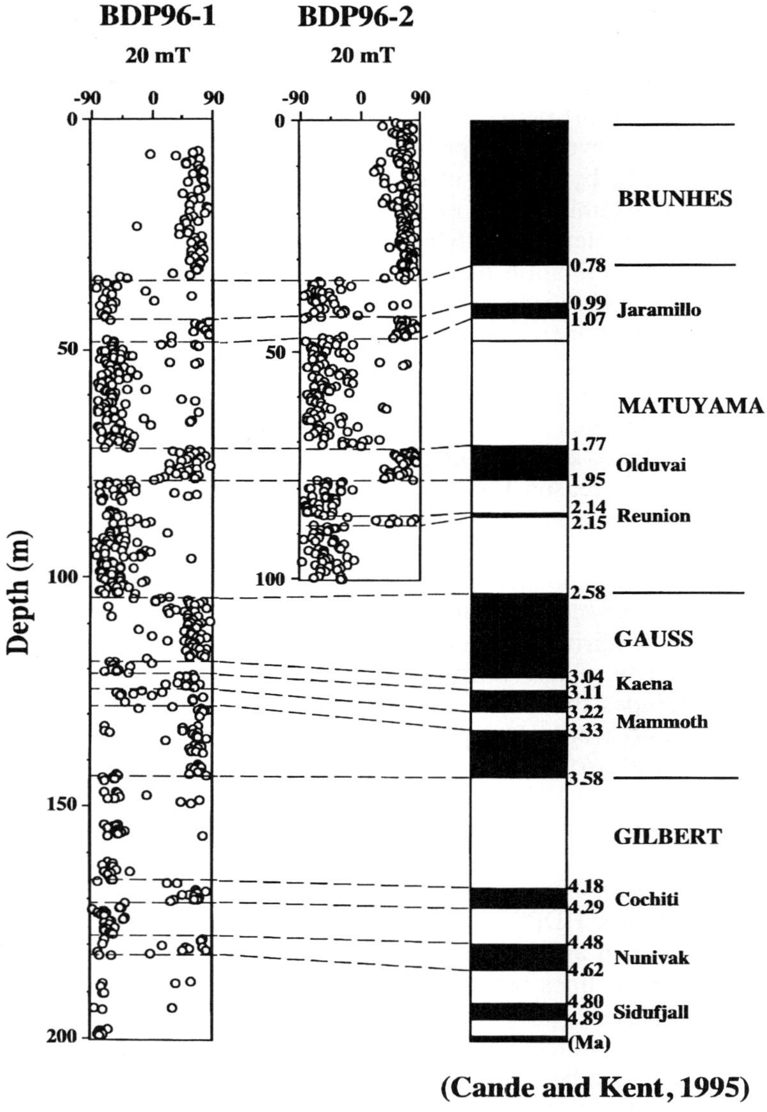

Fig. 3. Changes in inclination with the depth of discrete samples of BDP96 cores after 20 mT AF demagnetization. The geomagnetic polarity time scale of Cande and Kent (1995) is shown on the right for reference.

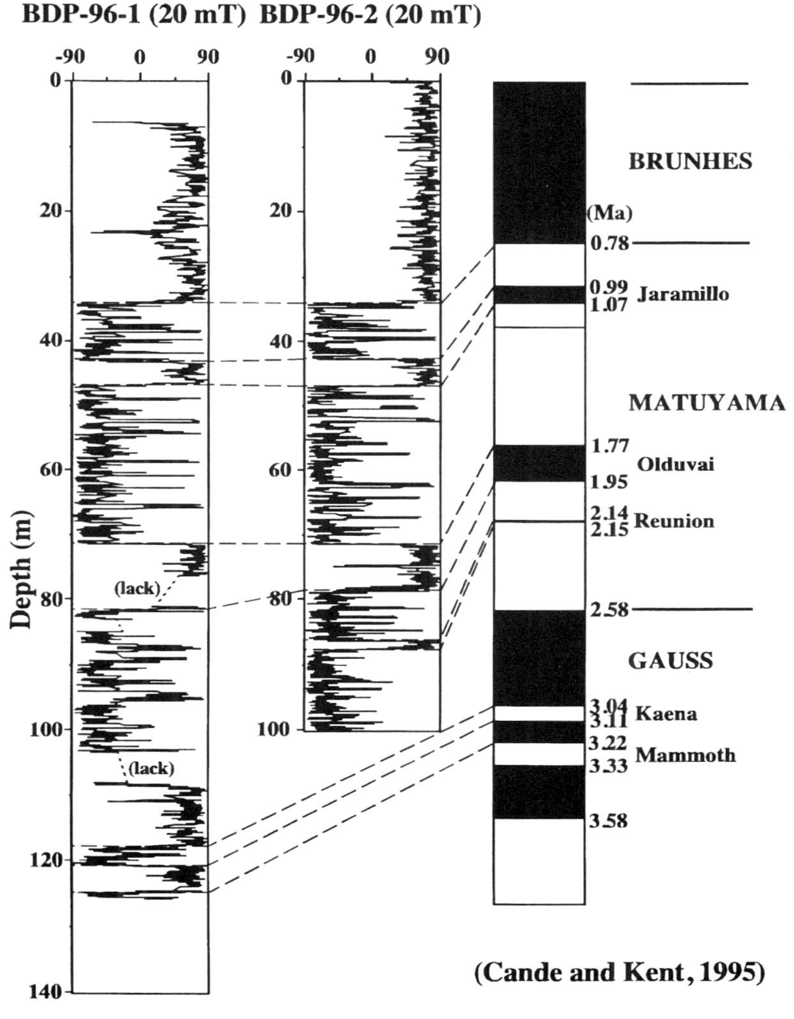

Fig. 4. Changes in inclination with depth of the QC samples of BDP96 after 20 mT AF demagnetization. The geomagnetic polarity time scale of Cande and Kent (1995) is shown on the right for reference.

also studied to investigate the sedimentary environment. The upper panel in Fig. 6 refers to the distribution of AMS in the St.18 short core near the BDP96 site. AMS revealed an oblate anisotropic fabric, that is, the maximum and intermediate axes are distributed in the horizontal plane, and the minimum axis lies in the vertical direction.

The lower panel shows the changes in the AMS parameter Max/Int-Int/Min in BDP96-2 with age, where the Max, Int, and Min values represent the degree of anisotropy of the maximum axis, intermediate axis, and minimum axis, respectively. This parameter is less than zero in most of the sequence, which means that the oblate AMS fabric is dominant. There does not seem to be any serious change in AMS features caused by the turbidite flow. The results of the AMS measurements corroborate the quiet sedimentary environment at Academician Ridge suggested by the constant sedimentation rate.

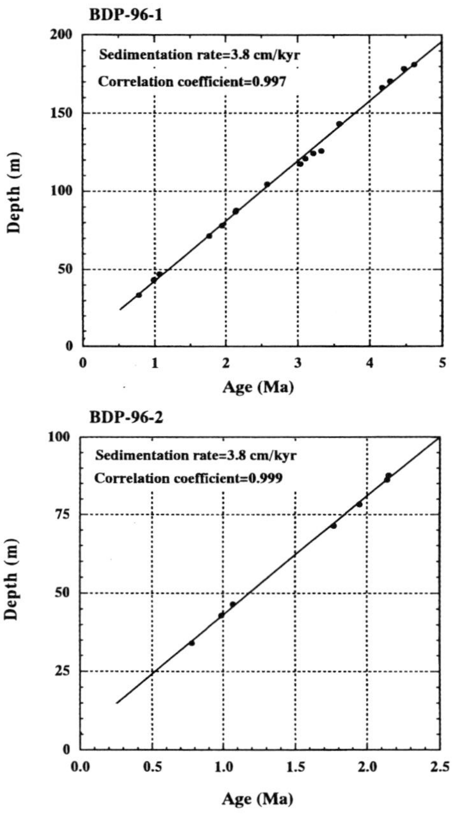

Fig. 5. Depth-age graphs of the polarity boundaries of geomagnetic events and epochs.

The upper graph is the diagram for BDP96-1, and the lower graph is for BDP96-2.

The geomagnetic polarity time scale by Cande and Kent (1995) has been used for reference.

Path-through measurement of QC samples

Path-through measurements were made on QC samples at 1 cm intervals and reveal its geomagnetic features in detail. We examined the QC data around the Brunhes/Matuyama (B/M) geomagnetic polarity boundary.

In Fig. 7, the inclination around the B/M boundary shows two normal/reverse transitions. That is, the polarity changes from Matuyama reverse to normal, then to reverse again, and finally back to Brunhes normal polarity. Such double transition phenomena at the B/M boundary

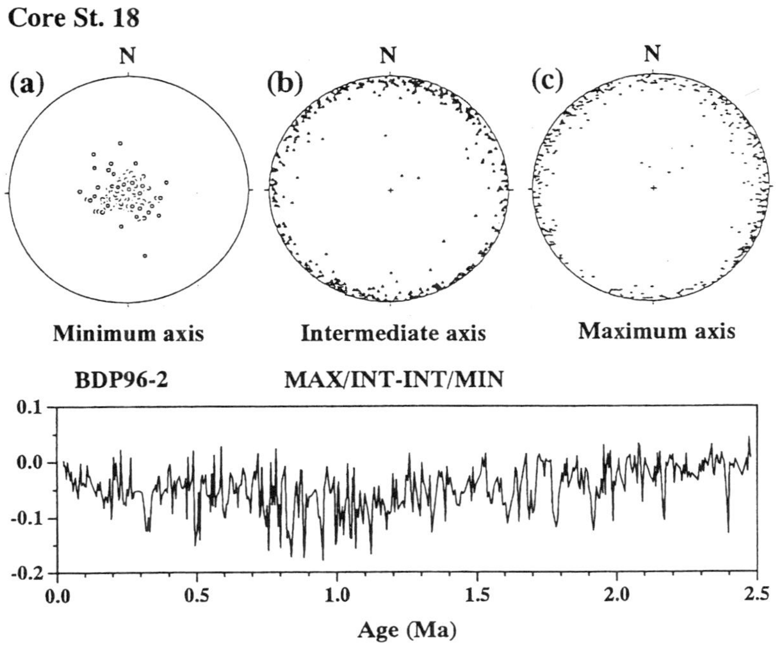

Fig. 6. The upper panel shows the distribution of the susceptibility anisotropy axes of the St. 18 sedimentary sequence. (a): minimum axis, (b): intermediate axis, (c): maximum axis. The lower panel shows the change in the AMS parameter (Max/Int-Int/Min) with age in BDP96-2.

have been reported in several studies (Jacobs, 1994). This QC study of BDP96 supports the existence of double polarity changes at the B/M boundary.

The greatest benefit of path-through long core measurements is that the data are continuous through minute measurement intervals, and further study will examine the possibility that QC data include formerly unknown geomagnetic events. However, when we analyzed the data around the region of core-breaks or disturbances, the path-through QC data yielded peculiar features. Therefore, path-through measurements on long cores should be used carefully by checking for disturbances in the core.

Changes in magnetic susceptibility as the environmental detector

Comparison with the content of diatom frustules and biogenic silica

Figure 8 shows the stratigraphic variation in magnetic susceptibility in short core St. 18 near BDP96. The figure also shows the fluctuations in the

Table 1 The age of geomagnetic polarity epochs, events, and the corresponding depth of BDP-96 cores. The geomagnetic polarity timescale of Cande and Kent (1995) is referred.

Polarity (Chron)	Boundary of Chron Subchron (Ma)		BDP-96-1 Discrete sample	BDP-96-2 Discrete sample
BRUNHES Normal			6.30m～	0.00m～
	B/M	(0.78)	33.48m/33.68m	33.96m/34.16m
	Jaramillo	(0.99)	43.27m/43.47m	42.88m/43.08m
		(1.07)	47.03m/47.23m	46.45m/46.65m
MATUYAMA	Olduvai	(1.77)	71.435m/71.615m	71.28m/71.48m
Reversed		(1.95)	78.025m/78.225m	78.20m/78.40m
	Reunion	(2.14)	86.955m/86.975m	86.07m/86.27m
		(2.15)	87.815m/87.835m	87.47m/87.87m
	Ma/Ga	(2.58)	104.27m/104.47m	
	Kaena	(3.04)	117.25m/117.45m	
GAUSS		(3.11)	120.61m/121.13m	
Normal	Mammoth	(3.22)	123.775m/124.525m	
		(3.33)	125.525m/126.04m	
	Ga/Gi	(3.58)	143.16m/143.36m	
	Cochiti	(4.18)	166.14m/166.34m	
GILBERT		(4.29)	170.40m/170.60m	
Reversed	Nunivak	(4.48)	178.48m/178.68m	
		(4.62)	181.14m/181.34m	

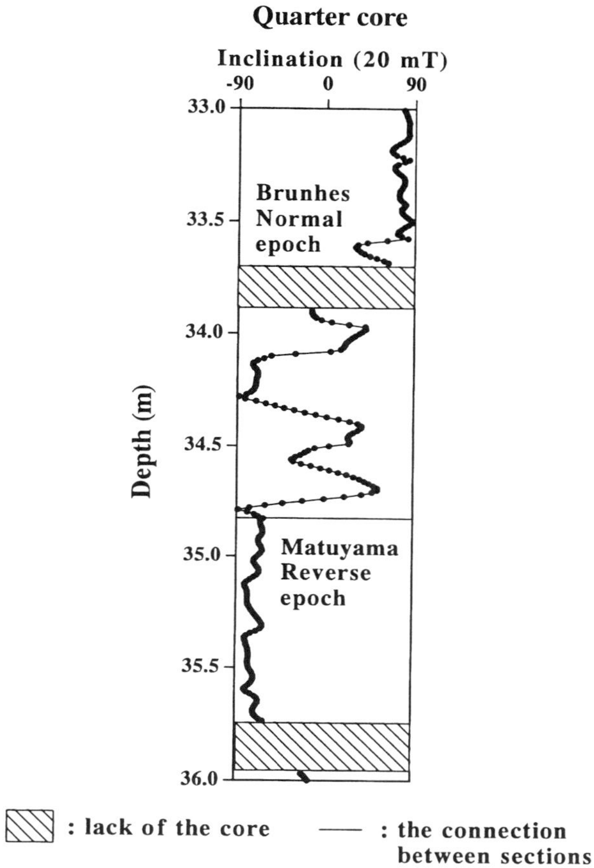

Fig. 7. Path-through inclination data of QC samples around the Brunhes/Matuyama boundary.

concentration of diatom frustules and fluctuations in the content of biogenic silica (bio-SiO_2), and the lithologic changes studied by Grachev et al. (1997) are also shown.

Magnetic susceptibility is low at depths where the sediment has high diatom frustule and bio-SiO_2 content. The fluctuations in diatom frustule and bio-SiO_2 content are an indicator of paleoclimatic change, and thus the fluctuations in magnetic susceptibility also serve as an indicator of paleoclimate. High susceptibility suggests a glacial (low diatom) period in Lake Baikal, and low susceptibility indicates an interglacial (high diatom) period.

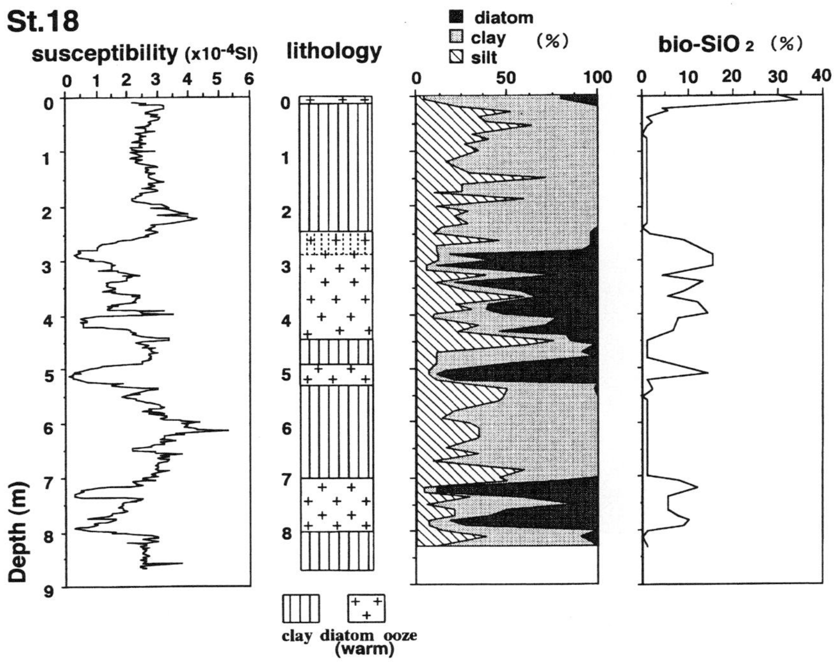

Fig. 8. Fluctuations in magnetic susceptibility, relative amount of diatom-clay-silt, and bio-SiO_2 content in core St.18 with depth at Academician ridge.

Comparison with iron content

Neutron activation analysis was conducted to study the fluctuations in iron content in the sediment of core St.18. Samples were collected at 10-cm intervals from 350 cm to 150 cm deep. In Fig. 9, the changes in iron content with depth are compared with the changes in magnetic susceptibility. There is a clear positive correlation between them, which suggests that the content of magnetic mineral is mainly responsible for the magnitude of susceptibility.

Around 280 cm deep (region-A), there was a distinct change in iron content, but the susceptibility changed little. Region-A corresponds to the boundary between glacial and interglacial sequences, and this may suggest intrusion of another mechanism on the correlation. Further study will be necessary to identify it.

Susceptibility change of BDP96

The upper panel in Fig. 10 shows the fluctuations in susceptibility with

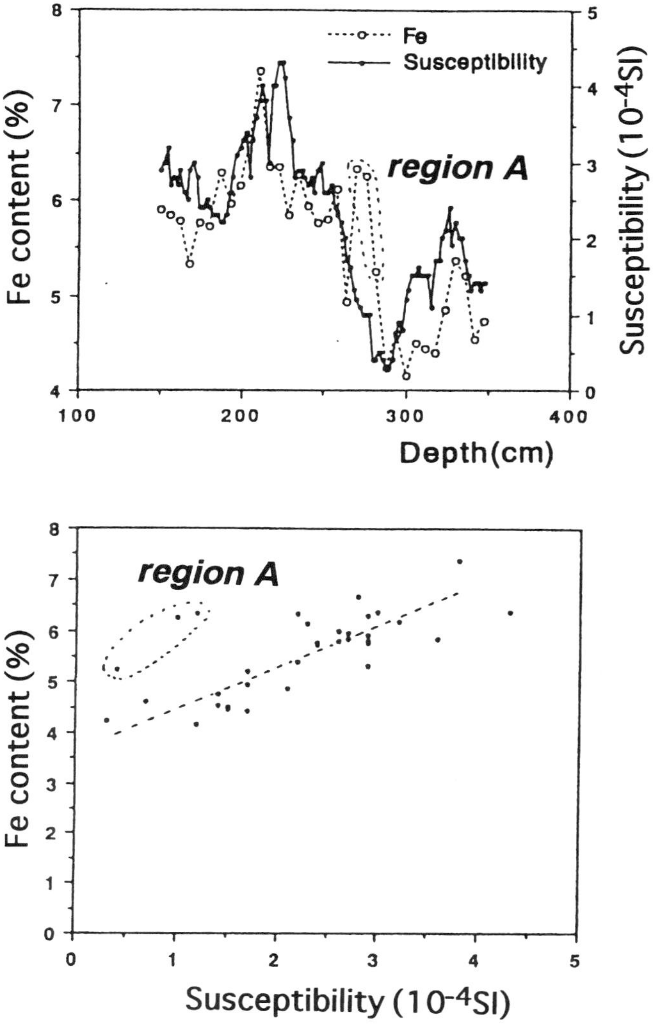

Fig. 9. Changes in the iron content and susceptibility of core St.18 from 350 cm to 150 cm deep. The lower figure shows the correlation between them (after Takamatsu, Sakai et al., 1997).

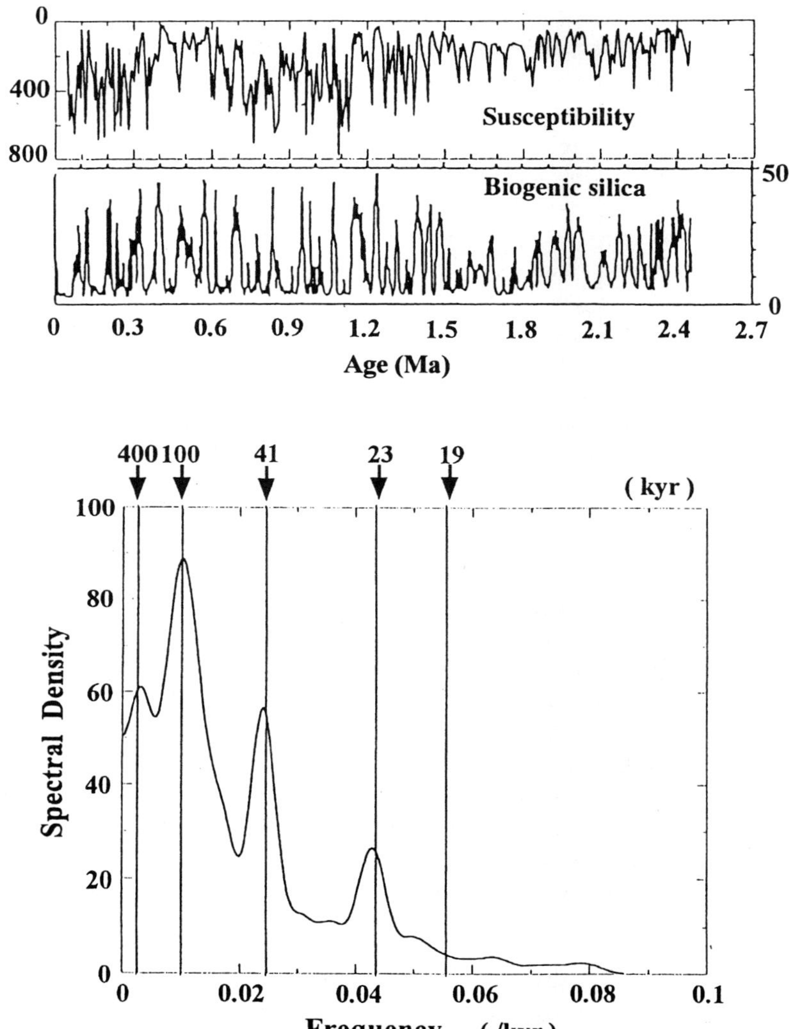

Fig. 10. Changes in susceptibility and biogenic silica with the depth of BDP96-2 (Williams et al., 1997) and the results of spectral analysis of susceptibility (Horii, 1999).

time of the BDP96-2 core (DC samples), and the middle figure shows the changes in the biogenic silica content of the sediment (Williams et al., 1997). An inverse correlation is seen between the two, similar to the correlation in Fig. 8. The lower panel shows the results of the spectral analysis of susceptibility, which reflects the distinct orbital Milankovitch cycle (Horii, 1999). These results indicate that the susceptibility changes in BDP96 are clearly related to global paleoclimatic changes.

The primary mechanism for the correlation between susceptibility and paleoclimate is thought to be as follows. The dilution of magnetic minerals during the interglacial period by the increase in biogenic mineral is responsible for the low susceptibility, and the increase in terrigenous flux with low biogenic mineral content during the glacial period causes the high susceptibility.

Susceptibility generally changes not only with fluctuations in the content of magnetic minerals, but with variations in species and the size of the magnetic minerals. The correlation between susceptibility and iron content in Fig. 9 suggests that the content of magnetic minerals is mainly responsible for the magnitude of susceptibility. We then examined the fluctuations in magnetic minerals with the changes in susceptibility by thermal demagnetization.

Thermal demagnetization

The thermal demagnetization analysis of isothermal remanent magnetization (IRM) was conducted on specimens prepared from the two regions: the specimens in group A taken from the interglacial period region where susceptibility was minimal, and the specimens in group B collected from the glacial period region with maximum susceptibility. The specimens in group A are specimen a (depth: 6.4 m), specimen c (14.19 m), and specimen e (19.71 m). The specimens in group B are specimen b (11.19 m), specimen d (16.41 m) and specimen f (42.93 m). These specimens were extracted from plastic cubes and coated with heat-resistant adhesives. After adequate drying for several days, IRM was achieved with a 0.2 T magnet. The thermal demagnetization experiment was carried out in a nitrogen atmosphere by stepwise heating from 100°C to 580°C in nine steps.

Figure 11 shows that the group A specimens with low susceptibility contain magnetic minerals whose magnetization drops at high temperatures (~580°C), whereas distinct decreases in the magnetization of the group B specimens with high susceptibility occurs at other temperatures around 350°C, in addition to 580°C. The same trend was observed in the experiments on the other specimens. These findings suggest that the differences in susceptibility between the glacial and interglacial periods may have been

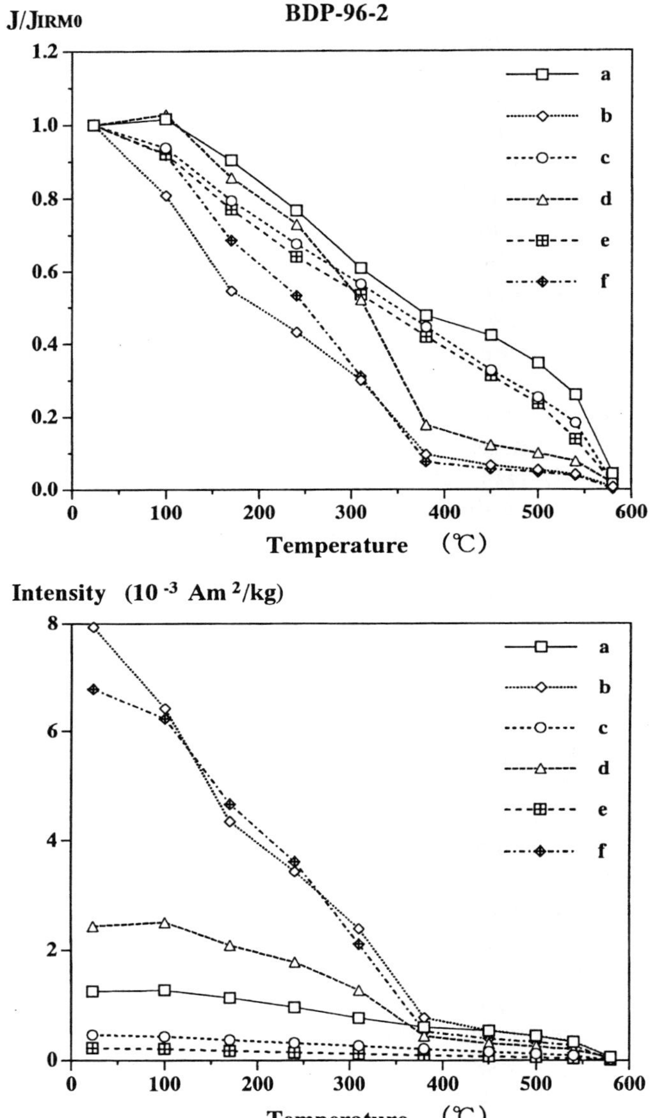

Fig. 11. Results of thermal demagnetization of BDP96 specimens. The lower panel shows the changes in intensity of remanent magnetization with temperature for each specimen. The upper panel shows the changes in relative remanent intensity normalized by the IRM intensities at 30°C (before heating).

associated with changes in the species and/or size of the magnetic minerals.

The supply of sediments in the lake may have two different origins, one being the terrigenous sediment from the river transportation system and the other being of biogenic origin. Since the Academician Ridge is of bathmetric high and away from the rivers, transpiration of terrigenous sediment brought by the river during the interglacial period is selected and limited. During the glacial period, the surface of the lake was covered with ice, and the terrigenous sediment may have been brought by the ice-rafting, giving rise to different magnetic minerals from the interglacial period.

Peck and King (1996) showed that the presence of magnetite could be traced to magnetotactic bacteria in Lake Baikal sediment. Magnetotactic bacteria have also been found in Antarctica (Funaki, private communication). The magnetotactic bacteria may be more active than other organisms (diatoms, etc.) in the glacial period, and magnetic minerals from the magnetotactic bacteria may be responsible for the remanent magnetization of sediment even in the glacial period. One interpretation of the differences between magnetic minerals in the glacial and interglacial periods is that sediment originated from ice-rafting contributes to magnetic mineral in the glacial period and that the magnetic minerals from magnetotactic bacteria are common to both periods.

Summary

Two BDP96 cores (BDP96-1 and BDP96-2) showed clear inclination reversals with depth. Comparison with the geomagnetic polarity timescale resulted in assignment of the sedimentary sequence of the 200 m long BDP96-1 core to the geomagnetic polarity epochs during the past 5 million years: the Brunhes, Matuyama, Gauss and Gilbert epochs. The sedimentation rate was estimated to be 3.8 cm/kyr by the least squares method based on the depth-age relationship. The fairly high correlation coefficient (0.997-0.999) of the depth-age relationship indicates that sedimentation at Academician ridge has been continuous in a quiet environment. This may be an important factor for the tectonic study of Lake Baikal.

Path-through measurements on quarter core samples around the B/M boundary showed the double polarity transitions. Path-through measurements are an effective means of investigating continuous magnetization of long cores, and it is necessary to examine disturbances in the core carefully.

The changes in magnetic susceptibility with time were inversely correlated with the changes in biogenic silica content, and spectrum analysis revealed clear Milankovitch orbital periodicities in the fluctuations in susceptibility. Susceptibility analysis makes it possible to study the paleoclimate, however, further study of the mechanism of the susceptibility changes in Lake Baikal associated with paleoclimate are needed.

Acknowledgements
We thank the BDP (Baikal Drilling Project) members from Russia, The United States, and Japan for their help in this study. We would like to express sincere gratitude to Professor M.I. Kuzmin, in particular, for support in the drilling and path-through study of the quarter core samples.

References

Baikal Drilling project II Members (1997) Continuous continental paleoclimate record for the last 4.5 to 5 million years revealed by leg II of Lake Baikal scientific drilling, EOS, 78(51), 597-604.

Cande S.C. and Kent. D.V. (1995) Revised calibration of the geomagnetic polarity timescale for the Late Cretaceous and Cenozoic, J. Geophys. Res., 100, B4, 6093-6095.

Grachev, M.A., Likhoshway, Ye.V., Vorobyova, S.S., Khlystov, O.M., Bezrukova, E.V., Veinberg, E.V., Goldberg, E.L., Granina, L.Z., Kornakova, E.G., Lazo, F.I., Levina, O.V., Letunova, P.P., Otinov,P.V., Pirog, V.V., Fedotov, A.P., Yaskevich, S.A., Bobrov, V.A., Sukhorukov, F.V., Rezchikov, V.I., Fedorin, M.A., Zolotaryov, K.V. and Kravchinsky, V.A. (1997) Signals of the paleoclimates of upper Pleistocene in the sediments of Lake Baikal, Russian Geology and Geophysics, 38, 957-980.

Grosswald, M. G. (1980) Late Weichselian ice sheet of Northern Eurasia, Quaternary Research 13, 1-32.

Horii, M. (1999) Paleomagnetic analysis during the past 2.5 million years by rock-magnetic measurement of sediments from Lake Baikal, Doctoral Thesis of Kanazawa University, 110 pp.

Jacobs, J.A. (1994) Reversals of the Earth's Magnetic Fields, Cambridge University press 187-192.

Peck, J.A. and King, J.W. (1996) Magnetofossils in the sediment of Lake Baikal, Earth and Planet. Sci. Lett., 140, 159-172.

Sakai, H., Nakamura, T., Horii, M., Kashiwaya, K., Fujii, S., Takamatsu, T. and Kawai, T. (1997) Paleomagnetic study with 14C dating analysis on three short cores from Lake Baikal, Bull. Nagoya Univ. Furukawa Museum, No.13, 11-22.

Williams, D.F., Peck, J., Karabanov, E.B., Pokopenko, A.A., Kravchinsky, V., King, J. and Kuzmin, M.I. (1997) Lake Baikal record of continental climate response to orbital insolation during the past 5 million years, Science, 278, 1114-1116.

Lake Baikal
K. Minoura (editor)
2000 Elsevier Science B.V.

Paleoclimatic signals printed in Lake Baikal sediments

Kashiwaya, K.[1]*, Tanaka, A.[2], Sakai, H.[3], and Kawai, T.[2]
[1]Department of Earth Sciences, Kanazawa University, Kakuma, Kanazawa 920-1192, Japan (kashi@kenroku.kanazawa-u.ac.jp)
[2]National Institute for Environmental Studies, Tsukuba, Ibaragi 350-0053, Japan (tanako@nies.go.jp; tkawai@nies.go.jp)
[3]Department of Earth Sciences, Toyama University, Toyama 930-8555, Japan (hsakai@sci.toyama-u.ac.jp)
(*corresponding author)

Abstract

Analyses of the physical properties (mean grain size and water content) and biogenic silica content in sediment cores (BDP96) from Academician Ridge in Lake Baikal have provided information on long-term fluctuations in environmental conditions, revealing that the continental interior has gradually cooled over the past 5 my with a characteristic periodicity of about 1.0 my. There are long periods around 1.0 my, 0.4 and 0.1 my in the datasets analyzed, which are related to the solar insolation. The 0.4 and 0.1 my are connected to eccentricity parameters (Milankovitch parameters). The 1.0-my period may also be related to the fluctuation in paleomagnetic intensity. Three intervals of cooling were found at about 2.6 - 2.8 my B.P., 1.7 - 2.0 my B.P., and 0.7 - 1.0 my B.P. These intervals correspond to the troughs in the 1.0 my period.

Introduction

Recent studies of Lake Baikal have helped to clarify the close relationship between climatic changes in continental interiors and global changes that are reflected in, for example, marine ^{18}O records (BDP members, 1995, 1997; Colman et al., 1995; Kashiwaya et al., 1997; 1998). Nevertheless, research on continental records has been limited in scope. In particular, longer and more detailed records from continental interiors are needed to understand the relationship between climatic factors such as terrestrial environments, oceanic conditions, and solar insolation. Lake Baikal is located in a crucial area for these studies (Short et al., 1991; BDP members, 1995), and recent studies of its sediments clearly indicate that paleoenvironmental changes in this part of Asia responded in a sensitive way to global climatic change and solar insolation (e.g., Colman et al., 1995; Kashiwaya et al., 1998). One of the major advantages to analyzing Lake Baikal sediments is that they comprise a long and continuous history, of unequaled scope, recording long-term environmental change in a continental interior. It is thought that sedimentation in the lake has been continuous since the Miocene, and that the entire lake remained uncovered by ice during the Pleistocene glacial periods (Grosswald, 1981).

The site selected for sampling these long records in the winter of 1996 was Academician Ridge, in central Lake Baikal (53°41'48"N, 108°31'00"E), a topographically isolated ridge with hemipelagic sediments and little direct fluvial input and turbidity flows. Two long cores (BDP96 Hole 1, BDP96 Hole 2) were obtained by Baikal Drilling Project members, consisting of American, German, Japanese, and Russian scientists.

In this report, we will discuss climatic signals recorded by mainly the physical properties of sediments in the two cores. The shorter core, BDP96 Hole 2, which is 100m in length, has 95% recovery and records continuous Pliocene-Pleistocene sedimentation over the past 2.5 million years. The data from this core are used here mainly to discuss comparatively short time periods. The longer core, BDP96 Hole 1, which is 200 m long, spans approximately the last 5.0 million years. It was 75% recovered and the upper 140-m segment records continuous Pliocene-Pleistocene sedimentation, so we utilized this upper part of the core (3.5 million years) for statistical analysis. The results from analysis of the upper 100m of this longer core are nearly the same as those from the shorter core, BDP96 Hole 2. Preliminary results from analyses of these cores were given by Kashiwaya et al. (1998, 1999b).

Data analyzed

Data dealt with here are mainly concerned with the physical properties (water content and mean grain size) of cores BDP96 Hole 2 and BDP96 Hole 1, while biogenic silica data from BDP96 Hole 1 (analyzed by A. Tanaka) are used for additional discussion. Subsamples for analyses of physical properties (water content, grain size, and grain particle density) were taken from each core at 20-cm intervals, about 500 subsamples from BDP96 Hole 2 and 800 from BDP96 Hole 1. Water content was determined by oven-drying, and grain size was measured using the laser reflection method (Shimadzu Sald 2000). Grain particle density was measured with an autopycnometer (Micromeritics Autopycnometer 1320). These physical properties (water content, grain size, and grain density) are closely related to biogenic (diatom) productivity and have been used as proxies for climato-limnological fluctuations in Lake Baikal sediments, especially for sediments on the Academician Ridge (Kashiwaya et al., 1999a). A close linkage between biogenic productivity and climatic change has also been noted here by Qiu et al. (1993), Carter and Colman, (1994). Colman et al.

(1995). and Grachev et al. (1997).

Age scaling

The age scale used here is based on magnetic epochs (H. Sakai et al., personal communication; ages after Cande and Kent, 1995); magnetic polarities and linear interpolation were used for age-scaling. The age scale was modified for core BDP96 Hole 2 by tuning the change in water content to 65°N July insolation (Berger and Loutre, 1991) and benthonic ^{18}O at ODP 677 (Shackleton et al., 1990), because the BDP96 Hole 2 data are dense overall and changes in water content are closely related to those in the ^{18}O values (Kashiwaya et al., 1998). Before these adjustments were made, however, some statistical tests were performed using only the magnetic polarity time scale. Three orbital parameters (a 100-ky period due to eccentricity parameters, a 40-ky period of obliquity, and a 20-ky period of precessional parameter) were found, suggesting that differences between the two age scales are too small to take into account in our discussion of longer overall trends. Age scales could not be obtained for the uppermost regions of both cores. Thus, the age scales were estimated using short cores obtained from the lake floor in 1997 from nearly the same geographical location.

Analytical results for the BDP96-2

We will first discuss results for the shorter core, BDP96 Hole 2. As has been shown (Kashiwaya et al., 1998), water content is high and the mean grain size is large for interglacial sediments, suggesting that the size of diatom tests and their gap were comparatively large during such periods, and show a large shift in fluctuations during the late Pleistocene. The results also show that there are shifts in fluctuations, at about 0.8 my B.P. and 1.7 - 2.0 my B.P., in the 100-ky band-pass-filtered curves, and at about 1.0 my B.P. in the 400-ky band-pass-filtered curves.

One result of spectral analysis (Barrodale and Ericsson, 1980) for periods of insolation longer than 70 ky (0 - 2.5 my B.P.) is shown in Figure 1, which shows distinct periods around 400 ky and 100 ky that are related to eccentricity parameters. We used the 65°N July insolation given by Berger and Loutre (1991) for our calculations. As mentioned above, there seem to have been changes (shifts) in the climato-limnological oscillations. Therefore, we will checked the magnitude of long periods related to eccentricity parameters, in order to clarify changes in solar insolation. The magnitude of these periods is assumed to be expressed as amplitude. Harmonic analysis is employed for the dominant periods that were obtained from spectral analysis. Four time domains (0 - 1.0 my B.P., 0.5 - 1.5 my B.P., 1.0 - 2.0 my B.P., and 1.5 - 2.5 my B.P.) were analyzed, and calculated results (Table 1) indicate no clear differences between the domains, although amplitudes around the 100-ky period are somewhat large for the 1.0 - 2.5 my B.P. interval, while amplitudes around the 400-ky period are somewhat large for the 0 - 1.5 my B.P. interval.

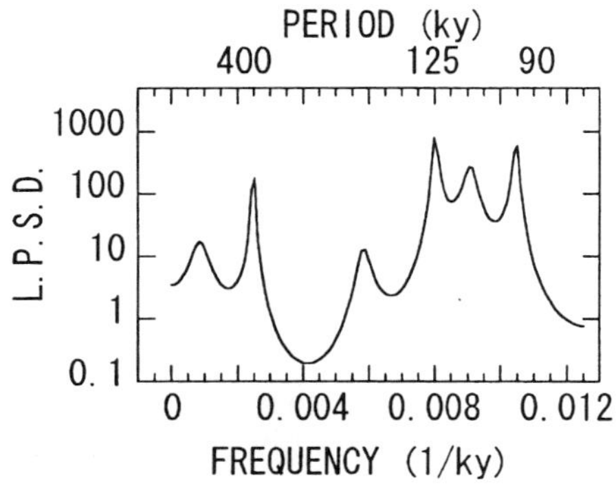

Figure 1. Spectral analysis for periods longer than 70 ky in the insolation (0 - 2.5 my B.P.).

Next, we will examine changes in amplitude of the climato-limnological oscillations (water content and mean grain size) over long periods. Equally spaced data (2,000-year intervals) are given with an interpolation method for statistical analyses. Spectral analyses of oscillations in the 0 - 2.5 my B.P. interval are shown in Figure 2. A 70- to 700-ky band pass filter was used for calculations, to make the 400-ky and 100-ky periods distinct. This figure shows a period around 200 ky, in addition to the 400-ky and 100-ky periods, that is related to the eccentricity parameter, and which may simply be a doubling of the 100-ky periods. Temporal changes in the amplitudes of the periods for the four time domains have been examined using harmonic analysis. The results are shown in Tables 2 and 3. For mean grain size, amplitudes around the 100-ky periods are comparatively large, and the 400-ky period is not clearly present in the most recent stage (0 - 1.0 my B.P.), while amplitudes around the 400 ky periods are large and those around the 100-ky periods are somewhat smaller in the older stage (1.0 - 2.5 my B.P.). Regarding water content fluctuations, amplitudes around the 100-ky periods are large for the recent stage (0 - 1.0 my B.P.), and gradually become small from the middle stage (0.5 - 2.0 my B.P.) to the old stage (1.5 - 2.5 my B.P.), while amplitudes around the 400-ky periods do not fluctuate significantly in any of the stages. These indicate that changes of the long periods in the climato-limnological environment do not always respond linearly to changes in insolation. It is well known that climatic oscillations related to the 100-ky periods respond non-linearly to the eccentricity parameters of insolation (e.g., Imbrie et al., 1993) and that climatic oscillations had large amplitudes in the late Pleistocene. Analytical results obtained here for the 400-ky and 100-ky periods also show their non-linear

Table 1. Amplitudes and phases from harmonic analysis of prevailing periods obtained from spectral analysis for insolation (65°N July). (a) 0 - 1.0 my B.P., (b) 0.5 - 1.5 my B.P., (c) 1.0 - 2.0 my B.P., and (d) 1.5 - 2.5 my B.P. DC refers to an average (w/m^2), and parentheses indicate errors.

(a) DC= 440.5

PERIOD (ky)	AMPLITUDE	PHASE
407.2	0.674 (0.0317)	184.3 (3.03)
212.5	0.041 (0.0315)	8.8 (26.03)
120.2	0.645 (0.0317)	111.7 (0.94)
108.9	0.339 (0.0316)	103.8 (1.63)
97.3	0.617 (0.0313)	9.1 (0.78)

(b) DC= 440.7

PERIOD (ky)	AMPLITUDE	PHASE
449.7	0.761 (0.0516)	108.5 (5.13)
209.8	0.250 (0.0537)	36.7 (7.03)
118.4	0.173 (0.0539)	16.5 (5.86)
107.5	0.498 (0.0546)	88.8 (1.89)
93.5	0.722 (0.0541)	35.5 (1.12)

(C) DC= 440.7

PERIOD (ky)	AMPLITUDE	PHASE
398.1	0.575 (0.0347)	214.7 (3.88)
178.8	0.108 (0.0350)	28.7 (9.29)
122.1	0.514 (0.0354)	7.9 (1.33)
106.8	0.424 (0.0358)	3.5 (1.45)
98.4	0.893 (0.0354)	77.9 (0.63)

(D) DC= 440.4

PERIOD (ky)	AMPLITUDE	PHASE
369.7	0.547 (0.0589)	340.9 (6.16)
201.1	0.225 (0.0575)	183.8 (8.19)
119.2	0.502 (0.0587)	59.9 (2.12)
103.7	0.285 (0.0585)	70.3 (3.37)
92.5	0.839 (0.0578)	81.1 (1.01)

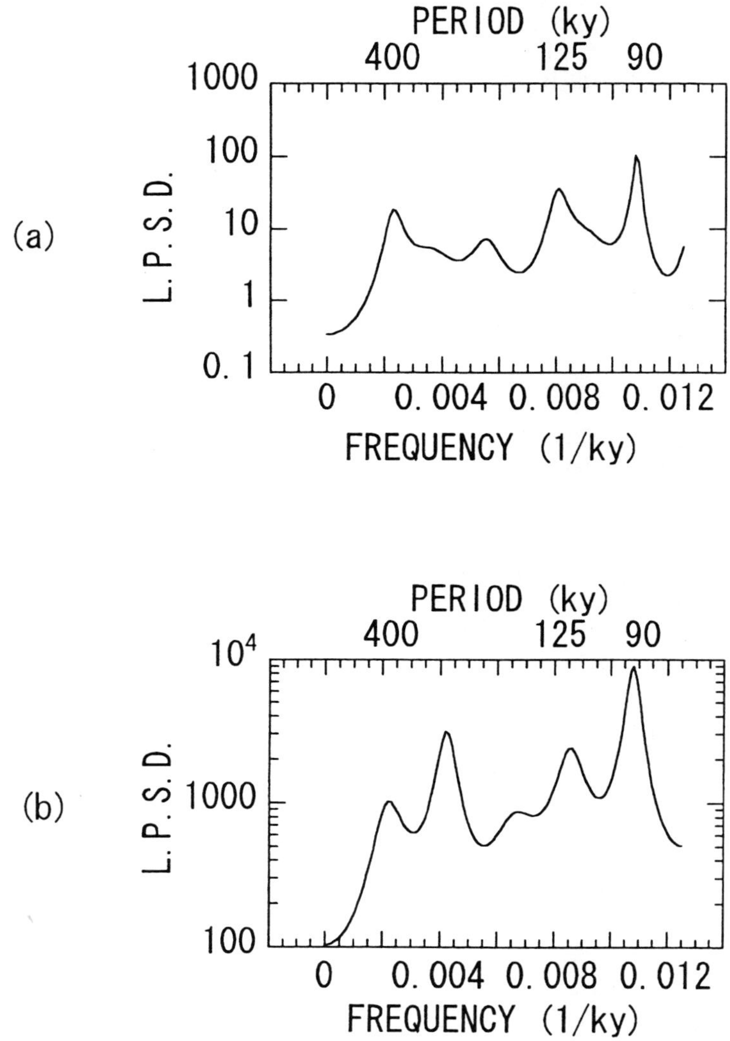

Figure 2. Spectral analysis for periods from 70 ky to 700 ky in (a) mean grain size and (b) water content of core BDP96-2 (0 - 2.5 my B.P.).

response to insolation. It is necessary to obtain additional detailed data and to thoroughly discuss the relationship among them, including phase lags, to clarify the causal mechanism.

Table 2. Amplitudes and phases from harmonic analysis of prevailing periods obtained from spectral analysis for grain size (mean) of core BDP96-2. (a) 0 - 1.0 my B.P., (b) 0.5 - 1.5 my B.P., (c) 1.0 - 2.0 my B.P., and (d) 1.5 - 2.5 my B.P. DC refers to an average (ϕ), and parentheses indicate errors.

(a) DC= 7.295

PERIOD (ky)	AMPLITUDE	PHASE
339.7	0.0746 (0.0123)	119.6 (9.01)
157.9	0.0748 (0.0134)	124.7 (4.47)
141.6	0.0962 (0.0135)	58.8 (3.20)
119.1	0.1652 (0.0126)	33.5 (1.45)
97.1	0.3862 (0.0127)	68.1 (0.51)
85.5	0.1527 (0.0127)	40.7 (1.13)
77.4	0.1145 (0.0125)	54.9 (1.35)

(b) DC= 7.265

PERIOD (ky)	AMPLITUDE	PHASE
413.7	0.0561 (0.0164)	59.3 (19.21)
147.9	0.0662 (0.0268)	124.1 (9.70)
137.9	0.2416 (0.0300)	92.4 (2.86)
126.9	0.2367 (0.0238)	103.6 (2.02)
108.5	0.1318 (0.0169)	20.7 (2.23)
94.9	0.1782 (0.0168)	66.8 (1.44)
80.4	0.2127 (0.0164)	76.6 (0.98)

(C) DC=7.109

PERIOD (ky)	AMPLITUDE	PHASE
370.3	0.1049 (0.0166)	254.8 (9.69)
156.5	0.0708 (0.0173)	155.4 (6.12)
129.1	0.1403 (0.0183)	37.6 (2.68)
115.6	0.3310 (0.0205)	78.6 (1.16)
108.2	0.1001 (0.0211)	107.1 (3.62)
95.5	0.1307 (0.0173)	27.7 (2.01)
84.9	0.1796 (0.0173)	13.3 (1.30)

(D) DC= 6.999

PERIOD (ky)	AMPLITUDE	PHASE
583.8	0.0754 (0.0127)	505.8 (15.70)
387.2	0.1590 (0.0126)	238.1 (4.87)
173.1	0.2453 (0.0125)	22.8 (1.40)
139.0	0.2307 (0.0126)	32.3 (1.20)
122.9	0.1735 (0.0123)	107.0 (1.39)
106.6	0.1842 (0.0129)	24.8 (1.19)
98.5	0.0943 (0.0127)	30.1 (2.15)

Table 3. Amplitudes and phases from harmonic analysis of prevailing periods obtained from spectral analysis for water content of core BDP96-2. (a) 0 - 0.5 my B.P., (b) 0.5 - 1.0 my B.P., (c) 1.0 - 1.5 my B.P., (d) 1.5 - 2.0 my B.P., and (e) 2.0 - 2.5 my B.P. DC refers to an average (%), and parentheses indicate errors.

(a) DC= 45.86

PERIOD (ky)	AMPLITUDE	PHASE
503.2	1.505 (0.444)	24.2 (20.62)
434.7	1.295 (0.421)	296.3 (21.71)
231.1	0.884 (0.198)	65.1 (8.66)
144.3	2.359 (0.200)	109.3 (1.95)
112.0	1.933 (0.200)	91.7 (1.98)
97.4	4.651 (0.196)	16.8 (0.65)
75.9	2.735 (0.195)	30.8 (0.86)

(b) DC= 44.46

PERIOD (ky)	AMPLITUDE	PHASE
429.7	1.563 (0.253)	248.5 (10.69)
335.8	1.769 (0.249)	214.0 (7.48)
214.3	2.383 (0.221)	110.8 (3.24)
128.1	1.493 (0.224)	33.0 (3.00)
108.5	1.999 (0.224)	68.7 (1.94)
95.6	2.952 (0.220)	9.8 (1.13)
79.9	1.969 (0.216)	50.0 (1.40)

(C) DC=43.42

PERIOD (ky)	AMPLITUDE	PHASE
365.1	2.219 (0.264)	87.0 (6.79)
321.5	1.258 (0.256)	125.7 (10.24)
232.7	2.013 (0.190)	54.6 (3.16)
204.8	1.415 (0.180)	159.4 (4.40)
119.9	2.455 (0.144)	80.7 (1.11)
93.7	2.111 (0.142)	18.1 (1.01)
84.9	2.205 (0.142)	54.8 (0.87)

(D) DC= 42.79

PERIOD (ky)	AMPLITUDE	PHASE
389.7	1.365 (0.116)	18.2 (5.36)
184.8	1.536 (0.246)	143.7 (4.39)
174.1	1.317 (0.234)	121.2 (4.72)
146.1	1.127 (0.124)	140.5 (2.47)
126.5	1.961 (0.117)	119.4 (1.20)
97.4	0.603 (0.116)	1.0 (3.02)

Summary for the core BDP96-2

Long-term changes in the climato-limnological environment inferred from physical properties of the 100-m core (BDP96-2) coincide with global climatic change without a notable time lag. Milankovitch parameters were also imprinted in the sediments over the past 2.5 my: the 400 ky period and the 100 ky period are both related to eccentricity parameters. Amplitudes around the 100-ky periods are large in the recent stage (0 - 1.0 my B.P.), gradually become smaller in the middle stage (0.5 - 2.0 my B.P.), and are comparatively small in the old stage (1.5 - 2.5 my B.P.), whereas amplitudes around the 400-ky periods do not fluctuate significantly at any stage.

Analytical results for the longer core BDP96-1

Next, we will present results for the longer core, BDP Hole 1, which includes the Pliocene. Long Pliocene-Pleistocene records are very valuable for detecting long-term changes and shifts in climatic conditions, and for discussing their causes, the influence of solar insolation and other factors. such as paleomagnetic conditions, on the climato-limnological environment in a deep continental interior. Preliminary results suggest that long-term climato-limnological fluctuations may be related to both solar insolation and paleomagnetic intensity (Kashiwaya et al., 1999b). Here, we will discuss some long-term fluctuations in climato-limnological environment and longer periods, including the 0.4-my (400-ky) period due to eccentricity parameters.

First, let us consider long-term trends in the climato-limnological environment over the past 5.0 my. Figure 3 shows the original data for mean grain size, water content and biogenic SiO_2 content. As shown previously (Grachev et al., 1997, Kashiwaya et al., 1998, 1999a), these parameters can serve as proxies for climatic conditions; large grain size, high water content and biogenic SiO_2 content indicate warm periods, and vice versa. Dotted lines in the figure indicate a calculated trend for each dataset, suggesting gradual cooling over the past 5.0 my, which coincides with the global tendency found in oceanic data.

As noted above, the upper part of the core (0-140m: 0 - 3.5 my B.P.) was utilized for statistical analysis, because data for the lower part of the core (140-200m, 3.5 - 5.0 my B.P.) are sparse. Equally-spaced data points for statistical analysis (5,000-year intervals) were obtained by interpolation. A high pass filter (- 750 ky period) (Ormsby, 1966) was used to check the trend and longer period in mean grain size, water content and biogenic SiO_2 fluctuations. The results are shown in Figure 4. The solid curves in the figure show filtered fluctuations, suggesting that a gradually cooling occurred over a long period of about 1.0 my, and that there were three troughs (peaks of cooling) at about 0.7 - 1.0 my B.P., 1.6 - 1.8 my B.P. and 2.6 - 2.8 my B.P. We will discuss these intervals later, because these may be related to changes in environmental regimes during the Pliocene-Pleistocene. Longer periods (- 350 ky) were checked with spectral and

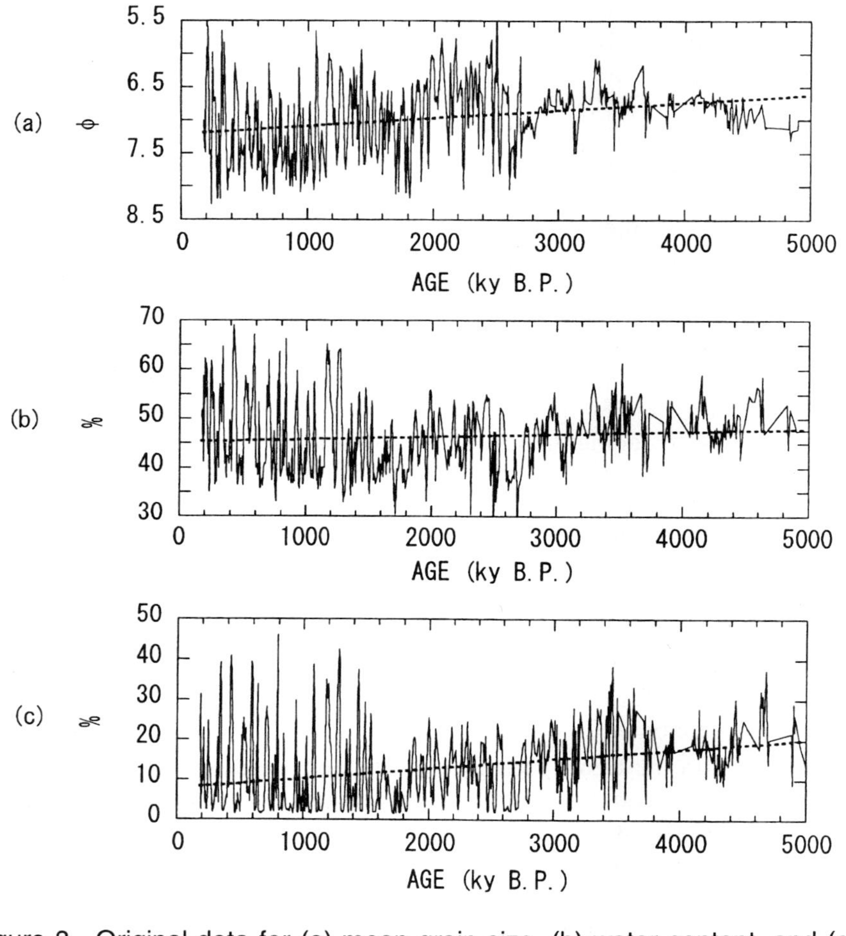

Figure 3. Original data for (a) mean grain size, (b) water content, and (c) biogenic SiO_2 content (thin solid line) of core BDP96-1. Dotted lines in the figure indicate calculated trends.

(a)
(b)
(c)

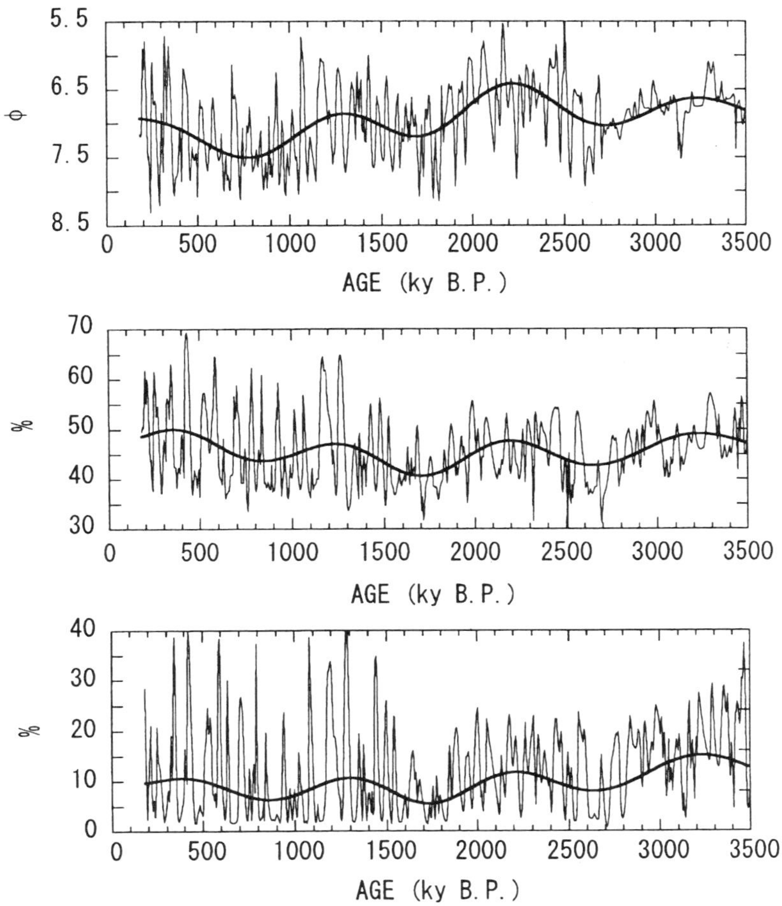

Figure 4. Interpolated datasets of (a) mean grain size, (b) water content, and (c) biogenic SiO_2 content (thin solid line) of core BDP96-1. Thick solid lines indicate filtered results (-0.75-my low-pass filter).

harmonic analyses. The results for the three datasets are shown in Table 4. As expected, strong periods around 1.0 my are found in all datasets. Periods related to eccentricity (410 ky) are also present in this interval, including the 500- and 700-ky periods.

Discussion of results for the BDP96-1 core

As mentioned above, strong periods of around 1.0 my occurred during

Table 4. Amplitudes and phases from harmonic analysis of prevailing periods obtained from spectral analysis for (a) mean grain size, (b) water content, and (c) biogenic SiO_2 content of core BDP96-1. DC refers to an average, and parentheses indicate errors.

(a) DC= 6.956 (ø)

PERIOD (ky)	AMPLITUDE	PHASE
993.0	0.266 (0.0117)	729.7 (7.00)
682.0	0.151 (0.0327)	444.1 (23.51)
653.6	0.127 (0.0323)	247.1 (26.52)
470.6	0.111 (0.0118)	339.8 (7.82)
414.7	0.096 (0.0114)	194.6 (8.01)

(b) DC= 45.70 (%)

PERIOD (ky)	AMPLITUDE	PHASE
1073.4	0.633 (0.0762)	926.8 (20.17)
890.3	3.016 (0.0989)	449.8 (4.10)
769.4	1.475 (0.0840)	331.1 (6.91)
529.6	2.650 (0.0818)	218.1 (2.68)
494.1	1.868 (0.0813)	32.1 (3.33)
369.7	0.245 (0.0536)	142.7 (12.89)

(C) DC=12.28 (%)

PERIOD (ky)	AMPLITUDE	PHASE
1157.4	1.288 (0.0768)	1081.0 (10.74)
844.6	2.170 (0.0954)	565.8 (5.67)
744.1	2.085 (0.0924)	510.5 (5.61)
496.9	0.971 (0.0755)	411.2 (6.07)
419.7	1.316 (0.0758)	339.7 (3.85)
346.2	0.669 (0.0744)	204.0 (6.17)

the past 3.5 my. The troughs appear to be cold intervals. The 0.7 - 1.0 my B.P. interval corresponds to the mid-Pleistocene change in climatic regime, the initiation of full glaciation in the Pleistocene (e.g., Maasch, 1988). The 1.7 - 2.0 my B.P. interval may be related to a large environmental change (i.e., the Tertiary-Quaternary boundary). The 2.6 - 2.8 my B.P. interval coincides with the beginning of cooling at about 2.7 my B.P. (e.g., Kukla et al., 1987), which is discussed for other datasets from the BDP core samples (e.g., Müller et al., 1998). As suggested previously (Kashiwaya et al., 1999b), other factors may have influenced climatic fluctuations. One of them is solar insolation. The period around 1.0 my has not been examined thoroughly, although discussion of a 0.4-my (400-ky) period has increased recently (e.g., Clemens and Tiedemann, 1997), and the 100-ky period is an important aspect of climatic change (e.g., Imbrie et al., 1993). One reason for this lack of discussion has been that long sampled intervals with high-resolution data were not available for discussion until now.

Here, we will discuss the long-term cycles of about 1.0 my and 0.4 my in duration. We used two numerical filters to clarify periodicity. One is a 0.75- to 1.5-my band pass filter and another a 0.35- to 0.50-my band pass filter. The results, applied to the three datasets (mean grain size, water

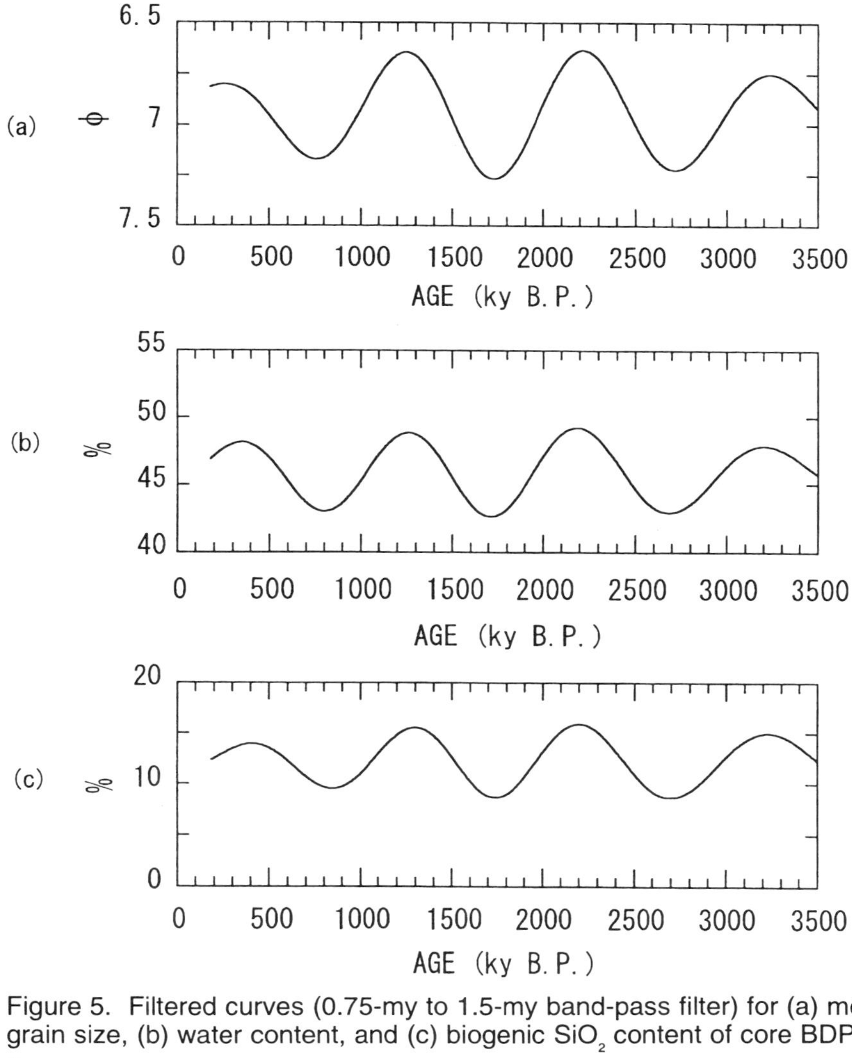

Figure 5. Filtered curves (0.75-my to 1.5-my band-pass filter) for (a) mean grain size, (b) water content, and (c) biogenic SiO_2 content of core BDP96-1.

content and biogenic SiO_2), are shown in Figures 5 and 6. Both figures show that the datasets synchronize with one another: in the 0.75- to 1.5-my band-pass-filtered datasets, they synchronously fluctuate with constant amplitude, while in the 0.35- to 0.50-my band-pass-filtered datasets, they

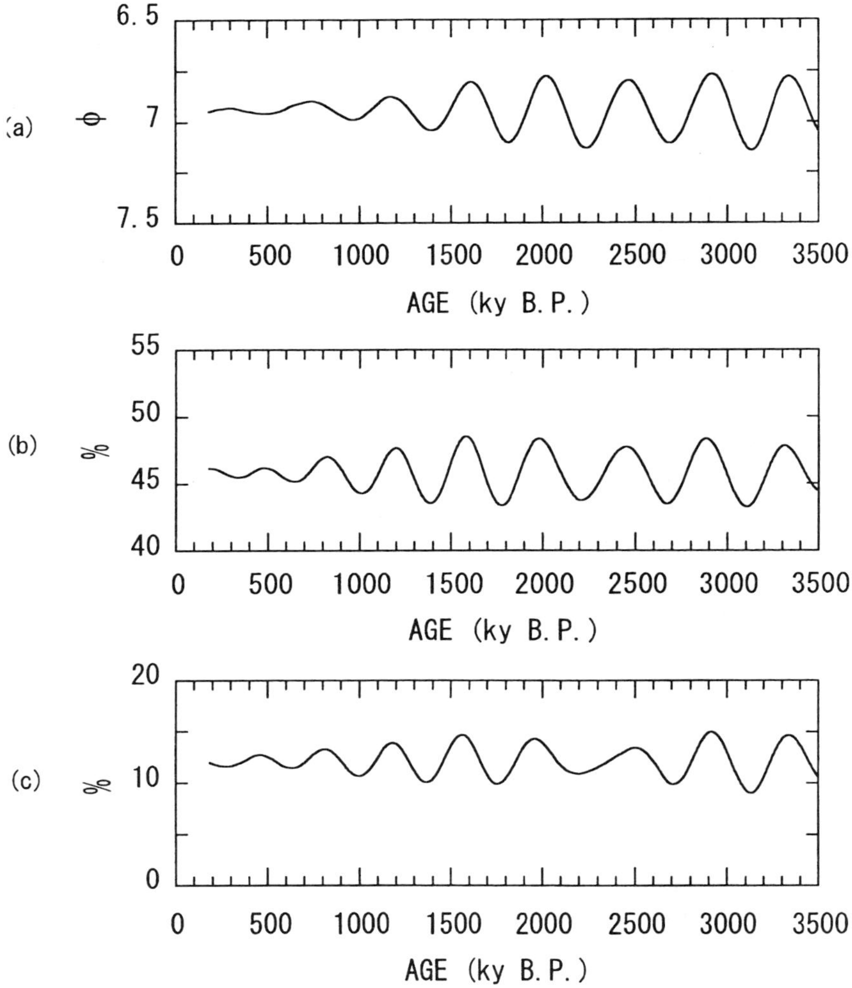

Figure 6. Filtered curves (0.35-my to 0.5-my band-pass filter) for (a) mean grain size, (b) water content, and (c) biogenic SiO_2 content of core BDP96-1.

begins to weaken (amplitudes become comparatively small) after 0.8 - 1.0 my B.P.

The same filters were also applied to solar insolation (65°N July insolation), and Figure 7 (a and b) shows the calculated results. There are only slight phase lags between the 0.75- to 1.5-my band-pass-filtered insolation

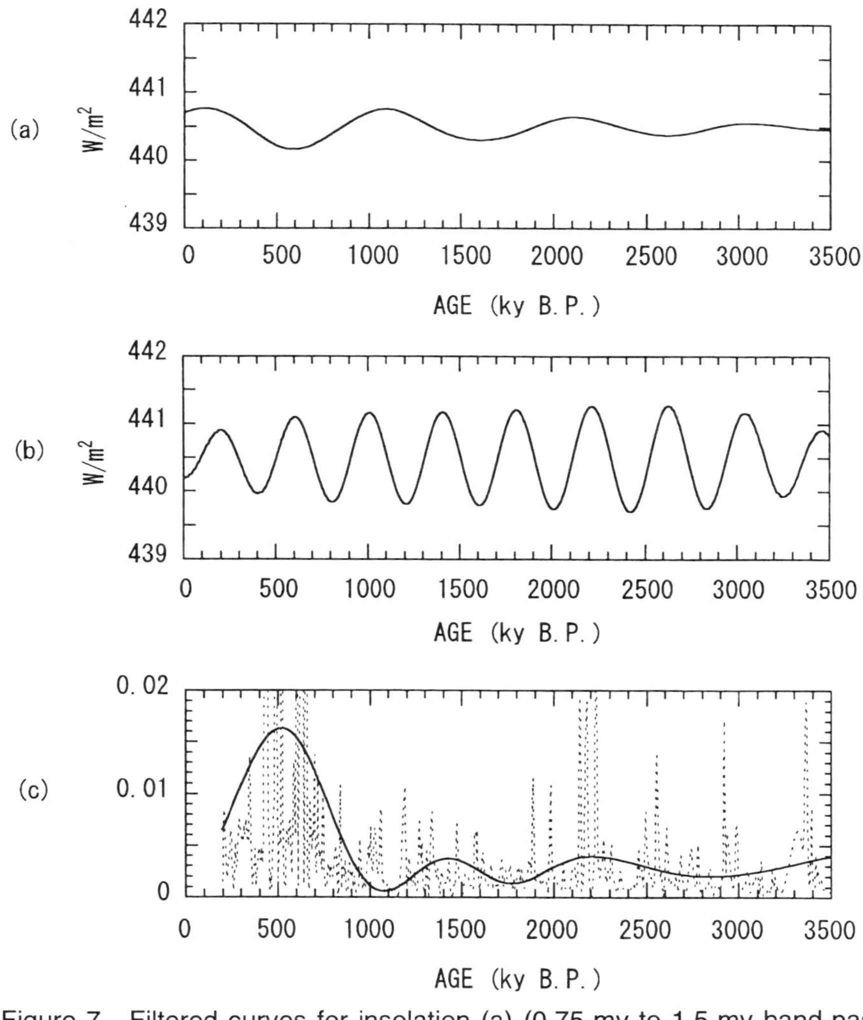

Figure 7. Filtered curves for insolation (a) (0.75-my to 1.5-my band-pass filter), (b) (0.35-my to 0.5-my band-pass filter) and (c) relative geomagnetic field intensity of core BDP96-1 (NRM/SIRM, dotted line) and filtered curve (thin solid line, 0.75-my to 1.5-my band-pass filter).

and datasets from BDP96 Hole 1, while about half of the phase lags are seen in the 0.35- to 0.50-my band-pass-filtered fluctuations. These suggest a close relationship between solar insolation and environmental changes from the Lake Baikal datasets, although causal mechanisms must be debated further. Another possible factor that may have influenced long-term environmental changes is geomagnetic field intensity (Kashiwaya et al., 1999b), although it is very difficult to properly obtain paleo-magnetic intensity. Here, the data obtained from the same core samples (BDP96 Hole 2) (NRM/SIRM) are assumed to reflect relative geomagnetic field intensity, shown in Figure 7c, which implies that it may have some relationship to other factors. This aspect of the research also awaits further data and detailed discussion.

Summary for the BDP96-1

Long-term fluctuations in environmental conditions represented by the physical and chemical properties of the 200-m core (BDP96 Hole 1) suggest that the continental interior has gradually cooled over the past 5 my and that the cooling was characterized by periodicity. There were long periods of cooling around 1.0 my and 0.4 my in the datasets analyzed, and both may be related to solar insolation. The 1.0-my period may also be related to a fluctuation in paleomagnetic intensity. Three intervals of cooling were found, at about 2.6 - 2.8 my B.P., 1.7 - 2.0 my B.P., and 0.7 - 1.0 my B.P.

Acknowledgements
The authors would like to thank the BDP Leg II members from Russia, the United States, Japan and Germany for collecting the BDP-96 sediment cores. We also thank Mr. M. Ryugo, who helped analyze the physical properties of the cores, and Dr. M. Horii and Mr. S. Nomura, who helped with the analysis of the paleomagnetic factors. Finally, we wish to thank all of our colleagues at the Hydro-geomorphological Laboratory, Kanazawa University, for their help and advice.

References

Baikal Drilling Project Members, 1995, Results of the first drilled borehole at Lake Baikal near the Buguldeika Isthmus. Russian geology and geophysics, 36(2), 3-32.

Baikal Drilling Project Leg II Members, 1997, Continuous continental paleoclimate record for the last 4.5 to 5 million years revealed by leg II of Lake Baikal scientific drilling. EOS, 78(51), 597-604.

Barrodale I. and R.E. Ericksson, 1980, Algorithm for least-square linear

prediction and maximum entropy spectral analysis, part 1. Theory. Geophysics, 45, 420-432.
Berger A. and M.F. Loutre, 1991, Insolation values for the climate of the last 10 million years. Quaternary Science Reviews, 10, 297-317.
Cande S.C. and D.V. Kent, 1995, Revised calibration of the geomagnetic polarity timescale for the Late Cretaceous and Cenozoic. Jour. Geophys. Res., 100, 6093-6095.
Carter S.J. and S.M. Colman, 1994, Biogenic silica in Lake Baikal sediments: results from 1990 -1992 American Cores. Jour. Great Lake Res., 20, 751-760.
Clemens S.C. and R. Tiedemann, 1997, Eccentricity forcing of Pliocene-early Pleistocene climate revealed in a marine oxygen-isotope record. Nature, 385, 801-804.
Colman S.M., J.A. Peck, E.B. Karabanov, S.J. Carter, J.P. Bradbury, J.W. King and D.F. Williams, 1995, Continental climate response to orbital forcing from biogenic silica records in Lake Baikal. Nature, 378, 769-771.
Grosswald M.G., 1980, Late Weichselian ice sheet of northern Eurasia. Quternary Research, 13, 1-32.
Grachev M.A., Ye.V. Likhoshway, S.S. Vorobyova, O.M. Khlystov, E.V. Bezrukova, E.V. Veinberg, E.L. Goldberg, L.Z. Granina, E.G. Kornakova, F.I. Lazo, O.V. Levina, P.P. Letunova, P.V. Otinov, V.V. Pirog, A.P. Fedotov, S.A. Yaskevich, V.A. Bobrov, F.V. Sukhorukov, V.I. Rezchikov, M.A. Fedorin, K.V. Zolotaryov and Kravchinsky, V.A., 1997, Signals of the paleoclimates of upper Pleistocene in the sediments of Lake Baikal. Russian. geology and geophysics, 38, 957-980.
Imbrie J., A. Berger, E.A. Boyle, S.C. Clemens, A. Duffy, W.R. Howard, G. Kukla, J. Kutzbac h, D.G. Martinson, A. McIntyre, A.C. Mix, B. Molfino, J.J. Morley, L.C. Peterson, N.G. Pisias, W.L. Prell, M.E. Raymo, N.J. Shackleton and J.R. Toggweiler, 1993, On the structure and origin of major glaciation cycles, 2. the 100,000-year cycle. Paleoceanography, 8, 699-735.
Kashiwaya K., T. Nakamura, N. Takamatsu, H. Sakai, N. Nakamura and T. Kawai, 1997, Orbital signals found in physical and chemical properties of bottom sediments from Lake Baikal. Journal of Paleolimnology, 14, 293-297.
Kashiwaya K., M. Ryugo, H. Sakai and T. Kawai, 1998, Long-term climato-limnological oscillation during the past 2.5 million years printed in Lake Baikal sediments. Geophysical Research Letters, 25, 659-663.
Kashiwaya K., M. Ryugo, M. Horii, H. Sakai, T. Nakamura and T. Kawai, 1999a, Climato-limnological signals during the past 260,000 years in physical properties of bottom sediments from Lake Baikal. Journal of Paleolimnology, 21, 143-150.

Kashiwaya K., H. Sakai, M. Ryugo, M. Horii and T. Kawai, 1999b, Long-term climato-limnological cycles found in a 3.5-million-year continental record. Journal of Paleolimnology. (to be submitted).
Kukla G., 1987, Loess stratigraphy in central China. Quaternary Science Review, 6, 191-219.
Maasch K.A., 1988, Statistical detection of the mid-Pleistocene transition. Climate dynamics, 2, 133-143.
Müller J., J. Kasbohm, H. Oberhäsli, M. Mellers and W. Hubberten, 1999, TEM analysis of smectite-illite mixed-layer minerals of BDP-96-1 - preliminary report. BBD symposium Proceedings. (in press)
Ormsby J.F.A., 1966, Design of numerical filters with applications to missile data processing. J. Assoc. Computer Mecha, 8, 440-466.
Qiu L., D.F. Williams, A. Gvorzskov, E. Karabanov and M. Shimaraeva, 1993, Biogenic silica accumulation and paleoproductivity in the northern basin of Lake Baikal during the Holocene. Geology, 21, 25-28.
Shackleton N.J., A. Berger and W.R. Peltier, 1990, An alternative astronomical calibration of the lower Pleistocene timescale based on ODP site 677. Trans. Royal Soc. Edinburgh: Earth Science, 81, 251-261.
Shackleton N.J., M.A. Hall and D. Pate, 1995, Pliocene stable isotope stratigraphy of site 846. Proc. O.D.P., Scientific Results, 138, 337-355.
Short D.A., J.G. Mengel, T.J. Crowley, W.T. Hyde and G.R. North, 1991, Filtering of Milankovitch cycles by Earth's geography. Quat. Res., 35, 157-173.

Lake Baikal
K. Minoura (editor)
2000 Elsevier Science B.V.

Glaciations of central asia in the late Cenozoic according to the sedimentary record from Lake Baikal

Karabanov, E. B.[1,2]*, Kuzmin, M. I.[2], Prokopenko, A. A.[1,3], Williams, D. F.[1], Khursevich, G. K.[1,4], Bezrukova, E. V.[5], Kerber, E. V.[2], Gvozdkov, A. N.[2], Gelety, V. F.[2], Weil, D.[6], and Schwab, M.[7]

[1] Baikal Drilling Project, Department of Geological Sciences, University of South Carolina, Columbia SC, 29208, USA, fax: (803)-777-6610, e-mail: ekarab@geol.sc.edu

[2] Institute of Geochemistry, Russian Academy of Sciences, Irkutsk, 664033, Russia, fax: (3952)- 46 4050, e-mail: karaban@chat.ru

[3] United Institute of Geology, Geophysics and Mineralogy, Russian Academy of Sciences, Novosibirsk, 630090, Russia

[4] Institute of Geological Sciences, NAS of Belarus, Minsk 220141, Belarus

[5] Limnological Institute, Russian Academy of Sciences, Irkutsk, 664033, Russia

[6] Alfred-Wegener-Institute for Polar and Marine Research, Box 120161, D-27515 Bremerhaven, Germany.

[7] GeoForschungsZentrum Potsdam, Project Area 3.3, Telegrafenberg, D-14473, Potsdam, Germany.

Correspondence should be addressed to E.B. Karabanov.

Abstract

This report describes the paleoclimatic record over the period of 5 million years based on variations in diatom abundance in the sediments of a 200-m core obtained from Lake Baikal. The data represent a long, continuous continental record of climate changes in Central Asia during the Late Cenozoic. The record shows the climatic cooling trend which started in Pleistocene and is superimposed on the short-term cyclic climatic variations controlled by the Earth's orbital parameters. The record also reveals the presence of the two cold episodes (each about 300 Ka long) at the time intervals 2.82-2.48 Ma and 1.75-1.45 Ma characterized by glaciation at their maximum phases. These cooling periods in Lake Baikal record were also registered as global coolings in other paleoclimate records of the Northern Hemisphere. The continental record of Lake Baikal contains the majority of climatic events found in marine records and demonstrates that continental regions of Asia responded to all major changes in the Earth's climate recorded in the long oxygen isotopic records.

Introduction

During the past decades the significant efforts have been put into obtaining the long continuous records of the Earth's climate. Marine sediments are the premiere climatic archive. However, in order to understand the functioning of global climate as a coupled ocean-atmospheric system the long continuous continental records are essential. Sedimentary archive of Lake Baikal was chosen to represent Siberia Central Asia, and as a result the international effort of the "Baikal Drilling Project" started in Siberian Branch of the Russian Academy of Sciences. Paleoclimate record for the last 5-million-year period was recovered by Baikal Project (BDP-Members, 1997). This is the most ancient continuous continental record obtained to date for Central Asia. Analysis of the Baikal records allows comparison of continental climatic events in Asia with global changes in the Earth's climate as recorded in marine, glacial and other continental records.

Materials and methods

In the winter of 1996, two boreholes were drilled in Lake Baikal from the ice-based platform at the water depth of 321 m. The coordinates of the drilling site were 53°41'48" N and 108°21'06" E. Core BDP96-1 was 200 m long, and the second core, BDP96-2, was 100 m long. Academician Ridge, the topographic high of the lake bottom (Fig. 1), was selected as the most suitable place for paleoclimatic research because of stable conditions of hemipelagic sedimentation and the relative isolation from any direct supply of coarse sediments from the coastal zone, bottom slopes and from the influence of the fluvial sediment supply.

In this article we use the data on diatom abundance in the 200-m BDP-96-1 core supplemented by data from the sediments of the upper 6-m interval of the twin BDP-96-2 core, which was not recovered in BDP-96-1. Diatom abundance was counted in the total of 700 samples using the semi-quantitative method based on comparing the smear slide observation data in light microscope with visual percentage comparison charts (Scholle, 1979; Terry and Chilingar, 1955) with an error factor of about 15%. These records served as the basis for the sedimentary climatic record of Baikal over the past 5 million years. The content of biogenic silica (produced by diatoms) in Lake Baikal sediments as expected exhibits remarkable correlation with smear slide diatom abundance data (BDP-Members, 1997). In this article we also use the new palynological and diatom species distribution data from the BDP-96-1 and BDP-96-2 drilling cores (Bezrukova et al., 1999; Khursevich et al., 1999).

Results and discussion

Lithologically the sediments recovered by drilling of 1996 were com-

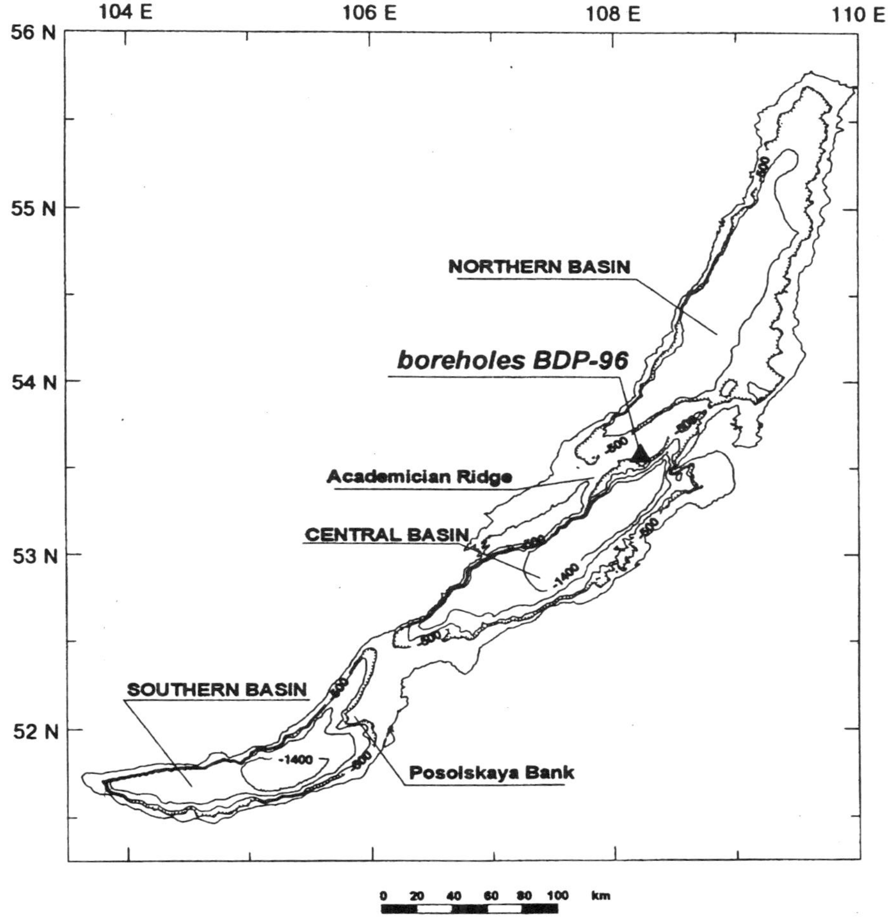

Figure 1. Map of Lake Baikal showing the location of the BDP-96 drilling site on Academician Ridge.

posed of alternating layers of biogenic diatomaceous ooze and terrigenous silty clay. Because the site was selected on the elevated Academician Ridge, the BDP-96 section does not contain turbidites typical of the deep basins of Lake Baikal. The content of diatom frustules varied from 0% to 85% sediment volume (Fig. 2B). According to the paleomagnetic studies, the age of the 200 m core at its base was slightly less than 5 Ma (Fig. 2A, B) (Williams et al., 1997; BDP-Members, 1998). The age model of the core based on 13 magnetic reversal/event boundaries indicates that sedi-

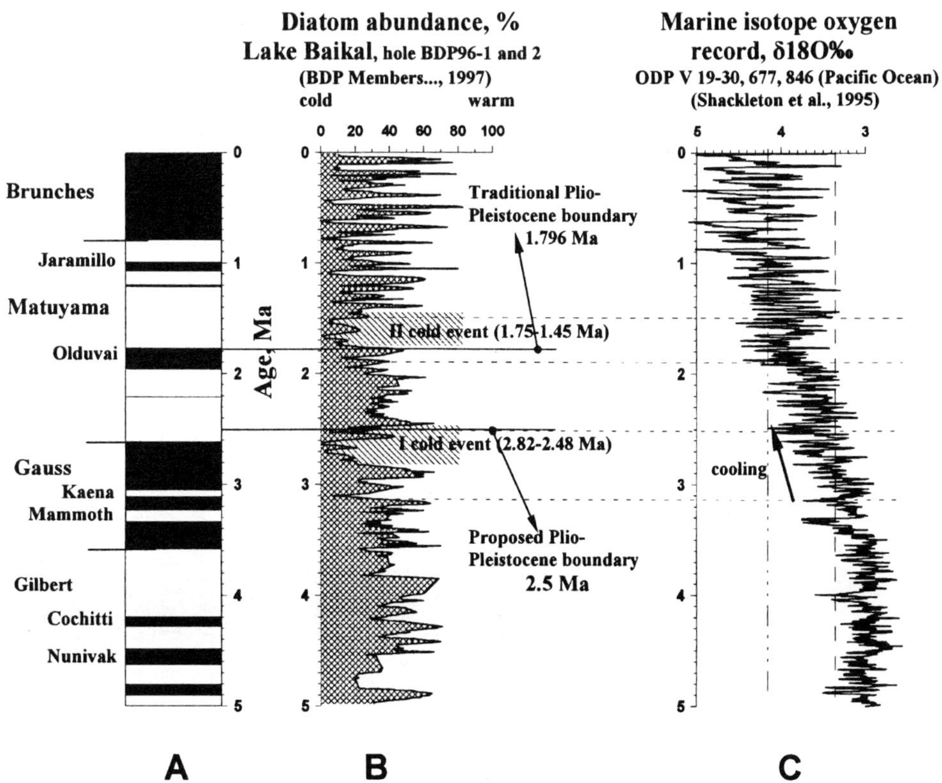

Figure 2. Correlation of the Baikal diatom record with marine oxygen isotopic record from Pacific.

A - Magnetostratigraphic scale for the Late Cenozoic (Cande and Kent., 1995). B - Baikal record of diatom abundance in cores BDP-96-1 and BDP-96-2 (five-point running average). Higher diatom abundance reflects warming, and vice versa. The general trend of decreasing diatom abundance towards the core top in response to cooling is evident from the shaded area. Hatching marks the zones of sharp decrease in diatom abundance, corresponding to cooling phases I and II in Lake Baikal record. C - The composite oxygen isotopic record reflecting global ice volume shows the general Plio-Pleistocene cooling trend. The vertical dashed line indicates modern ice volume. The vertical dotted-dashed line marks the average ice volume between the last glacial period and the Holocene. The Earth entered glacial climatic mode when oxygen isotopic record crossed this line at about 1.9 Ma. Arrow at the interval 3.1 - 2.5 Ma marks the transition to modern climate type. Horizontal dashed lines show that the cooling phases in Lake Baikal record correspond to the timing of major global climatic transitions.

ment accumulation at the BDP-96 site was continous, without hiati or redeposition. The average sedimentation rate at the drilling site was about 4 cm per 1000 years (BDP-Members, 1997; 1998; Williams et al., 1997).

The records show periodic variations in diatom abundance in Lake Baikal sediments (Fig. 2B). Such variations result from fluctuations in diatom plankton productivity due to the external climatic forcing controlled by cyclic changes in the Earth's insolation as a result of variations in orbital parameters (Colman et al., 1995; Williams et al., 1997). Biogenic silica content and diatom abundance were already shown to be good indicator of the relative warmer/colder climate fluctuations (BDP-Members, 1998; Williams et al., 1997). Decreases in the amount of diatom frustules correspond to colder episodes, whereas increases reflect warmings. The reduction in the amount of frustules to 0 – 1% indicates glacial conditions. This interpretation is corroborated by lithological and palynological data, as well as by absolute radiocarbon dating and by correlation age models (Colman et al., 1995; Williams et al., 1997; BDP-Members, 1998). The Baikal sediments recorded all glaciations that occurred during the Pleistocene and are reflected as global ice-volume buildups in the marine oxygen isotopic records (Williams et al., 1997). Climatic records from Lake Baikal reveal reliable and close correlation with marine isotopic records. In addition, similar to the marine records, the spectral characteristics of the Baikal records contain the frequencies corresponding to the Earth's orbital parameters. This has proven to be a good source for verification of the astronomic nature of diatom paleoclimatic signals from lake sediments (Colman et al., 1995; Williams et al., 1997).

The sedimentary record of the lake reveals that rhythmic variations in diatom abundance in Baikal occurred not only during the Pleistocene but throughout the entire Pliocene as well (Fig. 2B). The latter is not reflected in the available Siberian climatic curves (Arkhipov and Volkova, 1994; Williams et al., 1997), possibly because of insufficient resolution, although such cyclicity is well known from marine records (Shackleton et al., 1995). The frequency and the amplitude of variations in diatom abundance in the lower and upper parts of the core are different (Fig. 2B). The upper part of the record displays deep minima in frustule abundance corresponding to Quaternary glaciations*. This agrees with widely accepted concept concerning the onset of series of intense glaciation in the Northern Hemisphere since the Early Pleistocene (Gladenkov, 1978; Arkhipov and Volkova, 1994; Nikiforova, 1989).

* In this paper we accepted the age of the beginning of the Quaternary period and Pliocene-Pleistocene boundary to be 1.796 Ma according to the stratotype Pliocene-Pleistocene section (van Couvering, 1997), rather than 1.65 Ma, as accepted by the stratigraphic committee (Nikiforova, 1989). We do not refer to the Eopleistocene, and we distinguish Early, Middle,

and Late stages of the Pleistocene.

Lake Baikal diatom records of the long pronounced cooling events in southeast Siberia

The distribution of diatoms in BDP96-1 sediments (Fig. 2B) indicates that besides the short-term fluctuations as a result of astronomic factors, the average diatom abundance tends to decrease towards the core top. This reflects the trend towards colder climate of the Northern Hemisphere in the Late Cenozoic (Gladenkov, 1978; Nikiforova, 1989; Shackleton et al., 1995) as recorded in Asia. Although the Baikal diatom record begins only in the Early Pliocene, the cooling trend is evident in our records (Fig. 2B). Superimposed on this general cooling trend, there were two significant minima of diatom abundance (Fig. 2B), reflecting the pronounced cooling episodes. The first minimum occurred around the Gauss-Matuyama geomagnetic boundary, and the age of this interval is 2.82-2.48 Ma. The second minimum lies in the upper part of the Matuyama chron, and its beginning coincides with the Olduvai event. The age of the second interval is 1.75-1.45 Ma. The duration of both coolings was similar: 300 - 340 Ka. After the first cooling, the Baikal record shows the period of warm climate, comparable with Pliocene climate. Only after the second cooling phase did the intensity of cooling episodes reflected in Lake Baikal diatom abundance record reach the magnitude of Late Pleistocene glacial periods.

The abundance of diatom frustules was very low during these deep minima, at certain intervals reflecting the maximum cooling phases the amount of diatom frustules fell to zero, which is typical only for the glacial sediments of the Late-Middle Pleistocene, and not for the warm Pliocene. The sediments corresponding to these cold phases are analogous to the sediments of the Pleistocene glacial periods, i.e., they are composed of fine clay with textural elements of ice and probably iceberg rafting. Lithological evidence thus suggests the presence of glaciers around the lake at the time of the first and second cooling episodes, 2.82-2.48 Ma and 1.75-1.45 Ma. Diatom abundance indicates a significant climatic deterioration, and the ice- and iceberg-rafted detritus indicates the development of mountain glaciations in Siberia at the peaks of the cooling phases.

The first cooling event, 2.82-2.48 Ma

The results of palynological analysis of core BDP-96-1 (Bezrukova et al., 1999) provide evidence of critical changes in the composition and structure of vegetation in the Baikal region at the age boundary of about 2.5 Ma. These changes caused redistribution of the areas occupied by different wood assemblages. The areas covered by light-coniferous trees, as well as by deciduous and coniferous elements of moderately thermophilic flora (*Tsuga, Corylus, Quercus, Tilia*) diminished considerably. The dominant elements of the dendroflora were represented by dark-coniferous species of the taiga, such as *Pinus sibirica* and *Abies sibirica*, along with light-coniferous elements, such as *Larix* sp. Somewhat later, at the peak of the cold phase, the proportion of moderately-thermophilic elements in the wood vegetation diminished dramatically. At the same time, the area of open forestless steppe regions increased considerably.

The character of the variation of wood vegetation species and the entire structure of the vegetation cover of the region during the 2.82-2.48 Ma indicates a marked change in climatic conditions during this period. The dominance of dark-coniferous species, such as cedar and fir, provides evidence of a profound climate cooling. Considerable degradation of forests as well as the subsequent spread of the forest-steppe and steppe vegetation suggests reduction in precipitation. In some samples, referred to as the "maximum cooling phase", tree pollen is almost absent, indicating dramatic degradation of forests in the region. However, the absence of Arctic flora elements characteristic of the Pleistocene glacial epochs in the vegetation does not allow us to conclude that the cooling in the 2.82-2.48 Ma interval caused considerable glaciation of the region.

Significant changes in the diatom assemblage within the interval of the first cooling between 2.82-2.48 Ma is reflected in the species composition of diatom algae (Khursevich et al., 1999). The onset of cooling coincide with the last appearance datum (LAD) of *Stephanopsis* planktic diatom, with the first appearance datum (FAD) of the new genus *Tertiarius*. The algae of this genus are found only during the first cold interval, 2.82-2.48 Ma. A subsequent transition to warmer climatic conditions (as indicated by high diatom abundance) led to complete disappearance of diatoms of this genus and to the FAD of the yet another new genus, *Cyclotella - C. tempereiformica* and *C. distincta*. Such sharp and marked changes in the diatom assemblage at the high taxonomic level of geni indicates profound catastrophic changes which affected the plankton in Lake Baikal. Changes in diatom assemblage coincide with palynological evidence for the large-scale restructuring of the regional vegetation. Combined, this fossil evidence confirms the significance of regional climatic changes during the first cooling phase in Lake Baikal BDP-96-1 record.

The second cooling event, 1.75-1.45 Ma

At the start of the second cooling phase in the Baikal record, at about 1.75 Ma, significant changes in vegetation communities also took place (Bezrukova et al., 1999). The role of the arboreal species in that interval diminished, and that of herbaceous species increased, however, not reaching the high values of the first cooling phase. Nevertheless, after this cooling phase, Tsuga and moderately-thermophilic arboreal species disappeared from the vegetation of the region. The second cooling phase is also marked by the development of the forest-tundra type landscapes, which suggests that individual glaciers were developing in the mountains and likely exceeding the limited glaciation during the first cooling phase. Dramatic cooling at the end of the Olduvai event (1.8 Ma) is also shown by regional paleopedological data revealing strong cryogenic deformations indicative of the negative winter temperatures (Vorobyova et al., 1995; BDP-Members, 1997).

During the second cooling phase the diatom asseblage of the lake also underwent marked changes (Khursevich et al., 1999). During the this phase, the planktonic species *Cyclotella tempereiformica* and *C. distincta* characteristic of the previous warm period were replaced by the new diatom species of the same genus: *Cyclotella comtaeformica* et var. *spinata*. The latter were in turn replaced by the *Stephanodiscus majusculus* and *Aulacoseira* aff. *islandica* at the end of the cooling phase around 1.45 Ma. Although indicative of dramatic environmental changes, the changes in diatom flora at the boundaries of the second cooling phase were less significant than the changes of the first cooling phase 2.82 - 2.48 Ma BP, because they occurred at the speciation level and not at the genus level. Similar to lacustrine flora, terrestrial vegetation in Lake Baikal region did again undergo significant restructuring in response to the second cooling phase.

Correlation between Lake Baikal cooling phases and global cooling events

The cooling phases distinguished in the Baikal record correlate well with the global cooling events recorded in some of the marine and continental records. The first cooling phase (2.82 - 2.48 Ma) in the Baikal record, is roughly centered around the magnetic reversal of the Gauss-Matuyama boundary (Fig. 2A, B). According to Zagwijn (Zagwijn, 1996; 1997), the earliest glacial period in Northern Hemisphere, the Praetiglian, when a noticeable depletion of flora occurred around Gauss-Matuyama paleomagnetic reversal, and the forest/steppe boundary shift from the Netherlands far to the south. The beginning of the Middle Villafranchian and a sharp change in vegetation in southern Europe also corresponds to the Gauss-Matuyama paleomagnetic reversal (Nikiforova, 1989). The disappearance of warm-water assemblages and the appearance of cold-water assemblages of mollusks has occurred in the northern part of the Pacific and Atlantic oceans and in the Arctic seas at that time (Gladenkov, 1978). Significant cooling around this magnetic reversal was also observed in Northern Asia (Volkova and Baranova, 1980). The first prominent occurrences of iceberg-rafted detritus in North Atlantic and northern Pacific date back to 2.4-2.7 Ma (Gladenkov, 1978; Nikiforova, 1989; Shackleton et al., 1995). This cooling also produced the Elk Creek deposits, the most ancient moraine in North America, and the extinction of the thermophilic elements in the North American vertebrate fauna (Nikiforova, 1989). Thus, there was a sudden profound cooling at the Gauss-Matuyama magnetic reversal in Northern Hemisphere leading to regional glaciations and to dramatic changes in flora and fauna. The first cooling phase in the Baikal record within the interval 2.82-2.48 Ma appears to correlate with the global Pliocene cooling, and it's maximum peak could be correlated with the Praetiglian glaciation of Western Europe. In Central Asia this cooling was manifested in the large-scale restructuring of the vegetation cover, dramatic changes in Lake Baikal planktonic assemblages, and in lithologic structures indicative of the mountain glaciations in the Baikal region. It has to be noted, however, that according to the Baikal record, the first cooling phase started much earlier than the Praetiglian, which appears to correspond to cold maximum of this phase.

The second cooling phase dated as 1.75-1.45 Ma BP in the Lake Baikal record, corresponds to another pronounced global cooling found in many regions of the Northern Hemisphere at the Pliocene/Pleistocene boundary identified as the cold Eburonian period in Western Europe. The Eburonian is marked by significant changes in the floral and faunal composition in Western Europe (Zagwijn 1996). At the Pliocene/Pleistocene boundary the characteristic Arctic and northern boreal assemblages of mollusks appeared in Alaska, Iceland, in the Arctic and in northern boreal waters (Gladenkov, 1978). Around that time the forest-tundra and tundra elements of vegetation spread in Western Siberia indicating dramatic cooling (Volkova and Baranova, 1980). This cooling and the resultant changes in the composition of marine and on-shore flora and fauna served as the basis for distinguishing the upper boundary of the warm Neogene system at 1.796 Ma, followed by the cold glacial Quaternary epoch (Nikiforova, 1989; Berggren et al., 1995).

As shown by number of works, the significant climatic deterioration and profound changes in biota have occurred in the Northern hemisphere at about 2.5 Ma BP (Nikiforova, 1989; van Couvering, 1997; Zagwijn, 1996; Suc et al., 1997). That was the first time when Tertiary glaciations left their traces in the Northern Hemisphere (Nikiforova, 1989; van Couvering, 1997; Zagwijn, 1996; 1997; Suc et al., 1997). The importance of the 2.5 Ma boundary warrants the recently started discussion on lowering the Pliocene-Pleistocene boundary from 1.796 Ma to 2.5 Ma BP (van Couvering, 1997; Zagwijn, 1996; Suc et al., 1997).

In the diatom records from Lake Baikal both climatic benchmark episodes proposed as the Plio-Pleistocene boundary are well recognized as times of strong cooling accompanied by regional vegetation change and regional glaciation. At the same time, both the palynological data (Bezrukova et al., 1999) and the changes in diatom assemblages (Khursevich et al., 1999) indicate that changes caused by the first cooling phase were more profound than during the second cooling phase. Thus, the larger scale of environmental change in Central Asia reflected in the first cooling phase of the Lake Baikal record argues for lowering the age of Pliocene-Pleistocene boundary from 1.8 to 2.5 Ma BP, as proposed by Zagwijn and Suc (Zagwijn, 1996; Suc et al., 1997).

The proposed correlations of the first Baikal cooling phase with the Praetiglian and the second cooling phase with the Eburonian intervals of the West-European climate stratigraphic scale (Zagwijn, 1997) suggest that the warm interval between these coolings can be correlated with the warm Tiglian (see Figure 1 in G. Khursevich et al., same volume). Tiglian in Western Europe consisted of three warm intervals divided by two coolings, and the corresponding interval in the Baikal record also contains three coolings and two warmings. In addition, in Lake Baikal record smaller regular climatic fluctuations are observed, corresponding to the 41 Ka obliquity orbital cycle of the Earth.

According to Zagwijn (1996), the glacial Praetiglian largely lies above the Gauss-Matuyama boundary, while in the Baikal record, the majority of the first cooling corresponds to the Gauss epoch (see Figure 1 in G. Khursevich et al., same volume). This discrepancy between the Baikal and European records may be attributed to earlier cooling in Asia as a peculiar continental reaction of huge landmasses.

The age model of the European record (Zagwijn, 1996; 1997) based on paleomagnetic studies of continental deposits is still a point for discussion. The magnetic measurements of the Reuverian -Tiglian cross-section are not continuous. Also, a marked hiatus in the base of the Praetiglian coarse-grained sequence is observed in the Reuverian - Tiglian section (Zagwijn, 1996), and thus the lower boundary age of the cold Praetiglian might not be represented. The Praetiglian section contains additional episode of normal polarity, which is tentatively attributed to the Reunion I event (2.2 Ma) (Zagwijn, 1996; 1997). If the short episode of normal polarity turns out to be part of the Gauss chron then the Praetiglian glaciation would be shifted down into Gauss chron of normal polarity, as it is suggested by the Baikal record. The continuous sedimentary record of Lake Baikal (Williams et al., 1997) is preferable for constraining the timing of climatic events in the Late Cenozoic than the European record, which is based on the composite continental cross-section.

Comparison between the diatom abundance record from Lake Baikal and the marine isotope record

In addition to its climatic-stratigraphic relations with Europe and other regions of the Earth, the detailed Baikal paleoclimatic record offers a unique opportunity to compare the continental Pliocene-Pleistocene record from the center of the largest continent, Eurasia, with the detailed marine oxygen isotope records used for reference today. Practically all the long marine records clearly show the general climate deterioration trend in the Pliocene with a sharp cooling on the 3.1-2.5 Ma transition (Raymo, 1992; Shackleton et al., 1995). In the beginning of this period global ice volume began to increase, and by the end of this interval it reached the present level (Fig. 2C). The first cooling phase in the Baikal record between 2.82 and 2.48 Ma clearly corresponds to this global climatic transition. Moreover, the short cold episode distinguished by Raymo (1992) at the interval 3.1 - 3.2 Ma (Fig. 2C) is matched by pronounced drop in diatom abundance in Lake Baikal record at 3.14-3.07 Ma (Fig. 2A).

Within the interval of approximately 1.9-1.5 Ma the isotopic record (Raymo, 1992; Shackleton et al., 1995) points to one more critical change. Oxygen isotopes indicate that at about 1.9 Ma global ice volume crossed the threshold value of the average between global ice volumes of the last glacial and of the Holocene (Fig. 2C). This boundary actually marks the period when the Earth's climate entered the glacial mode. The second cooling phase in the Baikal record at the interval 1.75-1.45 Ma practically parallels the oxygen isotopic data. This cooling phase of Baikal is the response of the lake and its watershed to the beginning of the global glaciation on Earth.

Conclusion

Analysis of Baikal paleoclimatic records shows two profound cooling phases in Central Asia superimposed over the general Plio-Pleistocene cooling trend and over the regular rhythmic variations in the Earth's climate driven by changes in orbital parameters. The ages of these cooling phases were 2.82-2.48 Ma and 1.75-1.45 Ma, and the peaks of cooling during these phases were associated with regional glaciation, the earliest Late Cenozoic glaciations in Central Asia. The evidence of profound cooling phases in Lake Baikal paleoclimate record, which occurred concurrently with global ice volume changes reflected in marine oxygen isotopic records, suggests that continental regions of Asia experienced climatic changes similar to other regions of the Northern Hemisphere. The continuous Baikal record with its high resolution and robust age model allow the age of these global climatic events to be better constrained not only for Central Asia, but probably for the entire Eurasian continent. For instance, the first cooling phase dated in Lake Baikal record as 2.82 - 2.48 Ma BP appears to have started in the Gauss chron, significantly earlier than the presently accepted timing of the Praetiglian glaciation in Western Europe.

The dramatic changes in terrestrial and aquatic biota associated with the two cooling phases of 2.82 - 2.48 Ma and 1.75 - 1.45 Ma are recorded in the Lake Baikal sedimentary archive. The first cooling caused the most profound changes in Siberian terrestrial vegetation and in Lake Baikal planktonic assemblage on high taxonomic level, thus contributing the regional evidence to the current stratigraphic discussion on lowering the Quaternary to 2.5 Ma.

Acknowledgements
This work was financially supported by the Siberian Branch of the Russian Academy of Sciences as part of the program "Global changes of environment and climate" of the Ministry of Science and Technology of Russia, by Ministry of Geology of Russia, by US National Science Foundation (NSF), by International Continental Scientific Drilling Program (ICDP), by Science and Technology Agency (STA) of Japan, and by German Scientific Foundation (DFG). The authors would like to express their gratitude to all the participants of the Baikal Drilling Project involved in organizing and conducting the drilling operations at Lake Baikal, and in BDP-96 core description and sampling.

References
Arkhipov, S.A., and Volkova, V.S. (1994) Geological history, landscapes and climates of Pleistocene of Western Siberia. Novosibirsk, Nauka, 106 pp. (in Russian)

BDP-Members (1997) Continuous paleoclimate record of last 5 Ma from Lake Baikal, Siberia. EOS American Geophysical Union, Transactions, 78, 597-604.

BDP-Members (1998) Continuous record of climatic changes in Lake Baikal sediments during last 5 Ma. Russian Journal of Geology and Geophysics, 39, 139-165. (in Russian)

Berggren, W.A., Kent, D.V., Swisher, III C.C. and Aubry, M.-P. (1995) A revised Cenozoic geochronology and chronostratigraphy. In: Geochronology, time scales and global stratigraphic correlation, W. A. Berggren, D. V. Kent, C. C. Swisher III, and J. Hardenbol, Eds., SEPM, Tulsa, Oklahoma, 129-212.

Bezrukova, E.V., Kulagina, N.V., Letunova, P.P.and Shestakova, O.N. (1999) Evolution of vegetation and climate of Baikal region during last 5 million years accordingly the palynological investigation of lake Baikal sediments. Russian Journal of Geology and Geophysics, 5, 735-745. (in

Russian)
Cande, S.C., Kent, D.V., (1995) Revised calibration of the geomagnetic polarity time scale for the Late Cretaceous and Cenozoic. Journal Geophysics Research, 100, 6093-6095.
Colman, S.M., Peck, J.A., Karabanov, E.B., Carter, S.J., Bradbury, J.P., King, J.W., and Williams, D.F. (1995) Continental climate response to orbital forcing from biogenic silica record in Lake Baikal. Nature, 378, 769-771.
Gladenkov, Yu.B. (1978) Marine Cenozoic of the northern regions. Moscow, Nauka, 194 pp. (in Russian)
Khursevich, G.K., Karabanov, E.V., Williams, D.F. Kuzmin, M.M., and Prokopenko, A.A. (1999) Evolution of freshwater centric diatoms during the Late Cenozoic within the Baikal Rift Zone. (see pages ?? the same volume).
Nikiforova, K.V. (1989) The global climatic fluctuation and their displaying in Northern Hemisphere. Bulletin of commission of the Quaternary period investigation, 58, 37-48. (in Russian)
Raymo, M.E. (1992) Global climate change: a three million year perspective. NATO ASI series, Vol. 13, G.J. Kukla, E. Went, eds., Springer-Verlag, Berlin, Heidelberg, 207-223.
Ruddiman, W.F., and McIntyre, A. (1981) Oceanic mechanisms for amplification of the 23,000-year ice volume cycle. Science, 212, 617-627.
Shackleton, N.J., Hall, M.A. and Pate, D. (1995) Pliocene stable isotope stratigraphy of site 846. In: Proceedings of the Ocean Drilling program, Scientific Results, Vol. 138, N.G. Pisias, L.A. Mayer, T.R. Janecek, A. Palmer-Julson and T.H. van Andel, eds., College Satition, TX (Ocean Drilling Program), 337-355.
Scholle, P.A. (1979) A color illustrated guide to constituents, textures, cements and porosity of sandstones and associated rocks. AAPG Memories, 28, vii.
Suc, J-P, Bertini, A., Leroy, S., and Suballyova, D. (1997) Towards the lowering of the Pliocene/Pleistocene boundary to the Gauss-Matuyama reversal. Quaternary International, 40, 37-42.
Terry, R.D. and Chilingar, G.V. (1955) Summary of "Concerning some additional aids in studying sedimentary formations" by M.S. Shvetsov. Journal of Sedimentary Petrology, 25(3), 229-234.
Van Couvering, J.A. (1997) Preface: the new Pleistocene. In: The Pleistocene Boundary and the Beginning of the Quaternary, J.A. Van Couvering, ed., Cambridge University Press, Cambridge, xi-xix.
Volkova, V.S., and Baranova, Yu.P. (1980) Pliocene-Early Pleistocene changes of climate in Northern Asia. Russian Journal of Geology and Geophysics, 7, 43-52. (in Russian)

Vorobyova, G.A., Mats, V.D. and Shimaraeva, M.K. (1995) Late-Cenozoic paleoclimates in the Baikal region. Russian Journal of Geology and Geophysics, 36, 82-96. (In Russian)
Williams, D.F., Peck, J., Karabanov, E.B., Prokopenko, A.A., Kravchinsky, V., King, J. and Kuzmin, M.I. (1997) Lake Baikal record of continental climate response to orbital insolation during the past 5 million years. Science, 278, 1114-1117.
Zagwijn, W.H. (1997) The Neogene-Quaternary boundary in The Netherlands. In: The Pleistocene Boundary and the Beginning of the Quaternary, J.A. van Couvering, ed., Cambridge University Press, Cambridge, 185-190.
Zagwijn, W.H. (1996) Borders and boundaries: A century of stratigraphical research in the Tegelen-Reuver area of Limburg (the Netherlands). In: The dawn of the Quaternary. INQUA-SEQS-96. 16-21 June 1996, Kerkrade-the Netherlands. Volume of Abstract. T. van Kolfschoten and P. Gibbard, eds., Geological Survey of the Netherlands RGD. 2-9.

Lake Baikal
K. Minoura (editor)
2000 Elsevier Science B.V.

Palaeoclimatic changes from 3.6 to 2.2 Ma B.P. derived from palynological studies on Lake Baikal sediments

Demske, D.[1,2]*, Mohr, B.[1], and Oberhänsli, H.[3]
[1] Museum fuer Naturkunde, Humboldt Universitaet, Invalidenstr. 43, 10115 Berlin (Germany), fax: +49-30-2093-8868, e-mail: dieter.demske@rz.hu-berlin.de / barbara.mohr@rz.hu-berlin.de
[2] Alfred Wegener Institute for Polar and Marine Research, Research Department Potsdam, Telegrafenberg A43, 14473 Potsdam (Germany), fax: +49-331-288-2137
[3] GeoForschungsZentrum, Telegrafenberg C, 14473 Potsdam (Germany), fax: +49-331-288-1302, e-mail: oberh@gfz-potsdam.de
(*corresponding author)

Abstract:

The palynological record from BDP-96-1 drill cores (Academician Ridge, 321 m water depth) revealed late Pliocene development of mixed coniferous forests with a decline in associated broadleaved trees and hemlocks (*Tsuga*), followed by the expansion of open vegetation (*Artemisia*). The vegeta-tion and inferred climate changes in the Baikal region around 2.7 Ma B.P. (million years before present) are related to the intensification of northern hemisphere glaciations recorded as increases in ice-rafted debris in North Pacific (and North Atlantic) sediments.

Introduction

The position of Lake Baikal in the interior of the Eurasian continent provides a unique opportunity for reconstructing late Cenozoic vegetation history. Changes in the distribution of northern boreal taiga forests, southern Mongolian steppe elements, and steppe forests have implications for understanding past climatic changes in northeastern Eurasia during the Pliocene and Pleistocene epochs. Late Pliocene environmental changes in the northern hemisphere between 3.5 and 2 Ma B.P. (million years before present) are of special interest due to records of cooling and aridity between 3.5 and 3.0 Ma and after 2.7 Ma (Leroy, Dupont, 1994; Maslin et al., 1995; Kukla, Cilek, 1996; Han et al., 1997).

Materials and methods

The BDP-96-1 drill cores are composed of clay and diatom ooze in varying proportions, with silt, sand, and gravel in smaller, changing amounts. The provisional age model is based on palaeogeomagnetic reversals (BDP Members, 1997, 1998; Williams et al., 1997).

Sediment samples were taken from the cores at 50 cm intervals, representing a time resolution of 9-22 ka. Laboratory preparation included treatment with hydrochloric and hydrofluoric acid, followed by micro-sieving with 6-μm mesh. Acetolysed *Lycopodium clavatum* spores were used as spikes to calculate pollen and spore concentrations. The material was mounted in glycerine jelly with a paraffin seal. At least 300 grains (arboreal and non-arboreal pollen, AP+NAP) were counted, or when the concentration was very low, a minimum total of 100 grains was counted. Percentages are based on the AP+NAP sum, excluding aquatics and spores.

Pollen data and vegetation history

More than 100 different types of pollen and spores were identified. The sporomorph concentrations in the sediment varied in magnitude from 10^2 to 10^5 grains per cm^3, with the peaks tending to decrease in the upper part of the core. The preliminary zonation is based on percentage pollen data, resulting in five zones and two subzones (Fig. 1). The pollen spectra are dominated by bisaccate grains of coniferous trees, mainly pine (*Pinus*) and spruce (*Picea*). *Tsuga* pollen is abundant in the lower zones (I to III), while *Quercus* and *Ulmus/Zelkova* are rather abundant in zone I, and also frequent in zones II and V. Some broadleaved taxa (*Acer, Tilia, Juglans, Pterocarya* pollen) are confined to zones I and II, while *Betula* and *Alnus* are present throughout the section investigated. Grains of Cupressaceae (interpreted to be the *Juniperus*-type) are very frequent in pollen zone IV (subzone IV b). The total percentages of non-arboreal pollen vary considerably, with smaller peak values in zones I to II, higher peaks in zones III and IV and a maximum of about 50% in zone V.
According to the pollen data, the landscape around Lake Baikal was covered by mixed coniferous forests with pines (*Pinus* subgen. *Diploxylon* and subgen. *Haploxylon*), spruce (including *Picea* sect. *Omorica*), as well as firs (*Abies*), and up to ca. 2.6 Ma, by hemlock (*Tsuga*). Associated arboreal taxa of the coniferous forests included broadleaved trees, such as maple, linden, walnut, oaks and elms (*Acer, Tilia, Juglans, Quercus,* and *Ulmus,* pollen zones I and II). Around 3.4 Ma (zone I), forest communities with *Quercus* were important during climatically dry intervals, as they could partly occupy drier rocky sites with abundant *Lycopodium* and *Selaginella* (cf. Wang, 1961). Between 3.3 and 2.9 Ma (zone II) the admixture of hemlock in *Tsuga-Picea* forests was considerable, and since 3.0 Ma the importance of *Abies* has increased. Broadleaved taxa like *Quercus* and *Ulmus* by then played a minor role.

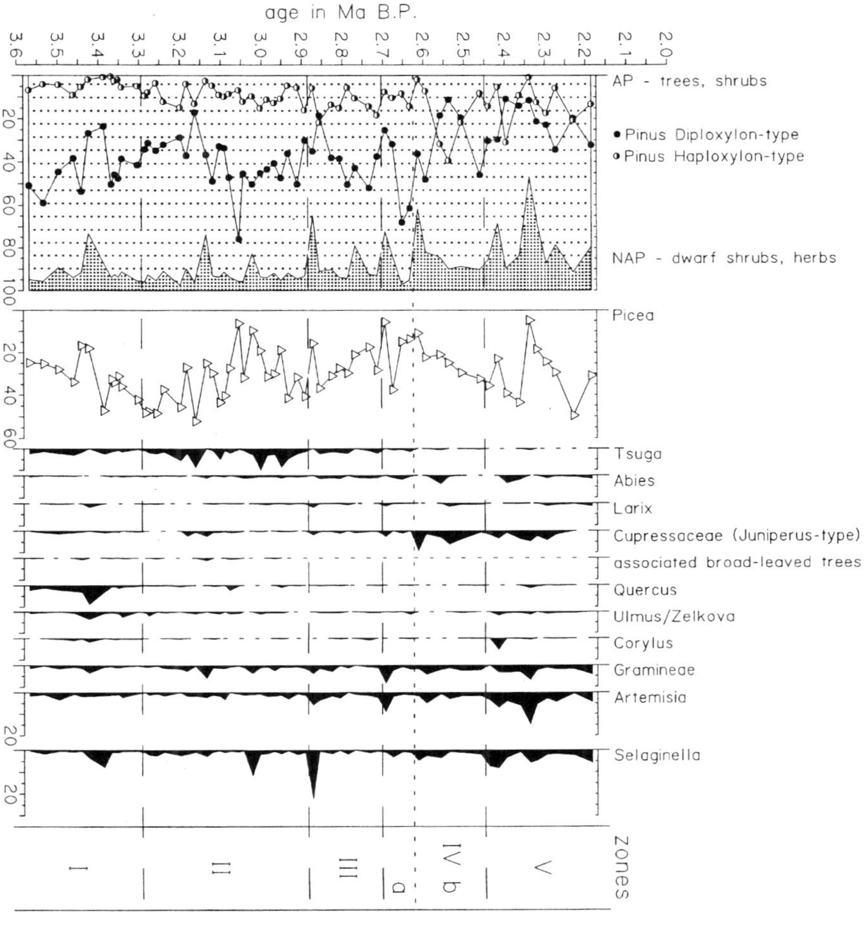

Fig. 1. Simplified percentage pollen and spore diagram for BDP-96-1 drill cores (2.2-3.6 Ma B.P.). Selected taxa are shown next to the summary diagram.

The significant decline of hemlock firs around 2.9 Ma (pollen zone III) was followed by slight dissemination of shrub alders (*Alnus fruticosa*-type), while at the same time steppe elements penetrated the landscape, as shown by maxima of *Artemisia* (from 2.9 to 2.8 Ma) and *Selaginella*. Increased percentages of non-arboreal taxa with *Artemisia* and Gramineae from 2.7 to 2.45 Ma (zone IV) point to an important step towards open vegetation (subzone IV a), while the diversity of forest vegetation was further reduced. Around 2.6 Ma *Tsuga* disappeared in the northern Baikal region, and juniper (*Juniperus*-type pollen) was able to spread (subzone IV b). After 2.45 Ma open vegetation and steppe communities became established in the landscape (zone V). The *Artemisia* maximum points to a very dry climate, whereas the slight increases in arboreal taxa, such as *Corylus, Quercus,* and *Ulmus,* may reflect fluctuations towards improved growing conditions.

Palaeoclimate

The vegetational succession (decline of hemlock, spread of steppe vegetation) outlined above reflects long-term variation in climatic conditions. Sharp peak percentages of certain pollen and spore types, mostly in low concentration samples, may reflect the cool/dry conditions of glacial interglacial. Exposed dry sites react readily with climatic signals, as suggested by expansion of cliff (*Selaginella*) and steppe (*Artemisia*) vegetation. Decreasing vegetation cover, as shown by increasing NAP/AP ratios, reflects drier conditions around 3.4 Ma, at 2.9 Ma and after 2.7 Ma. The recorded arboreal taxa reflect a more favourable, warmer climate prior to ca. 2.9 Ma and a late Pliocene cooling trend from 3 to 2.5 Ma accompanied by increasing dryness. These conclusions allow com-parisons with reconstructions of terrestrial vegetation in China (Han et al., 1997) and with marine records from the northwest Pacific (ODP Site 882), which show increased accumulation of ice-rafted detritus (IRD) inferred from data on magnetic susceptibility (Maslin et al. 1995). The minor increases in IRD around 3.4 Ma, suggesting a cool or dry climate, are related to distinct vegetational changes in the Baikal region, as evidenced by the spread of non-arboreal taxa and *Selaginella*, as well as by a *Quercus* maximum and a significant decrease in *Tsuga*, which is less drought-tolerant. The increasing dryness and cooling after 2.7 Ma, as inferred from the pollen record (zone IV), correspond to high IRD accumulation rates in the North Pacific and North Atlantic and reflect the dramatic intensification of northern hemisphere glaciations, whereas the fluctuations toward dry intervals after 2.45 Ma (zone V) may be related to somewhat lowered IRD deposition in the North Pacific.

References

BDP Members (1997). Continuous Paleoclimate Record Recovered for Last 5 Million Years. EOS, Transactions, American Geophysical Union, 78(51), pp. 597, 601, 604.

BDP Members (1998). Neprerywnaya zapis' klimaticheskikh izmeneniy v otlozheniyakh ozera Baikal za poslednie 5 millionov let (Engl. summ.: A continuous record of climate changes of the last 5 million years stored in the bottom sediments of Lake Baikal). Geologiya i Geofizika, 39(2), 139-156.

Han, J., Fyfe, W.S., Longstaffe, F.J., Palmer, H.C., Yan, F.H., Mai, X.S. (1997). Pliocene-Pleistocene climatic change recorded in fluviolacustrine sediments in central China. Palaeogeography, Palaeoclimatology, Palaeoecology, 135, 27-39.

Kukla, G., Cílek, V. (1996). Plio-Pleistocene megacycles: record of climate and tectonics. Palaeogeography, Palaeoclimatology, Palaeoecology, 120, 171-194.

Leroy, S., Dupont, L. (1994). Development of vegetation and continental aridity in northwestern Africa during the Late Pliocene: the pollen record of ODP Site 658. Palaeogeography, Palaeoclimatology, Palaeoecology, 109, 295-316.

Maslin, M.A., Haug, G.H., Sarnthein, M., Tiedemann, R., Erlenkeuser, H., Stax, R. (1995). Northwest Pacific site 882: The initiation of Northern Hemisphere glaciation. In: Rea, D.K., Basov, I.A., Scholl, D.W., Allan, J.F. (eds.), Proceedings of the Ocean Drilling Program, Scientific Results, 145, 315-327.

Wang, Chi-Wu (1961). The forests of China: with a survey of grassland and desert vegetation. Maria Moors Cabot Foundation, Publication No. 5. Harvard University, Cambridge, Massachusetts, 313 pp.

Williams, D.F., Peck, J., Karabanov, E.B., Prokopenko, A.A., Kravchinsky, V., King, J., Kuzmin, M.I. (1997). Lake Baikal Record of Continental Climate Response to Orbital Insolation During the Past 5 Million Years. Science, 278, 1114-1117.

Lake Baikal
K. Minoura (editor)
2000 Elsevier Science B.V.

TEM analysis of smectite-illite mixed-layer minerals of core BDP 96 Hole 1 : Preliminary Results

Müller, J.[1]*, Kasbohm, J.[2], Oberhänsli, H.[3], Melles, M.[1], and Hubberten, H. -W.[1]
[1] Alfred Wegener Institute for Polar and Marine Research, Potsdam, Telegrafenberg A43, 14473 Potsdam, Germany
[2] Department of Geological Sciences, University of Greifswald, Friedrich-Ludwig-Jahn-Str. 17a, 17487 Greifswald, Germany
[3]GeoForschungsZentrum Potsdam, Telegrafenberg, 14473 Potsdam, Germany
Correspondence should be addressed to jmueller@awi-potsdam.de,
fax: ++49-331-288-2137
(*corresponding author)

Abstract

First, TEM, STEM, and EDX analyses were carried out on four clay samples from Lake Baikal core BDP-96-1 which according to paleomagnetic data were formed ca. 3.29, 2.85, 2.78 and 2.42 Ma ago. The results showed that these methods are a useful and indispensable supplement to XRD measurements that provide additional information on the formation and transformation of clay minerals before deposition, and thereby assist in paleoclimatic interpretation. The illite-rich mixed-layer minerals in all of the samples generally haved a higher Fe content, which is probably displaced by Al during transformation into smectite. The smectite phase of all the samples could can be identified as montmorillonite. Whilst semiquantitative XRD analysis of all samples analysed showed little variation in interstratified illite-smectite [Sme > 40%] relative to illite content, EDX analysis showed that the clay minerals in the oldest sample (3.29 Ma) had a much higher substitution of Mg for Al in the octahedra sheets. In addition, a higher abundance of Smectite, ML10, and ML20 iswas observed in the oldest sample, and this may have beenbe the result of higher water discharge and intensive soil formation in the hinterland during this period. By contrast, the illite and illite-rich mixed-layer mineral content of the youngest sample (2.42 Ma) iswas distinctly higher. This is probably due to the predominant physical weathering conditions in the hinterland of a colder and dryer climate in the context of the major global cooling in the Late Pliocene epoch.

Introduction

X-ray diffraction (XRD) analysis of sediments from BDP-cores 93-1 and 93-2 has shown that clay minerals are a useful proxy for paleoclimate reconstruction in Lake Baikal (Melles et al., 1995; Oberhänsli et al., 1998; Yuretich et al., 1999). Semiquantitative analysis of interstratified illite-smectite [Sme > 40%], illite, chlorite, and kaolinite showed three different "clay facies": 1) high content of smectite, quartz, feldspar, low content of illite, chlorite, kaolinite; 2) high content of illite, chlorite, kaolinite, low content of smectite, quartz, feldspar; 3) high content of smectite, kaolinite, low content of illite, chlorite, quartz, feldspar; Melles et al., 1995). The high smectite content correlates well with the high biogenic silica, which is obviously a good indicator of warmer periods (Colman et al., 1995). However, the XRD analyses of clay minerals are semiquantitative and do not give such detailed information about the structure of the different clay mineral types. If we want to use the clay mineral distribution in Baikal sediments over time as an indicator for climate, a more precise description of the clay minerals themselves is needed. The description of grain forms, the mixed-layer mineral composition, and the structural formulae of the clay minerals help us to better understand the weathering conditions in the hinterland during formation. This information, supplementing the semiquantitative XRD-analysis, can be obtained with a transmission electron microscope (TEM) combined with a scanning electron microscope (STEM) and energy-dispersive X-ray analysis (EDX). For initial investigations to show the importance and necessity of these methods, we chose a time interval from BDP-96-1 for which a drastic climate change has already been identified. The relative diatom abundance in BDP-96-1 indicates a change from warmer to colder climate around 2.8 to 2.6 Ma (BDP Leg II Members, 1997; Williams et al., 1997), and the first pollen analyses indicate a change from warm-wet to cold-dry conditions in the hinterland starting around 3.0 Ma ago (Demske et al., 1999; Demske et al., this issue).

Materials and methods

A total of four samples from core BDP-96-1 were investigated by XRD, TEM, STEM, and EDX analyses. The samples were taken from depths of 92.36 m, 104.71 m, 106.58 m, and 120.11 m, which according to paleomagnetic data (BDP-Members 1998) have ages of 2.42 Ma, 2.78 Ma, 2.85

Ma, and 3.29 Ma, respectively. For XRD anayses the < 2 -μm fraction of each sample was isolated by gravimetric settling (Melles et al., 1995), and oriented mounts treated with ethylene-glycol were measured on a Philips PW 1820 automated powder diffractometer system with CoKα radiation (Ehrmann et al., 1992). Semiquantitative analysis of the major clay mineral groups interstratified illite-smectite [Sme > 40%], illite, chlorite, and kaolinite were made using integrated peak areas and the computer program "MacDiff" (Petschick et al., 1996; http://servermac.geologie.uni-frankfurt.de/HomePage.html). Relative abundances were determined after Biscaye (1964, 1965).

The preparation of samples for particle analysis on TEM was performed according to Henning & Störr (1986). TEM, STEM, and EDX analyses were performed on a JEM-1210 (120kV; 0.34 nm point resolution) from JEOL (Japan) with a LaB_6 Cathode, a LINK OXFORD EDX system and a GATAN MultiScanCamera with GATAN-Software DigitalMicrograph 2.5. We studied several TEM and secondary electron (SE) images of each sample as well as 13 to 41 EDX analyses of smectite-illite mixed-layer minerals. Based on the element analysis, the cation distribution in the clay minerals was calculated with regard to a charge of -22 per formula unit [$O_{10}(OH)_2$] (after Köster, 1977). The cation distribution allows estimation of the layer charge in the tetrahedra and octahedra sheets as well as the interlayer charge of the smectite-illite mixed-layer minerals. The charge distribution within the clay minerals studied was plotted in a Muscovite (M) – Celadonite (C) – Pyrophyllite (P) triangle after Köster (1977). From the P corner to the M/C side there was is an increasing interlayer charge, from the C corner to the P/M side an increasing octahedra charge, and from the M corner to the C/P side a decreasing tetrahedra charge. The cation distribution in the octahedra sheets of the clay minerals was plotted in a Mg-Fe-Al triangle diagram after Banfield et al. (1991).

The proportions of smectite and illite in the mixed-layer minerals were also estimated based on the Al^{IV}/Si ratio in the tetrahedra sheets according to the method of Srodon et al. (1992). Illite-dominated smectite-illite mixed layered minerals have a higher Al^{IV}/Si ratio in the tetrahedra sheet and vice versa. Estimation was performed in 10% steps.

Results

The results of the semiquantitative XRD analyses showed, relatively small differences in percentages between the interstratified illite-smectite [Sme >40%] and the illite in all samples (Table 1). TEM and the additional SE images of the four samples (Fig. 1) revealed representative sample spots. The clay minerals in the oldest sample (Fig. 1, g-h) have a xenomorphic flaky form, whereas the mixed-layer minerals in the three younger samples (Fig. 1, a-f) have more xenomorphic platy and edgy forms. The classification of the mixed-layer minerals (Srodon et al., 1992) and the mean cation distribution in the tetrahedra (IV) and the octahedra (VI) sheets, based on the EDX analysis are shown in Table 2. The highest amounts of smectites, ML10, and ML20 are found in the oldest sample (3.29 Ma). The youngest sample (2.42 Ma), on the other hand, contains higher amounts of illite and illite-rich mixed-layer minerals. The pure smectite phase in all four samples hasd a high Al content in the octahedra sheet, i.e., is a dioctahedral smectite and a nearly pure Si_4 tetrahedron. Thus, the smectite can be identified as a montmorillonite with the source of the layer charge primarily in the octahedra sheet (Moore and Reynolds, 1997, p. 155).

The charge distribution of the clay minerals plotted in the CMP triangle diagrams for each sample (Fig. 2, a-d) was quite similar in the samples with ages of 2.78 Ma and 2.85 Ma and plotwere plotted mainly near the P corner. In the youngest sample (2.42 Ma) the distribution liesay between the M and the P corner, whereas in the oldest sample (3.29 Ma) all clay minerals were plottedplot nearer to the C corner.

The cation distribution in the octahedra sheets of the clay minerals in the three younger samples was is quite similar (Fig. 2 e-h). In the in the oldest sample, however, the Mg content iswas distinctly higher in all clay minerals studied. In general, the four samples showed a trend to towards

Table 1. Results of XRD semiquantitative analysis of interstratified illite-smectite [Sme >40%], illite, kaolinite, and chlorite in the <2-μm fraction of BDP-96-1 sediments; relative aboundances after Biscaye(1964, 1965).

sample depth (m)	Sample age (Ma)	smectite (%)	Illite (%)	Chlorite and kaolinite(%)
92.36	2.42	32	36	32
104.71	2.78	33	33	34
106.58	2.85	30	36	34
120.11	3.29	37	28	34

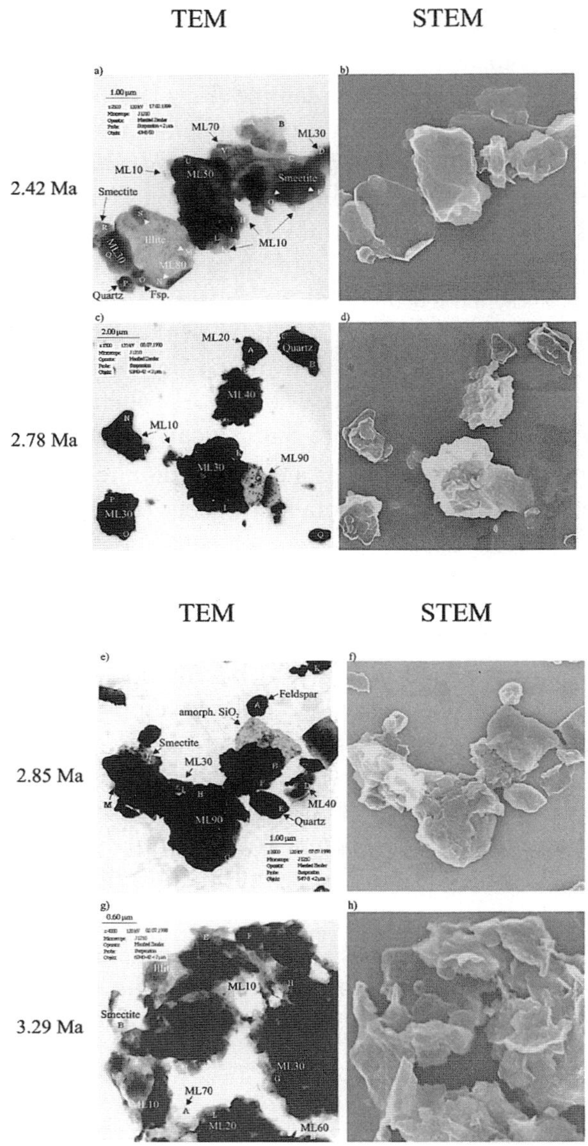

Fig. 1. Selected TEM(a, c, e, g) and STEM(b, d, f, h) micrographs of mixed-layer minerals of the <2-µm fraction of four BDP-96-1 samples, The letters in the TEM-image mark the positions of the EDX analysis.

increasing Fe content in the octahedra sheets as their age decreased that coincidinged with increasing illite content in the mixed-layer minerals.

Discussion

Table 2. Number of EDX measurements(n) of smectite, smectite-illite mixed-layer minerals (MLX with X% illite content), and illite in the <2-μm fraction of BDP-96-1 sediments, classification after Srodon et al. (1992), and mean cation distribution of the octahedra and the tetrahedra, based on charge-22[$O_{10}(OH)_2$].

Sample Age / Sample Depth	2.24 Ma / 92.36 m			2.78 Ma / 104.71 m		
	rel. %	Octahedra	Tetrahedra	rel. %	Octahedra	Tetrahedra
Smectite	25	$Al_{1,31} Fe_{0,50} Mg_{0,15}$	Si_4	0		
ML10	11	$Al_{1,44} Fe_{0,40} Mg_{0,12}$	$Al_{0,07} Si_{3,93}$	31	$Al_{1,51} Fe_{0,36} Mg_{0,13}$	$Si_{3,94} Al_{0,06}$
ML20	14	$Al_{1,31} Fe_{0,51} Mg_{0,18}$	$Al_{0,13} Si_{3,87}$	15	$Al_{1,43} Fe_{0,45} Mg_{0,09} Ti_{0,03}$	$Si_{3,83} Al_{0,17}$
ML30	9	$Al_{1,52} Fe_{0,40} Mg_{0,09}$	$Al_{0,24} Si_{3,76}$	31	$Al_{1,39} Fe_{0,41} Mg_{0,18} Ti_{0,02}$	$Si_{3,77} Al_{0,23}$
ML40	2	$Al_{0,99} Fe_{0,72} Mg_{0,29}$	$Al_{0,35} Si_{3,65}$	15	$Al_{1,39} Fe_{0,45} Mg_{0,16}$	$Si_{3,72} Al_{0,28}$
ML50	11	$Al_{1,25} Fe_{0,53} Mg_{0,22}$	$Al_{0,39} Si_{3,61}$	0		
ML60	5	$Al_{1,31} Fe_{0,61} Mg_{0,08}$	$Al_{0,48} Si_{3,52}$	0		
ML70	7	$Al_{1,50} Fe_{0,39} Mg_{0,11}$	$Al_{0,55} Si_{3,45}$	0		
ML80	5	$Al_{1,15} Fe_{0,47} Mg_{0,31} Ti_{0,07}$	$Al_{0,63} Si_{3,37}$	0		
ML90	2	$Al_{0,53} Fe_{1,06} Mg_{0,36} Ti_{0,05}$	$Al_{0,71} Si_{3,29}$	8	$Al_{0,51} Fe_{1,02} Mg_{0,47}$	$Si_{3,31} Al_{0,69}$
Illite	9	$Al_{0,57} Fe_{0,85} Mg_{0,49} Ti_{0,19}$	$Al_{0,89} Si_{3,11}$	0		
(n)	41			13		

Sample Age / Sample Depth	2.85 Ma / 106.58 m			3.29 Ma / 120.11 m		
	rel. %	Octahedra	Tetrahedra	rel. %	Octahedra	Tetrahedra
Smectite	8	$Al_{1,07} Fe_{0,64} Mg_{0,25} Ti_{0,04}$	Si_4	22	$Al_{1,07} Fe_{0,39} Mg_{0,53} Ti_{0,01}$	Si_4
ML10	8	$Al_{1,52} Fe_{0,31} Mg_{0,17}$	$Si_{3,95} Al_{0,05}$	25	$Al_{1,21} Fe_{0,35} Mg_{0,43} Ti_{0,01}$	$Si_{3,93} Al_{0,07}$
ML20	23	$Al_{1,30} Fe_{0,47} Mg_{0,23}$	$Si_{3,82} Al_{0,18}$	22	$Al_{1,11} Fe_{0,38} Mg_{0,50} Ti_{0,01}$	$Si_{3,82} Al_{0,18}$
ML30	8	$Al_{1,65} Fe_{0,17} Mg_{0,18}$	$Si_{3,80} Al_{0,20}$	9	$Al_{1,20} Fe_{0,33} Mg_{0,45} Ti_{0,02}$	$Si_{3,76} Al_{0,24}$
ML40	8	$Al_{0,95} Fe_{0,68} Mg_{0,31} Ti_{0,06}$	$Si_{3,66} Al_{0,34}$	3	$Al_{1,37} Fe_{0,31} Mg_{0,31} Ti_{0,01}$	$Si_{3,70} Al_{0,30}$
ML50	0			6	$Al_{1,22} Fe_{0,43} Mg_{0,33} Ti_{0,02}$	$Si_{3,63} Al_{0,37}$
ML60	23	$Al_{1,68} Fe_{0,24} Mg_{0,08}$	$Si_{3,54} Al_{0,46}$	6	$Al_{1,09} Fe_{0,50} Mg_{0,39} Ti_{0,02}$	$Si_{3,53} Al_{0,47}$
ML70	15	$Al_{1,69} Fe_{0,31}$	$Si_{3,44} Al_{0,56}$	3	$Al_{1,75} Fe_{0,25}$	$Si_{3,47} Al_{0,53}$
ML80	0			0		
ML90	8	$Al_{0,61} Fe_{1,06} Mg_{0,33}$	$Si_{3,28} Al_{0,72}$	0		
Illite	0			3	$Al_{0,58} Fe_{0,73} Mg_{0,62} Ti_{0,07}$	$Si_{3,22} Al_{0,78}$
(n)	13			32		

Smectite is mainly a product of soil formation produced by chemical weathering and can be generally viewed seen as an indicator of warm and wet climate conditions, whereas higher illite content is indicative of cold and dry conditions (e.g., Berner, 1971; Chamley, 1989). Because of the small differences (Table 1), the semiquanitative XRD analyses of smectite and illite in the four samples do not allow us to judge whether there are significant changes in the clay distribution, and thus in the weathering conditions. However, optical examination of the particles in the TEM and SE images showed that the clay minerals in the oldest sample haved undergone a high degree of chemical weathering because of their xenomorphic, flaky form, and their swelling structure. In contrast, the clay minerals in the three younger samples seemed to be more physically weathered due to their more xenomorphous platy habitus with sharper edges. Mixed-layer minerals, which often can not be distinguished on X-ray diffractograms (Reynolds, 1980), occurred frequently because of the rather weak chemical and structural linkage between successive layers in a given particle and mainly due to weathering or middle late diagenesis (Chamley, 1989). The higher Fe-content of the octahedra sheets of illites and illite-rich mixed-layer clay minerals of all samples, compared to the montmorillonites, suggests that the Fe is displaced by the less mobile Al during transformation from illite over the mixed-layer minerals to montmorillonite in soils.

Nearly all clay minerals studied in the oldest sample show a distinctly higher octahedra charge than in the three younger samples (Fig. 2 a-d), and clay minerals with a higher tetrahedra charge were are present and in the youngest sample (2.42 Ma). These results mirror the higher illite content in the sediment formed during a colder and dryer climate.

The cation distribution in the octahedra sheets showsed more Mg in the octahedra sheet in the oldest sample than in the three younger samples. This indicates that the cation substitution, especially in the octahedra sheet, is much higher in the mixed-layer minerals of the oldest sample. The rate of substitution depends mainly on the water discharge and soil temperatures of the hinterland (e.g., Schachtschabel et al., 1992). Chamley (1989) summarizes investigations on clay mineral transformations by stating that temperate climates coincide with the formation of poorly crystallized degraded 2:1 clay minerals, rather than to neoformation. Thus, the much higher substitution in the clay minerals studied in the oldest sample provides further evidence that the climate was warmer and wetter at 3.29 Ma

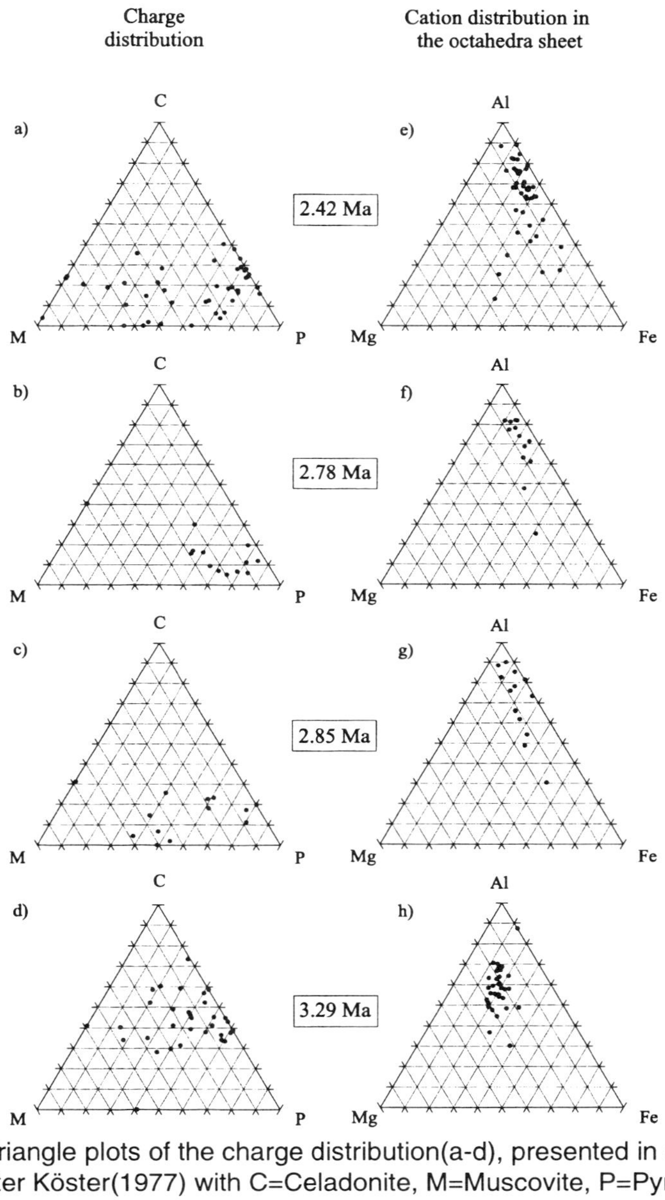

Fig. 2. Triangle plots of the charge distribution(a-d), presented in a CMP triangle after Köster(1977) with C=Celadonite, M=Muscovite, P=Pyrophyllite, and the cation distribution in the octahedra sheet of the smectite-illite mixed-layer minerals(e-h) in the <2-μm fraction of the four BDP-96-1 samples at approx. 2.42, 2.78, 2.85, and 3.29Ma.

than at 2.85, 2.78 Ma, and 2.42 Ma. This is in good agreement with pollen analysis on core BDP-96-1, which shows that a long term cooling trend accompanied by increasing dryness occurred in the Baikal area from 3.0 to 2.5 Ma (Demske et al., 1999, this issue). The distinctly higher content of illite and illite-rich mixed-layer minerals in the youngest sample (2.42 Ma) is obviously due to predominant physical weathering conditions and less intensive chemical weathering in a colder and dryer environment.

Conclusions

This study showed that XRD analysis of the clay fraction of Lake Baikal sediments only indicates a relative abundance of the major clay mineral groups. When used in combination with TEM, STEM, and EDX it enables closer examination of the pure smectite and smectite-illite mixed-layer minerals. We observed a much higher substitution of Mg for Al in the octahedra sheets and a shift in the charge distribution of the mixed-layer minerals of the oldest sample (3.29 Ma) relative to the three younger samples (2.85, 2.78, and 2.42 Ma). Thus, the TEM, STEM, and EDX analyses yieldsed important additional information on the formation and alteration of clay minerals and therefore improvesd our understanding of paleoclimatic and paleoenvironmental conditions in the Baikal area. They are an indispensable supplement to XRD analysis. Further investigations are under way.

Acknowledgements

Special thanks to Manfred Zander for his laboratory assistance and to Lutz Schirrmeister, Bernd Wagner, Birgit Hagedorn, and the two anonymous reviewers who provided important and helpful remarks on the manuscript. We are also grateful to Michael A. Grachev, whose devoted work on Lake Baikal has made international and interdisciplinary studies possible. We wish him all the best and a fast recovery. This study is part of the international Baikal Drilling Project and has been supported by the German Research Foundation (DFG Grant: HU 378/6-1). This paper is contribution No. 1636 of the Alfred Wegener Institute for Polar and Marine Research, Bremerhaven and Potsdam.

References

Banfield, J. F., B. F. Jones and D. R. Veblen (1991) An AEM-TEM of weathering and diagenesis, Albert Lake, Oregan: I. Weathering reactions in

the volcanics. Geochim. Cosmochim. Acta, 55, 2781-93.
Berner, R.A. (1971) Principles of Chemical Sedimentation. McGraw-Hill, New York, 270 pp.
BDP Leg II Members (1997) Continuous Continental Paleoclimate Record for the Last 4.5 to 5 Million Years Revealed by Leg II of Lake Baikal Scientific Drilling. Eos, 78 (51), 597-604.
BDP-Members (1998) A continuous record of climate changes of the last 5 million years stored in the bottom sediments of Lake Baikal. Geologiya i Geofizika, 39 (2), 139-156.
Biscaye, P. E. (1964) Distinction between kaolinite and chlorite in recent sediments by X-ray diffraction. The American Mineralogist, 49, 1281-1289.
Biscaye, P. E. (1965) Mineralogy and Sedimentation of Recent Deep-Sea Clay in the Atlantic Ocean and Adjacent Seas and Oceans. Geological Society of American Bulletin, 76, 803-832.
Chamley, H. (1989) Clay Sedimentology. Springer, Berlin New York Paris, 623 pp.
Colman, S. M., Peck, J. A., Karabanov, E. B., Carter, S. J., Bradbury, J. P., King, J. W. and Williams, D. F. (1995) Continental climate response to orbital forcing from biogenic silica records in Lake Baikal. Nature, 378, 769-771.
Demske, D., Müller, J., Eckert, C., Nowaczyk, N., Mohr, B., Oberhänsli, H., Hubberten, H.-W. and Melles, M. (1999) A sedimentological and palynological record of Lake Baikal at the Pliocene-Pleistocene boundary - A preliminary report. IPPCCE Newsletter, 12, 95-100.
Demske, D., Mohr, B., Oberhaensli, H. (2000) Palaeoclimatic changes from 3.6 to 2.2 Ma B.P. derived from palynological studies on Lake Baikal sediments. This issue.
Ehrmann, W. U., Melles, M., Kuhn, G. and Grobe, H. (1992) Significance of clay mineral assemblages in the Antarctic Ocean. Marine Geology, 107, 249-273.
Henning, K.-H. and Störr. M. (1986) Electron micrographs (TEM, SEM) of clays and clay minerals. Akademie-Verlag, Berlin, 350 pp.
Köster, H. M. (1977) The calculation of crystal-chemical formulas for 2:1 sheet-silicates in terms of measured interlayer charge and cation exchange capacity, and a representation of the charge distribution within the structure by means of triangular coordinates. Clay Minerals, 12, 45-54.
Melles, M., Grobe, H. and Hubberten, H.-W. (1995) Mineral Composition

of the Clay Fraction in the 100 m Core BDP-93-2 from Lake Baikal - preliminary results. IPPCCE Newsletter, 9, 17-22.
Moore, D. M. and Reynolds, R. C., Jr. (1997) X-Ray Diffraction and the Identification and Analysis of Clay Minerals. Oxford New York, Oxford Univ. Press, 378 pp.
Oberhänsli, H., Melles, M. and Hubberten, H.-W. (1998) Problems of stratigraphic age assignments to Lake Baikal sediments: consequences for paleoclimatic interpretations. IPPCCE Newsletter, 11, 48-58.
Petschick, R., Kuhn, G. and Gingele, F. (1996) Clay mineral distribution in surface sediments of the South Atlantic - sources, transport and relation to oceanography. Marine Geology, 130, 203-229.
Reynolds, R.C. (1980) Interstratified clay minerals. In: Crystal structures of clay minerals and their X-ray identification, G.W. Brindley and W. Brown, eds., Mineral. Soc., London, 249-303.
Schachtschabel, P., Blume, H.-P., Brümmer, G., Hartge, K.-H. and Schwertmann, U. (1992) Lehrbuch der Bodenkunde. Stuttgart, Enke-Verlag, 491 pp.
Srodon, J., Elsass, F., McHardy, W. J. and Morgan, D. J. (1992) Chemistry of illite-smectite inferred from TEM measurements of fundamental particles. Clay Minerals, 27 (2), 137-158.
Williams, D. F., Peck, J., Karabanov, E. B., Prokopenko, A. A., Kravchinsky, V., King, J. and Kuzmin, M. I. (1997) Lake Baikal Record of Continental Climate Response to Orbital Insolation During the Past 5 Million Years. Science, 278, 1114 -1117.
Yuretich, R., Melles, M., Sarata, B. and Grobe, H. (1999) Clay minerals in the sediments of Lake Baikal: a useful climate proxy. Journal of Sedimentary Research, 69 (3), 582-596.

Lake Baikal
K. Minoura (editor)
2000 Elsevier Science B.V.

Forest-desert alternation history revealed by the pollen-record in Lake Baikal over the past 5 million years

Kawamuro, K.[1]*, Shichi, K.[1], Hase, Y.[2], Iwauchi, A.[3], Minoura, K.[4], Oda, T.[4], Takahara, H.[5], Sakai, H.[6], Morita, Y.[7], Miyoshi, N.[7], and Kuzmin, M. I.[8]
[1]Forestry and Forest Product Research Institute, Japan
[2]Kumamoto University, Japan
[3]Avance Co., Ltd., Japan
[4]Tohoku University, Japan
[5]Kyoto Prefectural University, Japan
[6]Toyama University, Japan
[7]Okayama University of Science, Japan
[8]Institute of Geochemistry, SB, RAS

Corresponding author
Kimiyasu KAWAMURO
*Forest Environment division, Forestry and Forest Product Research Institute,
Tsukuba Norin Kenkyu Danchi-Nai, Ibaraki, 305 Japan
Fax: 81-298-73-1542,
E-mail: kawamuro@ffpri.affrc.go.jp

Abstract

We carried out the initial palynological study on the BDP96 Hole 1 core drilled by a Russian scientific drilling team in 1996. The pollen record from the core revealed a forest-desert shifting system during the past 5 million years in the region around Lake Baikal. Fifty-seven drastic alternations of forest taiga and subarctic desert appear in the periodic fluctuations in pollen content from the top to bottom of the entire core (192 m). During the Pleistocene, 23 forest-desert shifts coincide with marine $\delta^{18}O$ variations which are attributed to global glacial-interglacial cycles. In the Pliocene, two considerable forest retreats were observed for relatively long periods.

Introduction

The watershed of Lake Baikal constitutes the southern portion of the Siberian taiga and the northern part of the Mongolian steppe, and it is located at a high-latitude (51-56 N), far from the ocean. The paleoclimatic history of the Baikal region sensitively reflects past global changes, such as warm-cold and dry-moist oscillations of climate (Colman et al., 1995).

Paleovegetational reconstruction of the region is indispensable to understanding how the Siberian taiga forest responded in timing and magnitude to past global changes. However, there have been few palynological studies for paleovegetational reconstruction of the past million years in the region. We conducted our initial palynological study on lake sediment taken from the Baikal Drilling Project LegII Hole 1 in 1996 (BDP96 Hole 1).

In the winter of 1996, a Russian scientific drilling team successfully drilled on the submerged topographic high known as the Academician Ridge in 335 m of water (Baikal Drilling Project BDP96 (legII) Members, 1997). The BDP96 Hole 1 core recovered 93% of the upper 119 m and 61% of the 119 to 192 m subbottom (BDP96 Members,1997). Initial paleomagnetic results from the core show that the hemipelagic accumulation rate has been a nearly constant 4.0 cm per thousand years during the past 5 million years and that there are no signs of disconformities anywhere in the core (BDP96 Members, 1997, Williams et al., 1997).

The basic structure of the Plio-Pleistocene glacial-interglacial cycles for south-central Siberia over the last 5 million years (5 Ma) has already been determined from whole core magnetic susceptibility records, diatom abundance data, and biogenic silica (BDP96 Members, 1997,Williams et al., 1997). These paleoclimate proxies indicate that the climate was warm from 5 to 3 Ma, and from 2.5 to 1.8 Ma, and that the warm trend was punctuated by two strong cooling episodes at 2.8 to 2.6, and at 1.8 to 1.6 Ma. The paleovegetation changes obtained from the BDP96 core have not been described yet.

In this article we describe pollen records from the BDP96 Hole 1 core, and discuss a few points in regard to the outline relations between paleovegetation changes and the basic structure of the Plio-Pleistocene environment changes in the Baikal region.

Materials and methods

The Initial palynological study of the BDP96-1 core was conducted on a total of 560 samples, which on average represented 4 samples selected from each 2 m core.

The 1 cm^3 samples were collected from each 2.5 cm depth of core with a measuring spoon and placed into centrifuge tubes. Plastic microspheres, 50,000 grains per cubic centimeter, were added to determine the pollen concentration (Gordon and Can, 1986), and a KOH-HF-Acetolysis mixture method was used for pollen analyses. Heavy liquid separation with $ZnCl_2$ solution was also employed.

The age model for the BDP96-1 core was calculated by using age-depth relations based on the geomagnetic boundaries. Since the sedimentation rate was a constant average of 4.0 cm per thousand years (1ky), 4 samples per 2 m depth are equivalent on average to 4 dates per 50 k years that are available for reconstruction of vegetation changes corresponding to glacial-interglacial cycles.

Results and discussion

Pollen concentrations ranging from 0 to 10^5 grains per 1 cm^3 of sediment are shown. Alternate pollen increases and decreases occur periodically (fig.2 A). The major taxa during the pollen increasing periods are *Pinus* (including Diploxylon and Haploxylon types), *Picea*, and *Abies* pollen, which are thought to have originated from the main components of the taiga forest, such as pines, spruces, and firs (Miyoshi et al., 1999).

According to the results of pollen analyses of the surface gravity core (VER96-2) near the BDP96 Hole 1 site (Fig.1), pollen accumulation rates were calculated as 0-24 grain·cm^{-2}·$year^{-1}$ during the cold maximum of the late glacial time (21 to 14.4 k years BP), 10-444 grain·cm^{-2}·$year^{-1}$ during the last glacial (14.4 to 11.6 ky BP), and 80-1955 grain·cm^{-2}·$year^{-1}$ during the Holocene (9.2 to 3.5 ky BP)(Fig.1). Grichuk (1984) and Takahara et al. (1998) designate that montane subarctic desert and/or montane tundra with no vegetation cover expanded at the cold maximum of the late glacial, and that in the middle Holocene the montane taiga with coniferous and broad-leaved forest surrounding Lake Baikal developed. That is, a pollen accumulation rate of less than 24 grain·cm^{-2}·$year^{-1}$ indicates montane subarctic desert and/or montane tundra, and a rate of more than 80 grain·cm^{-2}·$year^{-1}$ reflects the forest development of the montane taiga, as at present (Fig.1). The accumulation rates of 24 and 80 grain·cm^{-2}·$year^{-1}$ from the surface gravity core could be converted into 6,000 and 20,000 grains·cm^{-3} of the BDP-96 core. Pollen content of less than 6,000 grains·cm^{-3} on the BDP-96 core implies montane subarctic desert and/or montane tundra similar to the vegetation type in the cold maximum of the late glacial, counts greater than 20,000 grains·cm^{-3} indicate development of montane taiga surrounding Lake Baikal in the Holocene. Pollen content between 6,000 and 20,000 grains·cm^{-3} represents a transitional vegetation cover, such as open forest with the prevalent grass. Changes between three types of vegetation cover are separated by pollen contents of less than 6,000 grains·cm^{-3} (desert), 6,000 to 20,000 grains·cm^{-3} (open forest), and more than 20,000 grains·cm^{-3} (taiga), represented in Fig.1 B. This shows 57 drastic alternations between forest taiga and subarctic desert during the past 5 million years.

The bold line in figure 1C shows a smoothed-out curve, which has the effect of grouping smaller fluctuations. The curve represents seven major

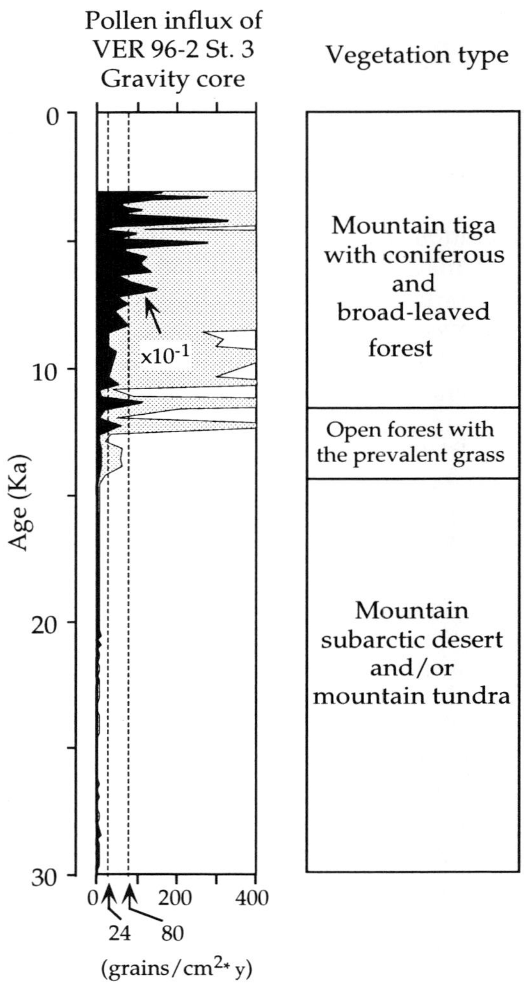

Fig. 1. Relationship between pollen influx of the VER96-2 St.3 core and vegetation type.

depressions, characterized by the relatively long periods dominated by desert and open forest from 5.1 to 4.36 Ma, 3.43 to 3.35 Ma, 2.8 to 2.7 Ma, 2.55 to 2.3 Ma, 1.8 to 1.7 Ma., 1.1 to 0.85 Ma, and 0.33 to 0.17 Ma. Some of them (1.8-1.7,1.1-0.85,0.33-0.17 Ma) coincide with the coldness of the alpha, beta, and gamma glacial, named for the peaks in the steppe/forest

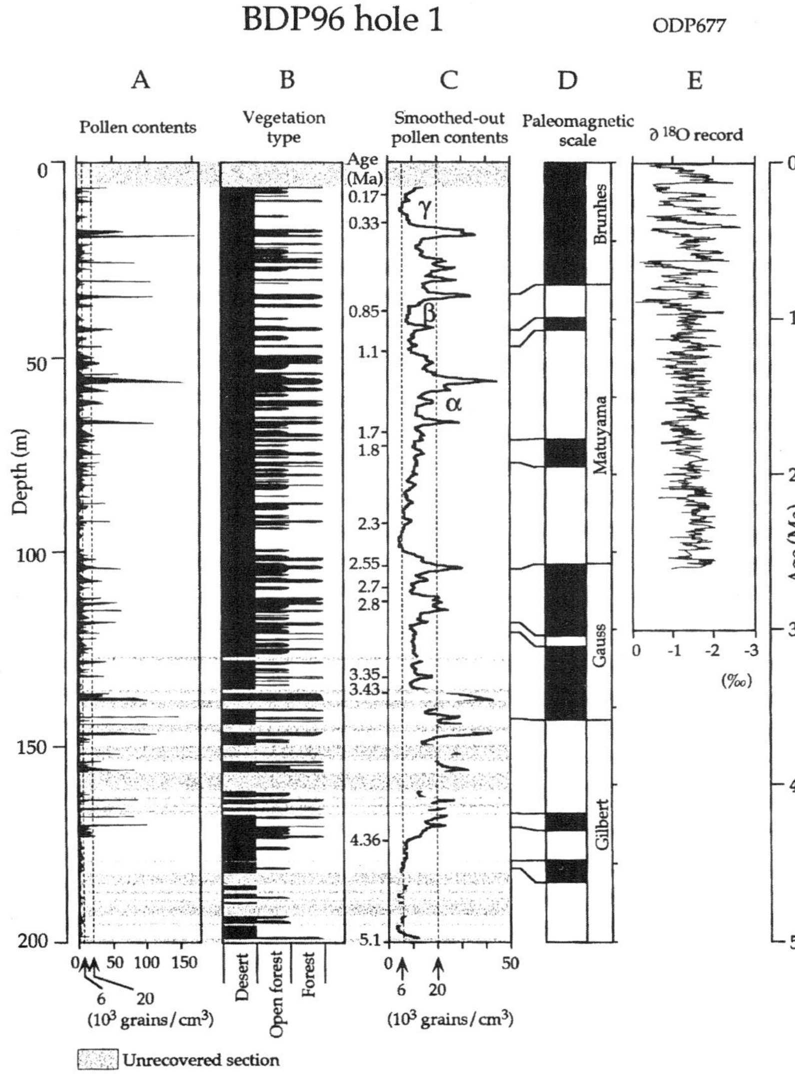

Fig. 2. A: Pollen content of sediment (grains per 1 cm³) plotted against depth of BDP96 Hole 1 core in Lake Baikal. There are some hiatuses at core depth 0-6.4 m,143-145 m, 155-160 m,180-186 m and 188-194 m. These hiatuses are represented by the unrecovered section in the figure. Age(Ma)-depth(m) relations based on the geomagnetic boundaries. B: Vegetation types are divided into forest, open forest and desert by pollen content . C: The bold line shows a smoothed-out curve and has the effect of grouping smaller fluctuations. D: Paleomagnetic scale calculated by age-depth relations. E: Change in the curve of benthonic $\delta^{18}O$ fluctuation at ODP677 (Shackleton et al., 1990)

index obtained from the Black Sea record (D.S.D.P.leg 42B)(Traverse,1978). The depressions from 2.8 to 2.7 Ma and 1.8 to 1.7 Ma correspond to a strong cooling episode obtained from the diatom record of the same BDP96 core (Williams et al., 1997). After 1.7 Ma., 23 forest-desert changes continue, almost perfectly corresponding to the fluctuations in diatom abundance, which are referred to glacial-interglacial cycles. From 2.7 to 1.8 Ma open forest and desert prevail. This indicates the general cooling trend for the Pleistocene glacial epoch. Furthermore, although Williams et al.(1997) suggested that climates were warm during most of the Gilbert and Gauss epochs before 2.9 Ma, based on diatom abundance, the pollen trend shows two distinctive barren periods, from 5.1 to 4.36 Ma, and from 3.43 to 3.35 Ma. The inconsistency between the pollen contents and diatom abundance before 2.9 Ma seems to be due to differences in productivity under terrestrial and aquatic conditions. That is, a warm, dry climate should bring on desert on land, whereas water productivity at the same time should be high.

In figure 1D, benthonic $\delta^{18}O$ fluctuation at ODP677 (Shackleton et al., 1990) is used for correlation with figure 1A. Major peaks in the forest-desert sifting curve after the Jaramillo event almost perfectly coincide with those of benthonic $\delta^{18}O$ variation, which are referred to as global glacial-interglacial cycles. During 2Ma to 1 Ma, before the Jaramillo event, the prevailing age of the forest is correlated with warming trend ages indicated by $\delta^{18}O$ values. Corresponding peaks in the two curves, however, are not visible as follows age.

Conclusion

The results obtained in this initial palynological study prove that the forest-desert alternations are closely related to global glacial-interglacial cycles during the Pleistocene. Considerable forest retreats are visible for relatively long periods during the late Pliocene, and a remarkable desert-prevailing period during the early and middle Pliocene preceded the forest-dominant period. The work in the Baikal area has also shown that these forest-desert alternations are more sensitive to global paleoclimate changes than those found in other continental region records (White et al., 1997, Hsü and Federico, 1980)

Acknowledgements
We thank M.A.Grachev, T.Kawai, D.Williams, and other members of Baikal Drilling Project Leg II in 1996. A.Traverse kindly reviewed the manuscript. This work was supported by the international cooperative project of STA Japan, the US NFS and Russian Academy of Sciences

References

Baikal Drilling Project BDP-96(LegII) Members, 1997, Cotinuous Paleoclimate Record Recovered for Last 5 Million Years, EOS 78,597-604

Colman S.M., J.A. Peck , E.B. Karabanov , S.J. Carter , J.P. Bradbury , J.W. King and D.F. Williams, 1995, Cotinental climate response to orbital forcing from biogenic silica records in Lake Baikal. Nature 378, 769-771

Gordon J.O., 1986, An alternative to exotic spore or pollen addition in quantitative microfossil studies. Can. J. Earth Sci. 23, 102-106

Gricbuk V.P., 1984, Late Quaternary environments of the Soviet Union, 155-200 (Univ. of Minnesota Press).

Hsu K.J. and G.Federico, 1980, Messinian event in the Black Sea. Palaeogeog. Palaeoclimatol. Palaeoecol. 29, 75-93(1979/1980)

Miyoshi N., 1999, (in preparation).

Shackleton N., J. Berger and W.R. Peltier, 1990, An alternative astronomical calibration of the lower Pleistocene timescale based on ODP site 677, Trans.Royal Soc. Edinburgh: Earth Science, 81, 251-262

Takahara H., S. Krivonogov, E. Bezrukova, N. Miyoshi, Y. Morita, T. Nakamura, Y. Shinomiya and K. Kawamuro, 1998, Vegetation history of the southeastern and eastern coasts of Lake Baikal. BICER, BDP and DIWPA Joint International Symposium on Lake Baikal, 102

Traverse A., 1978, Initial Reports of the Deep Sea Drilling Project, 42B, 993-1016

White J.M., T.A. Ager , D.P. Adam , E.P. Leopold , G. Liu , H. Jette and C.E. Schweger, 1997, An 18 million year record of vegetation and climate change in northwestern Canada and Alaska : tectonic and global climate correlates. Palaeogeog. Palaeoclimatol. Palaeoecol. 130, 293-306.

Williams D.F., J. Peck, E.B. Karabanov, A.A. Prokopenko, V. Kravchinsky, J. King and M.I. Kuzmin, Lake Baikal Record of Continental Climate Response to Orbital Insolation During the Past 5 Milllion Years. Science 278, 1114-1117

Lake Baikal
K. Minoura (editor)
2000 Elsevier Science B.V.

Vegetation history of the southeastern and eastern coasts of Lake Baikal from bog sediments since the last interstade

Takahara, H.[1]*, Krivonogov. S. K.[2], Bezrukova, E. V.[3], Miyoshi, N.[4], Morita, Y.[4], Nakamura, T.[5], Hase, Y.[6], Shinomiya, Y.[7], and Kawamuro, K.[7]
[1]Kyoto Prefectural University, Japan
[2]UIGGM SB, Russia
[3]Limnological Inst. SB, Russia
[4]Okayama University of Science, Japan
[5]Nagoya University, Japan
[6]Kumamoto University, Japan
[7]FFPRI, Japan

* Corresponding author: Hikaru Takahara, University Forests, Kyoto Prefectural University, 1-5 Shimogamo, Sakyo-ku, Kyoto 606-8522, Japan. Fax: +81-75-703-5680; E-mail: takahara@kpu.ac.jp

Abstract

Sediment samples from the last glacial and the Holocene were taken from eight bogs and two outcrops of peat sediment on the southeastern and eastern coasts of Lake Baikal and examined for fossil pollen, plant macro-fossils, and charcoal fragments. The chronology of each deposit is based on radiocarbon dating. During the last interstade (approximately 35-30 000 y B.P.), forests on the southeastern coast consisted mainly of spruce and birch, with Gramineae and *Artemisia*. Herbaceous plants such as Gramineae and *Artemisia* and shrubs such as willow, birch, and alder characterized the vegetation of the last glacial maximum. Between 12 000 and 11 000 y B.P., spruce expanded again in the coastal areas. The changes in spruce suggest that the Lake Baikal area was a glacial refugia of dark coniferous taiga. In the early Holocene, the spruce was replaced by fir and pine on the southeastern coast. On the eastern coast, the forests contained both spruce and fir. *Diploxylon* and *Haploxylon* pine forests have shared large areas with birch since 6000 y B.P. Charcoal analysis indicates that there were frequent fires during the glacial age and forest fires resulting from human activity in the most recent period.

Introduction

Lake Baikal is an ancient lake and contains a long record of the Earth's environmental changes, covering several million years in the sediment.

The Lake Baikal area is dominated by the continental high-pressure zone related to global climate. Knowledge of its vegetation history is very important for understanding the world's vegetation history. Unfortunately, in eastern Siberia, only a little pollen data is currently available for interpreting this history.

Long cores from the bottom of Lake Baikal are useful for palynological study of long-term changes in climate and vegetation. However, the palynological data contain vegetation history information for a long period and cover a huge area. On the other hand, mire sediments from coastal areas of Lake Baikal record the local vegetation history over a relatively short period, such as the period since the last glacial. Also, average sedimentation rates in Lake Baikal, estimated by studies on many piston cores and BDP cores, are very small. The whole Holocene corresponds to less than 1 m of the uppermost part of the cores, which is sometime disturbed by turbidites. The Holocene corresponds to less than 1 m of the uppermost part of the cores, which is sometime disturbed by turbidites. The Holocene mire sediments are stratigraphically sequential and have a thickness of several meters. To understand the history of the ecosystem around Lake Baikal, it is important to synthesize the results from cores from Lake Baikal and from mires in the coastal area. From this perspective, we conducted paleoecological studies of mire sediments from the coastal areas of Lake Baikal.

We examined sediment samples from eight bogs and two outcrops on the southeastern and eastern coasts of Lake Baikal for fossil pollen, plant macrofossils, and charcoal fragments. The chronology of each deposit is based on radiocarbon dating. Our results reveal the vegetation history of the taiga around Lake Baikal since the last glacial period.

This paper describes the geology and vegetation around the study sites and the regional vegetation history in the southeastern and eastern coastal areas of Lake Baikal. Detailed local pollen and plant macrofossil data for each site will be presented in other papers in future.

Study sites and geological features

The ten study sites were on the southeastern and eastern coasts of Lake Baikal (Fig. 1). We investigated the structure of the Quaternary sediments in this region. We gave special attention to determining the stratigraphic position and sedimentation history of geological materials from the Holocene and the Late Pleistocene, which we chose for detailed analysis. All the sites investigated are located on a plain. Eight sites were sampled by using hand-operated drills. Two sites were studied in outcrops of peat sediment along rivers.

The study sites along the southeastern coast were located mainly on the

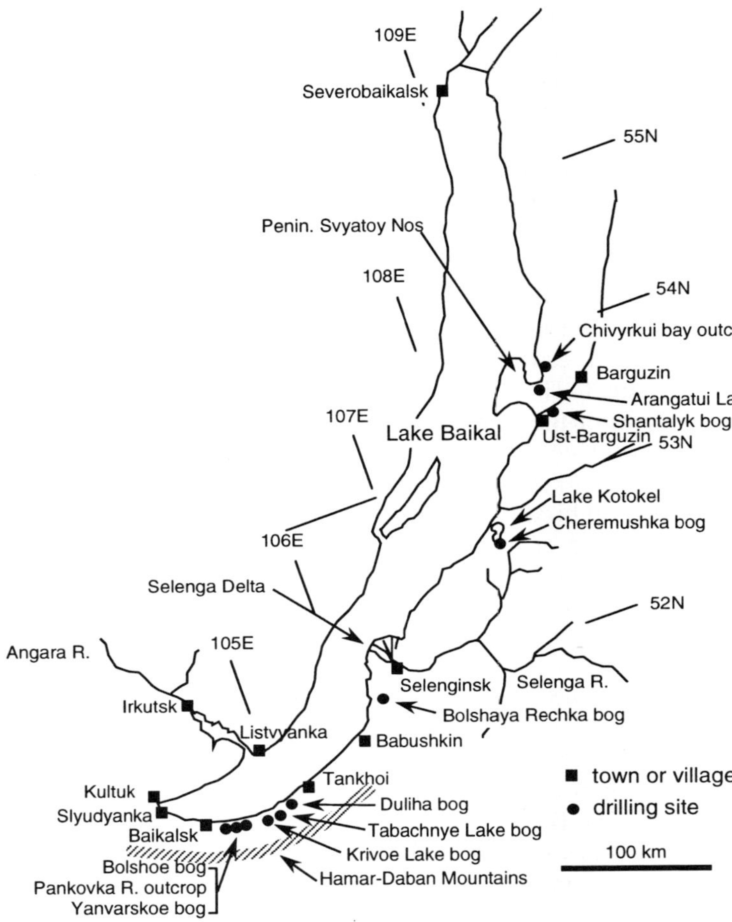

Figure 1. Map of Lake Baikal, showing the locations of the study sites.

Tanhoi Plain in the piedmont of the Hamar-Daban ridge. This plain is part of the Baikal depression bottom, which tectonic movement has connected with the uplift of the Hamar-Daban ridge (Imetkhenov, 1987, Mats, 1992).

The Late Pleistocene geological events are readily apparent. End moraines from the first glaciation in the Late Pleistocene are located on the plain opposite large trough valleys in the Hamar-Daban ridge. This end moraine was subsequently cut by fluvioglacial and river flow. Inset fluvioglacial boulders and pebble terraces 10-15 m high apparently correspond to the second glaciation in the Late Pleistocene. Near the end of the second glaciation, a loess-like layer formed in the interstream areas of the western part of the Tanhoi Plain. Glacial sediments, which can be referred to the last glaciation, were found only within the mountain area.

During the Holocene, sedimentation processes prevailed on the Tanhoi Plain, including swamping of the interstream areas, river activity, and formation of the low terraces of Lake Baikal. The most suitable places for Holocene research on the southeastern coast of Lake Baikal are the autotrophic sphagnum bogs located in interstream areas and small swampy depressions on the end moraine ridges.

On the eastern coast, geological materials related to the final stages of the development of Lake Kotokel were investigated. An extensive sphagnum bog is located on the southern coast of Lake Kotokel, between the villages of Kotokel and Cheremushka. The surface shape of the bed sediments was determined by measuring the thickness of the peat layer every 50 m along a 600-m transect across the bog.

Bogs develop in areas around Lake Kotokel, where the lake coast is retreating as a result of sand sedimentation by wave. At the study site, the modern coast has retreated about 600 m from the ancient abrasion cliffs. Between the cliffs and the coast there are three sandy pre-Holocene bars. The age of the most ancient coastline is 12 100 ± 60 y B.P. (Beta-115297) according to accelerator mass spectrometry (AMS) radiocarbon dating from lagoon sediments behind the second bar.

Conditions similar to the change in the Lake Baikal shoreline were also investigated on the neck of the Svyatoy Nos peninsula on the eastern coast of Lake Arangatui. The separation of this area from Lake Baikal, with the formation of the local lake and the first swamps, was pinpointed to 9400 ± 60 y B.P. (Beta-113968) by AMS radiocarbon dating.

The project study sites can be placed in six groups according to their geological position and the type of landscape:

1. Interstream autotrophic sphagnum bogs: Bolshoe, Yanvarskoe, and Duliha.
2. Bogs near large lakes with retreating shorelines: Chivyrkui Bay, Lake Arangatui, and Cheremushka.
3. Swampy depressions on the surface of end moraines: Lake Krivoe and Lake Tabachnye.
4. Swamped ancient erosion system: Bolshaya Rechka.
5. Swamped floodplain: Shantalyk.
6. Floodplain lake sediments: Pankovka.

The history of each site is determined by the geological situation, which has a significant influence on the structure of the local vegetation assemblage.

Vegetation around the study area

Southeastern area

The Hamar-Daban Mountains are on the southeastern coast of Lake Baikal. The taiga forests in this area consist mainly of evergreen conifers: *Pinus sibirica*, *P. sylvestris*, *P. pumila*, *Abies sibirica*, and *Picea obovata*. Relatively high precipitation in the Lake Baikal area, exceeding 1200 mm (Galaziy, 1993), supports the evergreen conifer forests.

Pinus sibirica can become quite tall, and it is one of the common taller trees on the plain and lower slopes of mountains. *Picea obovata* occurs mainly on flood plains near rivers. *Abies sibirica* stands extend up mountain slopes to the tree line. *Pinus pumila* thickets grow on rocky slopes and above the tree line. *Alnus fruticosa* (*Duschekia fruticosa*) thickets are also found above the tree line. This species also occurs in disturbed sites near villages. *Betula* species are common in all the forest types.

Sandy soil prevails around the Selenga delta. The annual precipitation in this area is 200-400 mm (Galaziy, 1993). *Pinus sylvestris* is predominant on sandy soil and grows along rivers with *Populus tremula* and *Betula* species.

Eastern area

On the eastern coast of Lake Baikal, between Lake Kotokel and the Svyatoy Nos peninsula, the annual precipitation is about 400 mm (Galaziy, 1993). *Pinus sylvestris*, *Larix sibirica*, and *Betula* are dominant in the forests around Lake Kotokel. *Abies sibirica* often occurs in the forests. On the hills near Ust-Barguzin, the forests are composed mainly of *Larix sibirica*, *Pinus sibirica*, *Abies sibirica*, and *Betula*. In addition, *Picea obovata* grows on the flood plains or wet sites near rivers. On sandbanks in the Svyatoy Nos peninsula, *Pinus pumila*, and *P. sibirica* form patchy communities. *P. sylvestris* also grows on sandbars.

Plant macrofossil analysis

The local vegetation can be readily identified from fossil seed and fruit assemblages. Rather large samples are required to obtain sufficient quantities for statistical analysis of plant macrofossils. Therefore, the cores were subdivided into 5-cm-thick samples. In our experience, such subsampling provides enough resolution to reveal patterns in local vegetation development (Krivonogov and Bezrukova, 1993).

The composition of seed and fruit complexes depends on a site's geographical and geological position, as well as on peculiarities of the local vegetation. As bog sediments were the main objects of our research, most of the remains found belong to bog plants. The type of bog is readily distinguished by the macrofossil assemblage. The vegetation of autotrophic sphagnum bogs can be differentiated by the occurrence of *Andromeda polifolia*, *Chamaedaphne calyculata*, *Betula* sect. *Nanae*, and *Oxycoccus palustris*. Eutrophic grassy bogs contain *Menyanthes trifoliata*, *Scheuchzeria palustris, Comarum palustre*, and species of *Carex*. In many of the study sites, aquatic vegetation existed during the initial stages of development of the lake, and is represented by *Nuphar candida*, *Nymphaea peltata, Trapa natans,* and species of *Potamogeton*, *Najas, Caulinia, Eleocharis, Ceratophyllum, Hippuris*, and *Myriophyllum*.

Plant macrofossil analysis indicated that Krivoe Lake, Cheremushka, and Arangatui were initially lakes, but that Bolshoe, Yanvarskoe, Duliha, Bolshaya Rechka, Shantalyk, and Chivyrkui began as grassy bogs.

Macrofossils of tree taxa are well preserved in the sediments from the Krivoe Lake bog. Each layer of sediment contains different proportions of seeds of the main forest taxa: *Larix sibirica*, *Abies sibirica*, *Pinus sibirica*, *P. sylvestris*, and *Picea obovata*.

Data on the development of local vegetation were taken into account when interpreting the palynological results.

Pollen and charcoal analyses

One-centimeter-thick subsamples of the cores were prepared by using standard KOH, HF, and acetolysis procedures for pollen analysis (Faegri *et al*., 1989). Charcoal fragments were measured in the pollen preparations on an image analysis system with the public-domain NIH *Imag*e program.

Correlations between the local pollen assemblage zones are shown in Figure 2. The correlations are based principally on the radiocarbon dates. Pollen data from Duliha and Cheremushka indicate the vegetation history since the last interstade and the last glacial maximum (LGM), respectively. We also obtained pollen data since the late glacial period or early Holocene from five sites: Krivoe Lake, Bolshaya Rechka, Shantalyk, Arangatui Lake, and Chivyrkui Bay. The results from Pankovka, Bolshoe, and Yanvarskoe correspond to the mid to late Holocene.

The results of the pollen analysis are also summarized in Figure 2. In the glacial period, herbaceous pollen such as Gramineae and *Artemisia* and shrub pollen such as *Salix*, *Betula*, and *Alnus* were dominant. *Picea* pollen increased in the late glacial and early Holocene and was followed by *Abies* pollen. The mid and late Holocene were characterized by the dominance of pollen from *Pinus* subgenera *Diploxylon* and *Haploxylon*, and *Betula*.

Vegetation history of the southeastern and eastern coasts

The vegetation history of the southeastern and eastern coasts of Lake Baikal since the last glacial period was reconstructed from pollen, plant macrofossil, and charcoal analyses, and radiocarbon dates for deposits

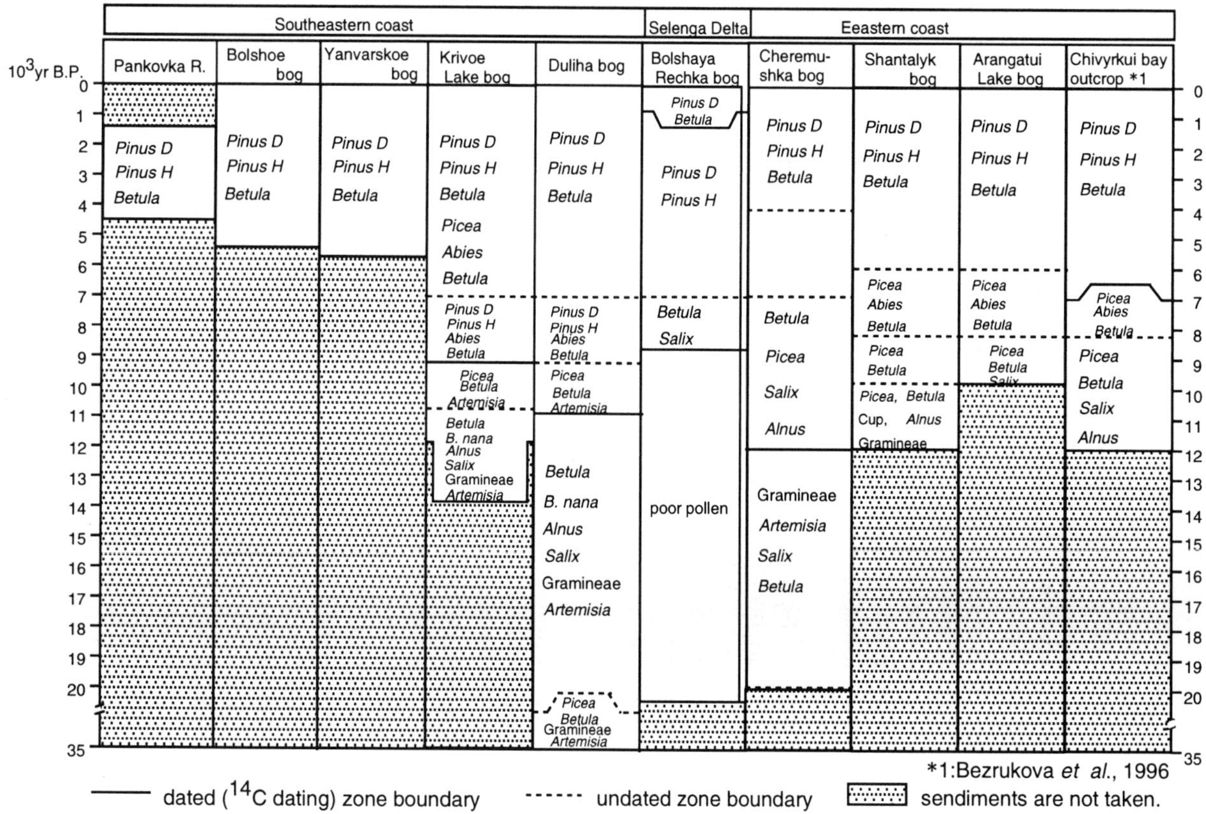

Figure 2. Correlations between the local pollen assemblage zones since the last glacial period on the southeastern and eastern coasts of Lake Baikal. *Pinus D*: *Pinus* subgenus *Diploxylon*; *Pinus H*: *Pinus* subgenus *Haploxylon*.

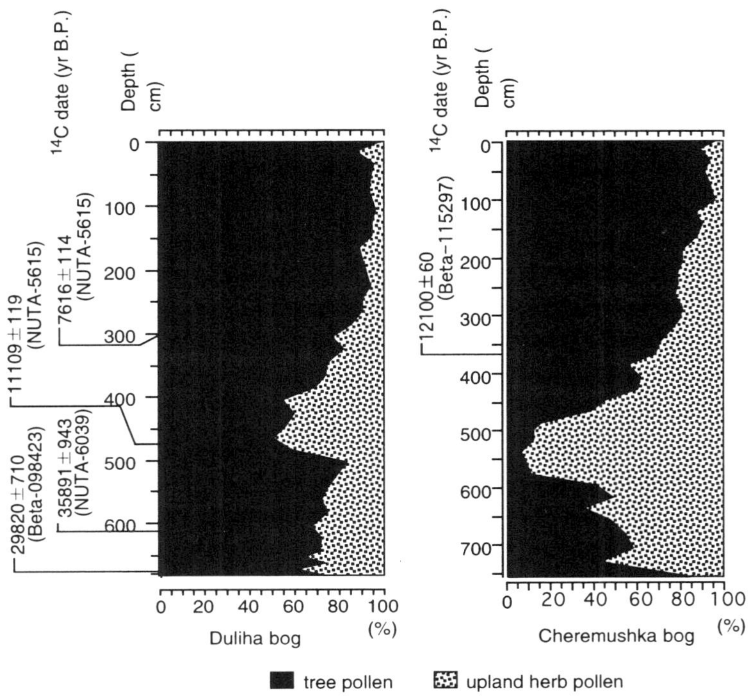

Figure 3. Ratio of tree pollen to upland herb pollen from Duliha and Cheremushka bogs.

from the ten sites.

Last glacial period

The period between approximately 35 000 and 30 000 y B.P. corresponded to the last interstade. Two radiocarbon dates are inverted at the bottom part of the sediment from Duliha bog (Fig. 3). The date at the lowermost level of the peat sediment appears too young, likely caused by the contamination of the bottom of the sediment with younger material when the core was taken. Pollen of this age from Duliha bog indicates that the forests were composed mainly of spruce and birch, along with herbaceous plants such as Gramineae and *Artemisia*. Herbaceous plants and shrubs characterized the vegetation during the LGM on both the southeastern (Duliha) and eastern (Cheremushka) coasts of Lake Baikal. The ratio of upland herb pollen to total pollen during the LGM was up to 40% at Duliha and 80% at Cheremushka (Fig. 3). Tree pollen assemblages from the LGM were composed of *Salix*, *Betula*, and *Alnus*.

Between 12 000 and 11 000 y B.P., spruce (*Picea*) was expanding again along the southeastern and eastern coasts of Lake Baikal. The spruce expansion occurred at different times from site to site. It occurred at about 11 000 y B.P. on the southeastern coast and at about 12 000 y B.P. on the eastern coast. However, more data are needed for detailed ages. An increase in *Picea* pollen during the late glacial is seen in cores from the bottom of Lake Baikal (Bezrukova *et al.*, 1992, Oda *et al.*, 1998, Miyoshi *et al.*, 1999). The spruce expansion was probably due to amelioration of climate in the late glacial. Also, in modern Siberia, spruce forests often occur on flood plains near rivers through the Siberian taiga. This fact indicates that wet sites probably prevailed on the coasts of Lake Baikal after the retreat of the mountain glaciers. However, it is also considered that the spruce did not form dense forests during the late glacial, because of the relatively high ratio of upland herb pollen in the pollen spectra (Fig.3).

Grichuk (1984) reconstructed refugia of dark taiga in western Siberia at the same latitude as Lake Baikal. The occurrence of spruce in the last interstade and its rapid expansion in the late glacial indicate that the Lake Baikal area was a full glacial refugia of dark taiga. Although a dry climate prevailed, spruce probably survived during the full glacial in the relatively humid conditions near the large lake.

Many charcoal fragments were detected in deposits from the LGM and the late glacial. This indicates that fires occurred frequently under the dry climate that prevailed from the LGM to the late glacial.

Holocene

In the early Holocene, fir (*Abies*) increased in forests everywhere. Fir and pine replaced spruce on the southeastern coast (Duliha and Krivoe Lake). On the other hand, spruce coexisted with fir in forests on the eastern coast (Cheremushka, Shantalyk, Arangatui Lake, and Chivyrkui Bay).

In the mid Holocene, *Diploxylon* and *Haploxylon* pines and birch occupied large areas around Lake Baikal. *Diploxylon* pine and birch pollen are long-distance dispersal types. For example, 20% - 30% of the pollen in fir stands where *Pinus sylvestris* does not occur in the upper part of the Hamar-Daban Mountains is *Diploxylon* pine pollen (Ogura et al., 1999). Therefore, *Diploxylon* pine pollen is over-represented in the fossil pollen spectra.

Many plant macrofossils were found in the peat from the Krivoe Lake bog. *Diploxylon* and *Betula* pollen are dominant in the pollen spectra. In contrast, seeds of *Pinus sibirica* characterize the assemblage of plant macrofossils.

Peterson's isopoll map (Peterson, 1993) of *Pinus sibirica* indicated that it expanded northward through the Holocene in western and eastern Siberia.

Our observed pine expansion in the Lake Baikal area is consistent with this.

In some sites on the southeastern coast, the concentration of charcoal fragments, which indicates fires, increases sharply in the upper sediments. This increase in charcoal probably reflects forest fires caused by human activity.

We are extending our study to the northeastern coast and northern area of Lake Baikal, and expect to publish the vegetation history of the light taiga in the northern area of Lake Baikal in the near future.

Acknowledgements

This work has been supported by the Science and Technology Agency of Japan and the Russian Foundation for Basic Research grant No. 97-05-64278. The fieldwork was promoted by the Siberian branch of the Russian Academy of Science and RFBR. We acknowledge Mikhail Grachev for giving us opportunities for the fieldwork and Nadezhda Cherepanova for arranging our fieldwork. We also thank Hideaki Noi, Akihide Takehara, and Takashi Uchiyama for their help in the fieldwork, and Kyoko Tanida, Madoka Ota, and Toshiyuki Fujiki for their help in the laboratory.

References

Bezrukova E. V., P. Letunova and E. Karabanov, 1992, Palynological investigations of Holocene deposits of Baikal, In: International Project on Paleolimnology and Late Cenozoic Climate No. 6, eds. Horie S., Universitätsverlag,, Wagner, Innsbruck, 59-68.

Bezrukova E. V., V. D. Mats, P. P. Letunova, T. Nakamura and S. Fujii, 1996, Holocene peat bogs in Prebaikalia as an object of paleoclimatic reconstructions. Russian Geology and Geophysics, 37, 78-92.

Faegri K., P. E. Kaland and K. Krzywinski (1989). Textbook of Pollen Analysis, 4th Edition. Wiley, New York, 328 pp.

Galaziy G. I., 1993, Baikal Atlas. Nauka, Moscow, 160 pp. (in Russian).

Grichuk V. P., 1984, Late Pleistocene vegetation history, In: Late Quaternary Environments of the Soviet Union, eds. Velichko, A. A., University Minnesota Press, Minneapolis, pp.155-178.

Imetkhenov A. B., 1987, Late Cenozoic sediments of the Lake Baikal coast. Nauka, Novosibirsk, 150 pp. (in Russian).

Krivonogov S. K. and E. V. Bezrukova, 1993, On the history of sedimentation, development of vegetation and climate of the Upper Chara basin in the end of the Late Pleistocene and the Holocene. Russian Geology and Geophysics, 34, 195-206.

Mats V. D., 1992, The structure and development of the Baikal Rift Depression. BICER, Irkutsk, 70 pp.

Miyoshi N., T. Takeuchi, H. Kataoka, K. Ueda, Y. Morita, K. Kawamuro, H. Takahara, Y. Hase, Y. Inouchi, T. Oda and K. Minoura, 1999, Pollen analysis of upper sediment (VER94/5-St. 21) in Lake Baikal. Jpn. J. Palynol. 45: (in press).

Oda T., T. Sato, H. Takahara, K. Minoura, Y. Hase, N. Miyoshi and T. Nakamura, 1998, Vegetation history since the last glacial in the watershed of Lake Baikal, In: Scince of Global Environmental Change - Baikal Drilling Project, eds. Matsumoto G. I., K. Kashiwaya and K. Minoura, Kokon Shoin, Tokyo. pp.137-144 (In Japanese).

Ogura, A. H. Takahara, S. K. Krivonogov, E. V. Bezrukova, Y. Morita, Y. Shinomiya and K. Kawamuro, 1998, Pollen-tree abundance relationship from Hamar-Daban Mountains, in the southeastern area of Lake Baikal. Abstract of BICER, BDP and DIWPA Joint International Symposium on Lake Baikal (November 5-8), Yokohama, Japan, 83.

Peterson G. M., 1993, Vegetation and climate history of the western former Soviet Union, In: Global Climate Since the Last Glacial Maximum, eds. Wright H. E. Jr. *et al.*, University Minnesota Press, Minneapolis, pp.169-193.

Lake Baikal
K. Minoura (editor)
2000 Elsevier Science B.V.

Estimation of paleoenvironmental changes in the Eurasian continental interior during the past 5 million years inferred from organic components in the BDP96 1 sediment core from Lake Baikal

Matsumoto, G. I.[1]*, Kosaku, S.[2], Takamatsu, N.[3], Akagi, T.[2], Kawai, T.[4], and Ambe, Y.[2]

[1]Department of Environmental Information Science, School of Social Information Studies, Otsuma Women's University, Tama, Tokyo 206-8540, Japan.

[2]Faculty of Agriculture, Tokyo University of Agriculture and Technology, Fuchu, Tokyo 183, Japan.

[3]Department of Chemistry, Faculty of Science, Toho University, Funabashi, Chiba 274-8510, Japan.

[4]National Institute for Environmental Studies, Onogawa, Tsukuba, Ibaragi 305-0053, Japan.

Correspondence: Genki I. Matsumoto (Otsuma Women]s University)
TEL 042-339-0088 FAX 042-339-0044 E-mail genki@otsuma.ac.jp

Abstract

Organic components in a BDP96/1 sediment core (length 192 m, age of core bottom 5.2 Ma) from the Academician Ridge (53°41'48"N, 108°21'06"E, water depth, 321 m) in Lake Baikal were studied to estimate paleoenvironmental changes in the interior of the Eurasian continent during the past 5.2 million years. Low TOC content (0.21-1.56%) indicated low primary productivity in the drainage basin of the lake throughout the sedimentation. The TOC values fluctuated largely within a short period of time (less than 20 kilo years) reflecting short-term climate changes. The similarity of the TOC profile and marine $\delta^{18}O$ records revealed that the climate changes in the interior of the continent occurred simultaneously with changes in the marine environment. The TOC profile is a useful proxy for global climate change. TOC and allocthonous organic matter data indicated a fairly warm climate from 4.3 to 2.8 Ma and from 2.0 to 1.5 Ma, and a cool climate from 2.8 to 2.2 Ma and from 1.5 Ma to the present. Allochthonous organic matter is mainly supplied by inflow of river water in warm periods. The decrease in relative abundance of n-C_{27} alkane and n-C_{26} alkanoic acid from about 4 Ma to the present suggests a decrease in the contribution of *Populas* spp. and *Salix* spp. in the drainage basin of the lake, while the increase in relative abundance of n-C_{31} alkane from about 4

Ma to the present probably reflects an increase in herbaceous plants due to aridification of the climate.

Introduction

The increase in greenhouse gasses, such as carbon dioxide, methane, chlorofluorocarbons, and nitrous oxide, in the atmosphere as a result of human activity is causing global warming. Studies on long-term paleoenvironmenal changes are important for estimating the future influence of the global warming induced by human activity, since the earth's climate has changed considerably in the past. Information on long-term paleoenvironmental changes is obtained by analysis of marine and lacustrine sediment cores and by analysis of ice cores from Antarctica and Greenland. However, very little is known about the long-term paleoenvironmental changes in continent interiors. Lake Baikal is one of the oldest (30 million years old), largest freshwater lakes in the world, and is located in southern Siberia on the Eurasian continent. The sediment layer of the lake is more than 7,000 m at its thickest point and can be expected to provide a record of paleoenvironmental changes in the interior of the Eurasian continent during the past 30 million years [e.g., Nikolaev *et al.*, 1985; Baikal Drilling Project BDP-96 (Leg II) Members, 1997]. Organic components have been supplied by living and dead organisms, and they reflect changes in the distribution of organisms as well as changes in the paleoenvironment in the drainage basin.

The Baikal Drilling Project (BDP) has been carried out by scientists from Russia, the United States, Japan, and Germany as a means of elucidating long-term paleoenvironmental changes in the Lake Baikal basin and in the interior of the Eurasian continent. In this study we investigated the total organic carbon (TOC), total nitrogen (TN), hydrocarbons, fatty acids, and sterols in a BDP96/1 sediment core in relation to BDP, and compared the results with marine $d^{18}O$ records.

Materials and Methods

Samples

The sampling methods are described in detail elsewhere [Baikal Drilling Project BDP-96 (Leg II) Members, 1997]. During January - March 1996, BDP96/1 (192 m) and BDP96/2 (100 m) sediment cores were taken from the Academician Ridge (53°41'48"N, 108°21'06"E, water depth 321 m) of Lake Baikal in southern Siberia. Core recovery in the upper 100 m and 100-192 m of the BDP96/1 sediment core was approximately 100% and 70%, respectively.

Analytical methods

The TOC and TN content of the BDP96/1 sediment core was determined with a Fisons NCS NA 2500 automatic elemental analyzer after treatment with 6M hydrochloric acid to remove carbonate-carbon. Hydrocarbons, fatty acids, and sterols were analyzed by the methods of Matsumoto et al. (1979, 1982) and Matsumoto and Watanuki (1992). Briefly, after saponification with 0.5M potassium hydroxide/methanol (80°C, 2h), organic components were extracted with ethyl acetate. The ethyl acetate extracts were chromatographed on a silica gel column (160 mm x 6 mm i.d.), and hydrocarbon and fatty acid-sterol fractions were obtained by elution with hexane and ethyl acetate/benzene (7/3), respectively. Fatty acids were methylated with diazomethane, an aliquot of the methyl ester fraction was trimethylsilylated (TMS) with 25% *N,O*-bis(trimethylsilyl acetamide) acetonitrile solution, and, sterol TMS derivatives were obtained. Hydrocarbons, fatty acid methyl esters, and sterol-TMS derivatives were analyzed with a JEOL JMS Automass 150 gas chromatograph-mass spectrometer equipped with a fused silica capillary column (DB5, 30 m x 0.25 mm i.d., film thickness 0.1 mm). The analytical uncertainty was within ±10%.

Results and Discussion

Primary production and allochthonous organic matter

Paleomagnetism of the BDP96/1 sediment core showed that the sedimentation rate was about 4 cm/kilo years (kys) with almost constant sedimentation. The age of the core bottom was estimated to be 5.2 Ma [Baikal Drilling Project BDP-96 (Leg II) Members, 1998, Sakai *et al.*, unpublished].

TOC content is a marker of biomass and/or primary productivity in the drainage basin and reflects climate changes, i.e., a high TOC content means a warm climate, and a low TOC content means the opposite. The TOC and TN content in the sediment core ranged from 0.21% to 1.56% and from 0.049% to 0.23%, respectively (Fig. 1). These values were very low and indicated low primary productivity in the drainage basin of the lake throughout the course of sedimentation (5.2 million years). The TOC values fluctuated greatly within a short period of time (less than 20 kys), reflecting short-term climate changes.

Although 6.3 m of the core top was thought to have been lost during drilling of the BDP96/1 sediment core [Baikal Drilling Project BDP-96 (Leg II) Members, 1998], comparison of the lithological and physical property profiles of the sediment core showed that the top 8 m had not been recovered at the time of drilling (Minoura *et al.*, unpublished). Comparison of the TOC profile of the sediment core and marine $\delta^{18}O$ data (SPECMAP, NOAA Paleoclimatology Program, 1997) for the past 780 kys also showed an estimated 7.9 m loss of the core top. The TOC data revealed the cycles of glacial and interglacial periods in the last 780 kys.

To facilitate comparison of long-term climate changes in the BDP96 sediment core and marine climate records, weighted mean values (5%) of the TOC and TOC/TN levels are shown in bold solid lines together with the marine $\delta^{18}O$ records compiled by Masuda (1991) in Figure 1. The overall changes in the TOC content of the sediment core were similar to those of the marine $\delta^{18}O$ records. The TOC profile is a useful proxy for global climate changes, and thus these findings demonstrated that the climate changes in the interior of the Eurasian continent occurred simultaneously with the changes in the marine environment. The data indicated that fairly warm climate conditions prevailed from 5.0 to 2.8 Ma, and cool climate conditions from 2.8 to 2.0 Ma. The major cooling at around 2.8 Ma could be explained by the onset of Northern Hemisphere cooling influenced by closing of the Isthmus of Panama in late Neogene times and the uplift of the Himalayan Mountain chain (Keigwin, 1982; Masuda, 1991; Driscoll and Haug, 1998).

The TOC/TN weight ratios of vascular plants are generally high, and those of plankton are low. Thus, the TOC/TN weight ratios reflect the relative contribution of vascular plants (allochthonous organic matter) and plankton (autochthonous organic matter) in the lake sediment. The TOC/TN weight ratios of the sediment core ranged from 2.1 to 13.2, except for an extremely high value of 17.0 at a depth of 194.8 m. The TOC content was well correlated with the TOC/TN weight ratios, with a correlation coefficient of +0.79 (n=696). This shows that the high TOC layers were mainly due to the contribution of allochthonous organic matter.

We calculated the relative contribution of autochthonous and allochthonous organic matter by using the following equations.

$$X + Y = 1.0$$
$$2.1X + 13.2Y = TOC/TN$$

Here, we assumed that the lowest TOC/TN weight ratio of 2.1 represented 100% contribution by autochthonous organic matter (X), and that the highest TOC/TN weight ratio, 13.2, represented 100% contribution by allochthonous organic matter (Y). The pattern of changes in allochthonous organic matter was similar to the pattern in the TOC results (Fig. 1). The increase in allochthonous organic matter in the warm periods can be

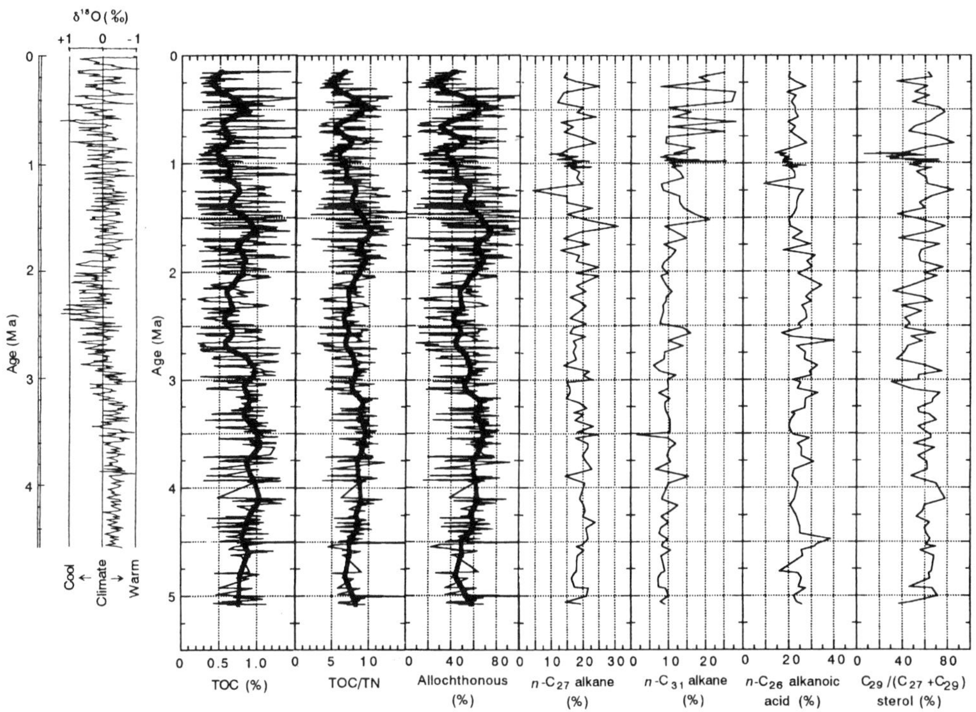

Fig. 1. The vertical distribution of TOC content, TOC/TN weight ratios, allochthonous organic matter, and selected biomarker compounds in the BDP96/1 sediment core, compared with the $\delta^{18}O$ records in marine sediments compiled by Masuda (1991). The n-C_{27} and n-C_{31} alkane values are percentages of total alkanes (n-C_{15} - n-C_{35}). The n-C_{26} alkanoic acid values are percentages of total n-alkanoic acid (n-C_{20} - n-C_{34}). C_{27} sterol: cholesterol and cholestanol. C29 sterol: 24-ethylcholesterol and 24-ethylcholestanol.

explained by the increased inflow of river water.

Molecular fossil information

Normal-alkanes ranging from n-C_{15} to n-C_{35}, with a predominance of odd-carbon numbers, were found in the sediment core together with pristane, phytane, and squalane. The major hydrocarbons were n-C_{25}, n-C_{27}, n-C_{29}, and n-C_{31} alkanes, and/or squalane. Long-chain n-alkanes (C_{20}-C_{35}) originating in vascular plants accounted for 94-100% of the total n-alkanes (C_{15}-C_{35}) and were well preserved in the sediment core. Normal-alkanoic acids ranging from n-C_{10} to n-C_{34} were detected, with a predominance of even-carbon numbers. A blank experiment, however, showed that a considerable portion of the short-chain n-alkanoic and alkenoic acids (C_{12}-C_{19}) represented contamination during the experimental procedures. Thus, only long-chain n-alkanoic acids (C_{20}-C_{34}) were considered for discussion. Sterols (C_{27}-C_{29}) were detected, with preponderance of C_{27} and C_{29} stenols and stanols.

Normal-C_{27} alkane was the predominant hydrocarbon in all sediment layers. This finding is consistent with the abundance of *Betula* spp., *Salix* spp., *Populas* spp., and some *Pinus* spp. in the drainage basin of the lake (Matsumoto et al., unpublished). The percentages of n-C_{27} alkane decreased from about 4 Ma to the present, while n-C_{31} alkane increased abruptly from about 1 Ma to the present (Fig. 1). Normal-C_{26} alkanoic acid, which is abundant in *Populas* spp. and *Salix* spp., tended to decrease from about 4 Ma to the present, as in the case of n-C_{27} alkane. Normal-C_{31} alkane is abundant in some herbaceous plants, such as *Agrosis clavata* (Matsumto et al., unpublished), suggesting that herbaceous plants increased in the drainage basin of the lake due to aridification of climate.

The C_{29} /(C_{27}+C_{29}) sterol ratio (%) is a marker of allochthonous organic matter, because C_{29} sterols (24-ethycholesterol and 24-ethylcholestanol) mainly originate in vascular plants, while C_{27} sterols (cholesterol and cholestanol) are mainly derived from plankton (e.g., Nishimura, 1977; Matsumoto et al., 1982, unpublished). The ratios ranged from 40% to 80%, and were generally similar to the range for allochthonous organic matter calculated from the TOC/TN weight ratios (Fig. 1).

Conclusions

Organic components in the BDP96/1 sediment core from the

Academician Ridge in Lake Baikal were studied to elucidate paleoenvironmental changes in the Eurasian continent interior over the last 5.2 million years. Primary productivity in the drainage basin of the lake during the period was low. The TOC profile is a useful proxy for global climate change, and the TOC and allocthonous organic matter data showed a fairly warm climate from 4.3 to 2.8 Ma and from 2.0 to 1.5 Ma, and a cool climate from 2.8 to 2.2 Ma and from 1.5 Ma to the present. Biomarkers suggested a decrease in the contribution of *Populas* spp. and *Salix* spp. in the drainage basin of the lake from about 4 Ma to the present and an increase of herbaceous plants due to aridification of climate.

Acknowledgements

The authors thank Misses R. Hinata, M. Kiyono, M. Nishi, K. Yamamura, and Y. Yoshino for the biomarker analyses. We thank the Leg II Scientific Drilling Team, led by D. Lykov, and A. Goroglad, Captain M. I. Kazakov of the BDP support ship the R/V Ulan Ude and many colleagues for participation in the logistics, description, and sampling, etc. This research was supported by the Science and Technology Agency of Japan, US National Science Foundation grants EAR94-13957 and EAR96-14770, the Russian Ministry of Science, the Alfred Wegner Institute (AWI no.1297), the Baikal International Center for Ecological Research, and the Japanese Association for Baikal International Research Programs.

References

Baikal Drilling Project BDP-96 (Leg II) Members, 1997, Continuous paleoclimate record recovered for last 5 million years. EOS, 78 (51), 597-604.

Driscoll N. W. and G. H. Haug, 1998, A short circuit in thermohaline circulation: A cause for Northern Hemisphere glaciation, Science, 282, 436-438.

Keigwin L. D., 1982, Isotopic paleocenography of the Caribbean and east pacific: Role of Panama uplift in late Neogene times. Scicene, 217, 350-353.

Masuda F., 1991, Evaluation of the present on the history of paleoclimatic change. J. Geology, 100, 976-987 (in Japanese).

Matsumoto G. I. and K. Watanuki, 1992, Geochemical features of organic components in an extremely acid crater lake (Yugama) of Kusatsu-Shirane Volcano in Japan. Geochem. J., 26, 117-136.

Matsumoto G., T. Torii and T. Hanya, 1979, Distribution of organic constituents in lake waters and sediments of the McMurdo Sound region in the Antarctic. Mem. Natl Inst. Polar Res., Spec. Issue, 13, 103-120.

Matsumoto G., T. Torii and T. Hanya, 1982, High abundance of algal 24 ethylcholesterol in Antarctic lake sediment. Nature, 299, 52-54.
Nishimura M., 1977, Origin of stanols in young lacustrine sediments. Nature, 270, 711-712.
Nikolaev V. G., L. A. Vanyakin, V. V. Kalinin and V. Y. Milanovsky, 1985, The sedimentary section beneath Lake Baikal. Internat. Geol. Rev., 27, 449-459.
NOAA Paleoclimatology Program, 1997, Paleoceanographic Data. File 1. file 17 (http://www.ngdc.noaa.gov/paleo/data.html).

Lake Baikal
K. Minoura (editor)
2000 Elsevier Science B.V.

Paleoenvironmental changes in the Eurasian continent interior inferred from chemical elements in sediment cores (BDP96/1, BDP96/2) from Lake Baikal

Takamatsu, N.[1]*, Matsumoto, I. G.[2], Kato, N.[3], and Kawai, T.[4]

[1]Department of Chemistry, Faculty of Science, Toho University, Funabashi, Chiba 274-8510, Japan
[2]Department of Environmental Information Science, School of Social Information Studies, Otsuma Women's University, Tama, Tokyo 206-8540, Japan
[3]Department of Chemistry, Toho University School of Medicine, Ohta-ku, Tokyo 143-8540, Japan
[4]National Institute for Environmental Studies, Tsukuba, Ibaragi 305-0053, Japan

Correspondence: Nobuki Takamatsu
Tel & Fax: 047-472-5135, E-mail takamatu@chem.sci.toho-u.ac.jp

Abstract

Chemical elements in sediment cores (BDP96/1, BDP96/2) from the Academician Ridge of Lake Baikal in southern Siberia were studied to estimate paleoenvironmental changes in the interior of the Eurasian continent. The terrigenous element content (Al, Ti, and K, etc.) was inversely correlated with diatom abundance, which is a proxy for paleoclimate change. Principal component analysis of data sets for 28 element/aluminum weight ratios in the BDP96/1 core showed that the first principal component scores increased abruptly at 3.4 Ma. The inorganic element content of the BDP96/1 core indicated significant climate change at 2.5 Ma (onset of Northern Hemisphere cooling) and at 3.5 Ma (onset of the aridification of Asia) in the Eurasian continent interior. Spectral analysis of K distribution in the BDP96/2 core showed that the relative variance of the frequency of eccentricity increased from 2.5 Ma to the present. The contributions of frequency of precession and obliquity in the Eurasian continent interior were much greater than those of marine climate records (SPECMAP).

Introduction

Lake Baikal is located in the great Baikal Rift of southern Siberia (51° 30'~56° 00' N, 104~110° E) and is the world's deepest (1632 m; Weiss *et al.*, 1991) and most voluminous freshwater lake (23,000 km^3; Kozhov, 1963). Lake Baikal sediments contain a long, continuous stratigraphic record since the Miocene epoch (Hutchinson *et al.*, 1992). Short *et al.* (1991) suggested that the energy balance model based on the Milankovitch theory indicated that the temperature responses of this region may have been as high as 14°C during the past 800 kilo years (ky). Lake Baikal sediments therefore provide the best opportunity to estimate climate change in the central Eurasian continent interior and to make comparisons with the marine climate records.

Between January and March 1996, the Nedra drilling team of Russia succeeded in obtaining two long cores (BDP96/1 and BDP96/2) from a topographic high region known as the Academician Ridge (53°41'48"N, 108°31'00"E, water depth 321 m) in Lake Baikal. Because the Academician Ridge is isolated from direct fluvial and down slope sedimentation, the cores have no major hiatuses or disconformities [(Baikal Drilling Project BDP-96 (Leg II) Members, 1997)]. The paleomagnetic evidence showed that the recovered sediments span 5.2 Ma (BDP96/1 core) and 2.5 Ma (BDP96/2 core), respectively, with a constant sedimentation rate of approximately 4 cm/ ky.

The purpose of this study was to estimate the paleoenvironmental changes in the Eurasian continent interior during the past 5.2 million years (My) from the levels of inorganic chemical elements in the BDP96/1 core, and to examine the relative contribution of the frequency of precession, obliquity, and eccentricity by spectral analysis of K distribution in the BDP96/2 core.

Materials and Methods

The sampling methods are shown in detail elsewhere [Baikal Drilling Project BDP-96 (Leg II) Members, 1997]. We analyzed the inorganic element content of 142 samples of the BDP96/1 core at approximately 1.4 m intervals, and 447 samples of the BDP96/2 core at intervals of approximately 20 cm. A 100 mg powdered subsample of each sediment sample was digested with a mixture of nitric acid (3 mL), hydrogen peroxide (2 mL), and hydrofluoric acid (4 mL) in a Teflon vessel. The vessel was then exposed for 90 min to microwaves generated by a magnetron in a microwave-assisted decomposition apparatus (OI. Analytical Co. Ltd.). After drying the digested sample solution, the residue was dissolved in 1 mL of nitric acid and transferred to a 100 mL volumetric flask with distilled water. All acids and reagents used in this experiment were of very pure grade (TAMAPURE-AA-100, Tamakagaku Kogyo Co. Ltd). After appropriate dilution, the content of the elements (Li, Be, B, Al, Sc, Ti, V,

Cr, Mn, Co, Ni, Cu, Zn, Ga, Ge, As, Rb, Sr, Mo, Cs, Ba, W, Pb, Th and U) in the sediment samples was determined with an inductively coupled plasma-mass spectrometer (Perkin Elmer ELAN 5000) by using rhodium as the internal standard. Total-Fe_2O_3 content was determined by a colorimetric method. Calcium oxide, MgO, Na_2O, and K_2O content was determined by atomic absorption spectrometry. The precision and accuracy of all of the element data were assessed by comparison with the results of analysis of JLk-1 (lake sediment) issued by the Geological Survey of Japan. The uncertainly of precision and accuracy for all elements was less than 10%.

Results and Discussion

Distributions of chemical elements

Depth along the BDP96/1 and BDP96/2 cores was converted into age by using an age scale based on paleomagnetic epochs (Sakai *et al.*, unpublished data), except for ages less than 38 m of BDP96/2 core, which were determined by using the tuning data (Kashiwaya *et al.*, 1998).

The profiles of some typical selected elements in the BDP96/1 core are shown in Fig. 1. The content of alkali metals (Na and K) and alkaline earth metals (Ca and Mg) increased gradually from 5.2 Ma to the present, with repeated major fluctuations after about 3 Ma, while the aluminum content was relatively high. The Ti/Al weight ratios were almost constant throughout the core, indicating that these elements were well preserved in silicate minerals under chemical weathering. The content of terrigenous element (Al, Ti, V, and W, etc.) was inversely correlated with the diatom frustule content of the core, suggesting that they can also be serve as proxies for major cooling or warming episodes of the southern Siberia.

Abnormally high As, Mo, Ni, and Co content was occasionally observed (Fig. 1). These elements are known to be sensitive to changes in the redox potential of sediments. Their high content may reflect low redox conditions caused by chemical reactions such as sulfate reduction after sedimentation (T. Takamatsu, 1998).

Principal component analysis

We applied principal component analysis to our data sets to examine whether the inorganic chemical elements of the BDP96/1 core reflect geological events, such as the onset of Northern Hemisphere cooling, which is known to have occurred at the time of the Matsuyama/Gauss reversal at 2.5 Ma (Shackleton *et al.*, 1984).

The ratios of 28 elements to Al in the BDP96/1 core were used as the explanatory variables to compensate for the dilution effect by diatom frus-

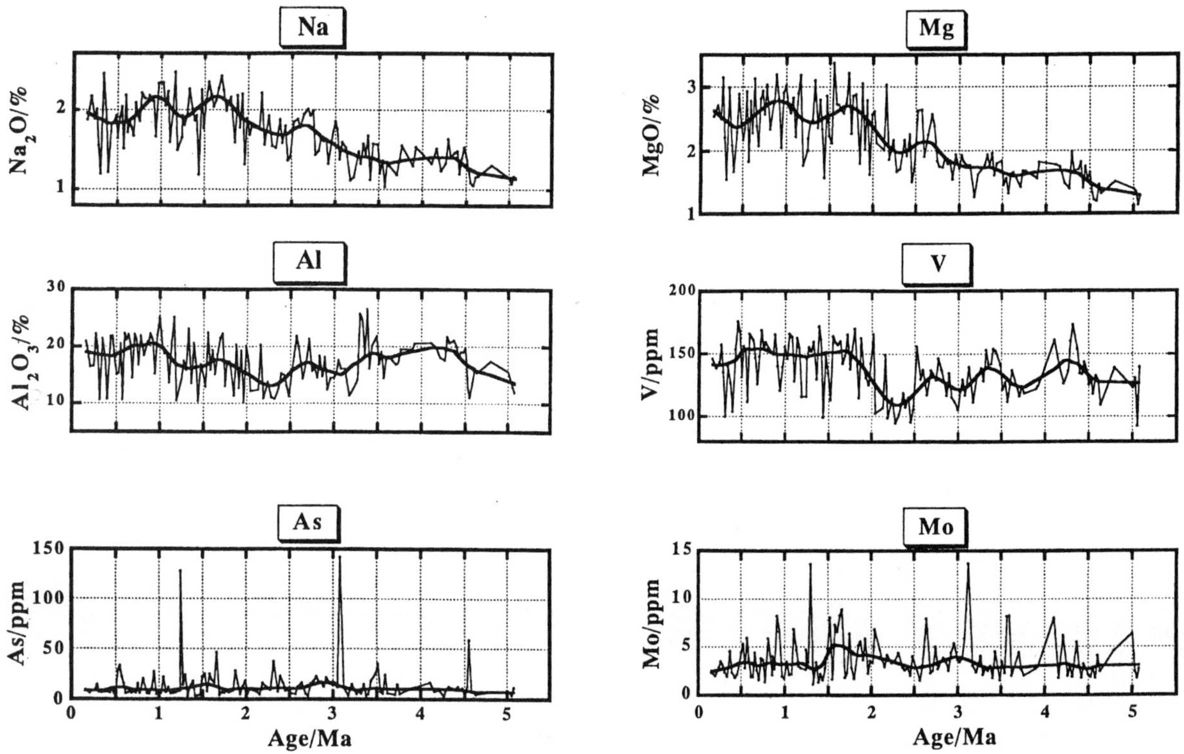

Fig. 1. Variations in the distribution of some typical elements in the BDP96/1 core during the past 5 million years.

tule content. The first principal component scores were relatively low, from 5.2 to 3.5 Ma, with narrow fluctuations, but they increased abruptly at 3.4 Ma, with wide fluctuations. They decreased slightly at 2.8 Ma but increased again at 2.5 Ma (Fig. 2). Mats (1993) reported that rapid rifting began in Lake Baikal at about the time of early-late Pliocene boundary at 3.5 Ma, and the aridification of Asia, which began at 3.6 Ma, is probably related to the uplift of the Tibetan Plateau and Himalayas (e.g., Raymo and Ruddiman, 1992; Ding *et al.*, 1997). The inorganic element content of the BDP96/1 core reflects the significant climate changes at 2.5 Ma (onset of Northern Hemisphere cooling) and at 3.5 Ma (onset of the aridification of Asia) in the Eurasian continent interior.

Spectral analysis

We performed spectral analysis of the K distribution in the BDP96/2 core at 500-ky intervals over the past 2.5 Ma (except for 0-1.0 Ma) to estimate changes in the contribution of precession, obliquity, and eccentricity cycles (Fig. 3). The 41-ky obliquity band was dominant from 2.5 to 1.5 Ma, whereas the 100-ky cycle was stronger from 1.5 Ma to the present. The 23-ky and 1.9-ky precession frequencies were distinct for the past 1000-ky (including Brunhes chron 800-ky), although the frequencies were

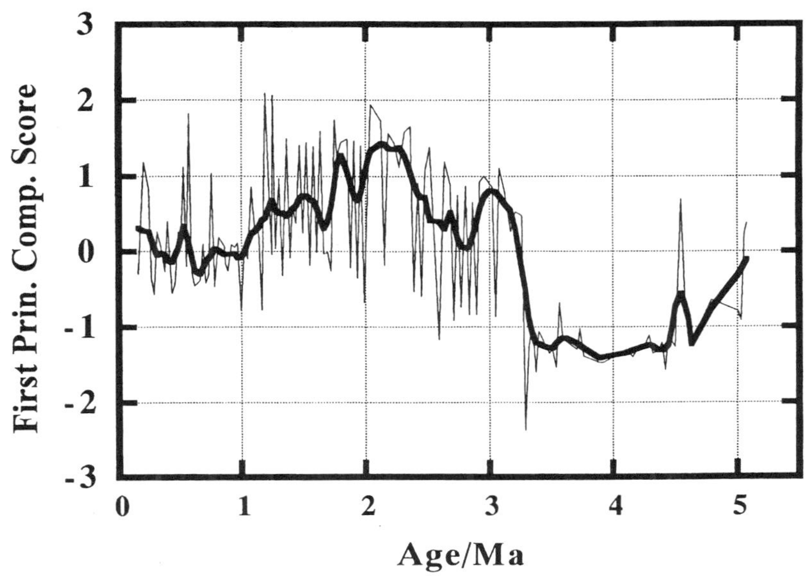

Fig. 2. Variation of the first principal component score obtained by principal component analysis of data sets for 28 element/aluminum weight ratios in the BDP96/1 core.

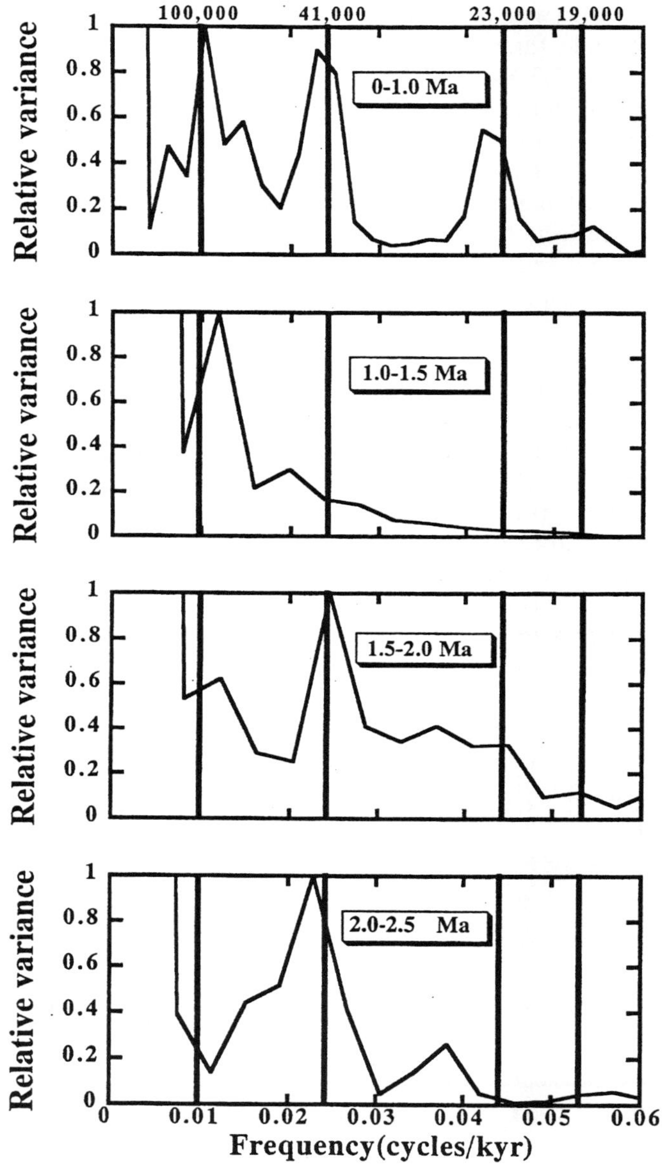

Fig. 3. The power spectra of K distributions in the BDP96/2 core over the past 2.5 million years.

not very large from 2.5 to 1.5 Ma. Kashiwaya *et al.* (1997) found from spectral analysis of the water content and grain size of the BDP96/2 core that a precession frequency of 23-ky was high in the past 500 kys. The relative variance for K for the last 1000ky was distinctly different from that of SPECMAP (NOAA Paleoclimatelogy Program, 1997) over the past 800 ky (Fig. 4), reflecting the fact that Lake Baikal is located at a relatively high latitude and in the central Eurasian continent interior where strong obliquity and precession bands occur.

Conclusions

Al and Ti were found to be terrigenous elements over the period studied. Terrigenous element distributions were inversely correlated with diatom abundance, especially those younger than 2.5 Ma. Principal component analysis for data sets of 28 element/aluminum weight ratios in the BDP96/1 core revealed that the abrupt change that started at 3.5 Ma was associated with the uplift of the Tibetan Plateau and Himalayas, and that the second occurrence of high values at 2.5 Ma may be related to the onset of Northern Hemisphere cooling. The spectral analysis of the K distribution in the BDP96/2 core shows that the 41-ky obliquity cycle is strong from 2.5 to 1.5 Ma, while the variance in the 100-ky eccentricity band, the 41-ky obliquity, and 23-ky precession are large from 1.0 Ma to the present.

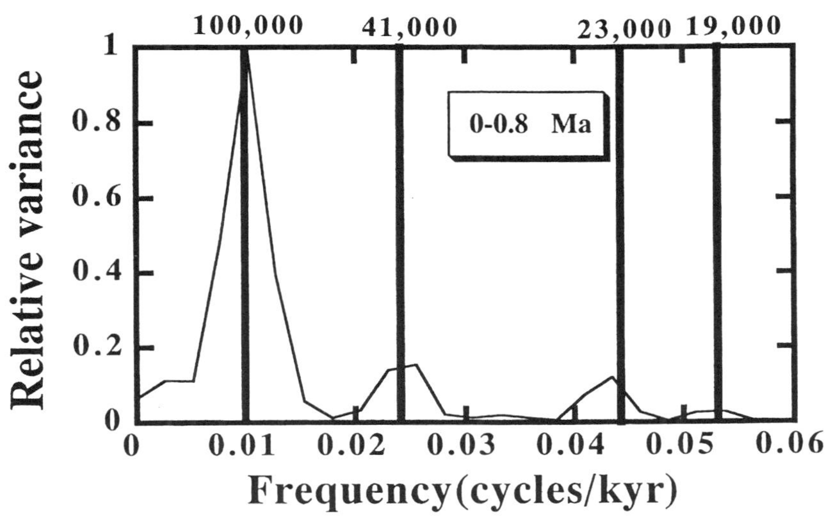

Fig. 4. The power spectrum of SPECMAP data for the last 800 kilo years.

These results correlate well with the late Pleistocene global climate change in the central Eurasia continent interior.

Acknowledgements
The authors thank Messrs T. Kawamura and J. Yamamoto and Misses M. Kodama, K. Kondo and A. Oshitanai for their chemical analyses. We thank the Leg II Scientific Drilling Team, led by D. Lykov, and A. Goroglad, Captain M. I. Kazakov of the BDP support ship Ulan Ude and many colleagues for participation in the logistics, description, sampling, etc. This research was supported by the Science and Technology Agency of Japan, US National Science Foundation grants EAR94-13957 and EAR96-14770, the Russian Ministry of Science, the Alfred Wegner Institute (AWI no.1297), the Baikal International Center for Ecological Research, and the Japanese Association for Baikal International Research Programs.

References

Baikal Drilling Project Leg II Members (1997): Continuous paleoclimate record recovered for last 5 million years. EOS, **78**(51), 597-604.

Ding, Z., Rutter, N. W. and Liu, T. (1997): The onset of extensive loess deposition around the G/M boundary in China and its paleoclimatic implications, Quat. Internat., **40**, 53-60.

Hutchinson, D. R., Golmshtok, A. J., Zonenshain, L. P., Moore, T. C., Scholz, C. A., and Kligord, K. D. (1992): Depositional and tectonic framework of the rift basins of Lake Baikal from multichannel seismic data. Geology, **20**, 887-890.

Kashiwaya, K., Ryugo, M. (1998): Long-term climato-limnological oscillation during the past 2.5 million years printed in Lake Baikal sediments. Geophy. Res. Lett., **25**, 659-662.

Kozhov, M. (1963): Lake Baikal and its life. W. Junk. The Hague, Netherlands, 344 p.

NOAA Paleoclimatelogy Program (1997): Paleooceanographic Data. File 1. File 17 (http://www.ngdc.noaa.gov/paleo/data.html).

Raymo, M. E., Ruddiman, W. F., Backman, J., Clement, B. M. and Martinson, D.G. (1989): Late Pliocene variation in northern hemisphere ice sheet and North Atlantic deep water circulation. Paleoceanography, **4**, 413-446.

Shackleton, N. J., Backman, J., Zimmerman, H., Kent, D. V., Hall, M. A., Roberts, D.G., Schnitker, D., Baldauf, J. G., Desprairies, A., Homrighausen, R., Huddlestun, P., Keene, J. B., Kaltenback, A.J., Krumsiek, K. A. O., Morton, A.C., Murray, J. W. and Westberg-Smith, J. (1984): Oxygen isotope caliblation of the onset of ice-rafting and history of

glaciation in the North Atrantic region. Nature, **307**, 620-623.
Short, D. A., Mengel, J. G. Growley, T. J. Hyde, W. T. and North, G. R. (1991):Filtering of Milankovitch cycles by Earth's geography. Quat. Res., **35**, 157-173.
Takamatsu, T. (1998) : Scince of Global Environmental Change -Baikal Drilling Project. G. I., Matsumoto, K. Kashiwaya and K. Minoura (eds), Kokon Shoin, Tokyo. 137-144 pp. (In Japanese). Weiss, R. F., Carmack, E. C. and Koropalov, V. M. (1991) Deep-water renwal and biological production in Lake Baikal. Nature, **349**, 665-669.

Lake Baikal
K. Minoura (editor)
2000 Elsevier Science B.V.

A new preparation method for qualitative and quantitative analysis of fossil sponge spicules by light microscopy

Eckert, C.[1]*, Veinberg, E. V.[2], Kienel, U.[1], and Oberhänsli, H.[3]
[1] Alfred Wegener Institute for Polar and Marine Research, RD Potsdam, Telegrafenberg A43, 14473 Potsdam, Germany
[2] Limnological Institute Irkutsk, Siberian Branch of RAS, Ulan-Batorkaya str. 3, 664033 Irkutsk, Russia
[3] GeoForschungsZentrum Potsdam, Telegrafenberg, 14473 Potsdam, Haus C, Germany
*Correspondence should be addressed to eckert@awi-potsdam.de, fax: ++49-331-288-2137

Abstract

In this paper we present a gravity settling method for preparation of siliceous freshwater sponge spicules. Application of the new technique makes qualitative and quantitative analyses of fossil sponge assemblages more reliable and reproducible. Paleoecological interpretation limnic environments also become more accurate. The method has been tested on sediments from the Lake Baikal, but it can also be applied to other fresh water lake sediments lacking carbonate.

Introduction

The uniqueness of Lake Baikal and its sediments, compared to other lakes, is documented by the diversity of its living flora and fauna and of its fossil assemblages. The distribution pattern of the spicules of siliceous sponges and diatom frustules holds particular promise for ecological interpretations, and both are well preserved and present in large quantities in the sediments. At the Academician Ridge the sediments consist of 10% to 90% biogenic silica by weight. Diatom frustules are the main constituents and are therefore preferentially used for biostratigraphic analysis. The sponge spicules are a candidate for a new tool for stratigraphic assignment in the sediments deposited in Lake Baikal. Dr. Y. Masuda (Kawasaki Medical School, Okayama, Japan) and Prof. S.M. Efremova (State University of St. Petersburg, Russia) initiated an extensive, ongoing investigation of living siliceous freshwater sponges, and at the beginning of the 90s began to revise the classification of the endemic species of Lake Baikal (Efremova, Veinberg and Masuda, unpublished results).

Since 1993 a revival of the idea of Martinson (1936a,b, 1955), who investigated the fossil spongiofauna of Central Asia in the 30s, has become evident and has been actively promoted by one of the co-authors (Veinberg E.V., 1997). However, a new method of preparation had to be developed to assure correct identification of individual sponge spicules and thereby guarantee reliable and reproducible quantitative analyses. The new method we developed and discuss in this paper replaces the preparation technique based on smear slides, which is no longer appropriate for the new quality demands, and it is based on the preparation method used by Zielinski (1993) for diatom analysis. The new method allows determination of specific sponge spicules in sediments and provides the basis for a reproducible quantitative analysis of sponge spicule assemblages.

Veinberg *et al.* (1999) for the first time presented evidence that sponge spicules are indeed an excellent biostratigraphic tool and that they are valuable paleoecological indicators of changing environmental conditions in Lake Baikal.

Material

The sediments investigated were sampled from a gravity core collected during a cruise by the research vessel "Vereshagin" in summer 1997. The core was 480 cm long. Sediments from the Academician Ridge have a relatively homogeneous composition, with grain size fractions of <2 μm, 2-32 μm, and >32 μm accounting for 25 - 50%, 30 - 60% and 10 - 30%, respectively, by weight.

Clay minerals are major components of Baikal sediments, and biogenic silica in the form of diatoms and sponge spicules is a minor constituent. The >32-mm grain size fraction has the following average composition: quartz, up to 45%, feldspar, up to 40%, heavy minerals, up to 15%, by weight. The organic carbon content (pollen, plant remains, and humic acids) averages. 0.5 % by weight, and never exceeds 3 % by weight. The sediment is carbonate-free.

Sample preparation

A 5 g sample of freeze dried sediment is oxidised with 100 ml hydrogen peroxide (3% concentration) mixed with a drop of conc. NH_4OH, and the sample is disaggregated on a shaker for 12 hours. After disaggregation, the samples are washed through a 32 mm sieve. This method of fractionation is most suitable, because sponge spicules only rarely reach a length of 100 mm. The influence of a few adhesive clay particles, which can interfere with identification under a light microscope, can be ignored. The <32-mm fraction is stored for subsequent sedimentological analysis, whereas the

>32-μm fraction is carefully rinsed off the sieve into a 50 ml polyeturan (PE) bottle with a cover or into other containers that can be sealed and stored until used for the following preparation.

Several Petri dishes (diameter 4.9 cm), according to the number of samples, are placed on a horizontal surface (Fig. 1). Two cover glasses are placed into each dish, and the dishes are filled approximately 2/3 full with a gelatine solution (0.06 g of gelatine dissolved in 700 ml distilled water). After shaking the PE bottle, a 1ml aliquot of from the well in which the >32-μm fraction was suspended is pipetted onto the gelatine solution so that the solution is uniformly distributed over the entire Petri dish. The Petri dish with the sample solution is allowed to stand for at least 2 hours to permit even gravitational settling of the suspended material onto the glass holders. Prepared filter strips are then inserted into the Petri dish with one end touching the bottom of the dish and the other end touching the surface outside the Petri dish. Capillary attraction draws the water from the dish to the horizontal surface, from which it is removed with a pipette or absorbing paper.

A sedimentation 'stairway' (Fig. 2), designed by Zielinski (1993), makes this preparation step much easier. The arrangement of the stairs permits simultaneous preparation of up to 50 samples. If fewer samples are to be processed at the same time we recommend using a sedimentation "lattice" (Fig. 3). To make it more efficient we attach a plastic or metallic lattice to a deep black (settling process can be better observed) plastic basin

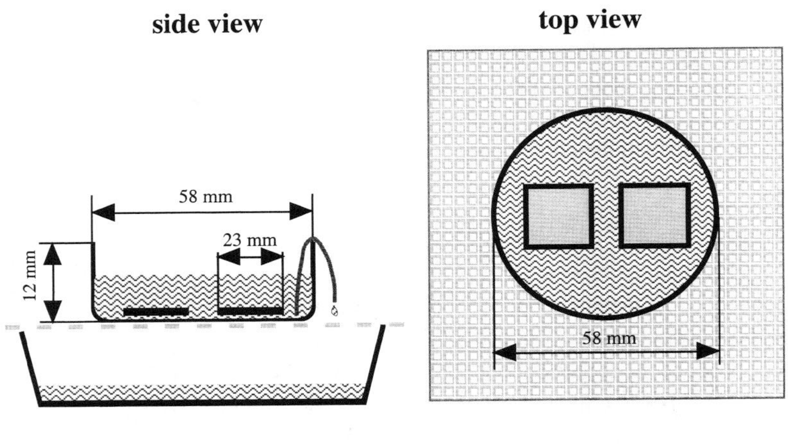

Fig. 1: Schematic diagram of the Petri dish with cover glasses for sample settling. The basin and the lattice are not in scale.

(approx. size: 40 x 30 x 5 cm) and place the Petri dishes on the grid. The gelatine solution is drained from the Petri dish through the lattice to the bottom of the plastic basin with absorbent paper strips until the Petri dish is dry and an equally dispersed sample (>32 μm fraction) has settled on the dry cover glasses.

The next step is carried out in the hood (Fig. 5). A hot plate covered with aluminium foil is heated to approx. 80°C, and a labelled (use a water-proof pencil) glass slide the size (or double the size) of the cover glass is placed on the hot plate and coated evenly with Canada balsam. The dry

Fig. 2: The sedimentation "stairway" after Zielinski (1993). The material is plexiglass.

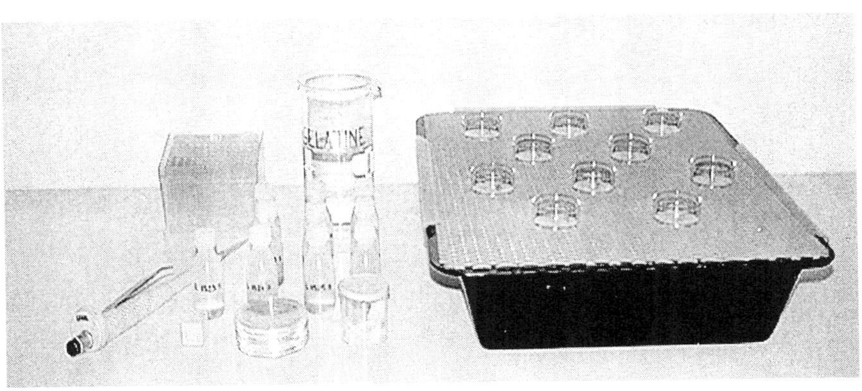

Fig. 3: The sedimentation "lattice" used for preparation of a smaller number of samples.

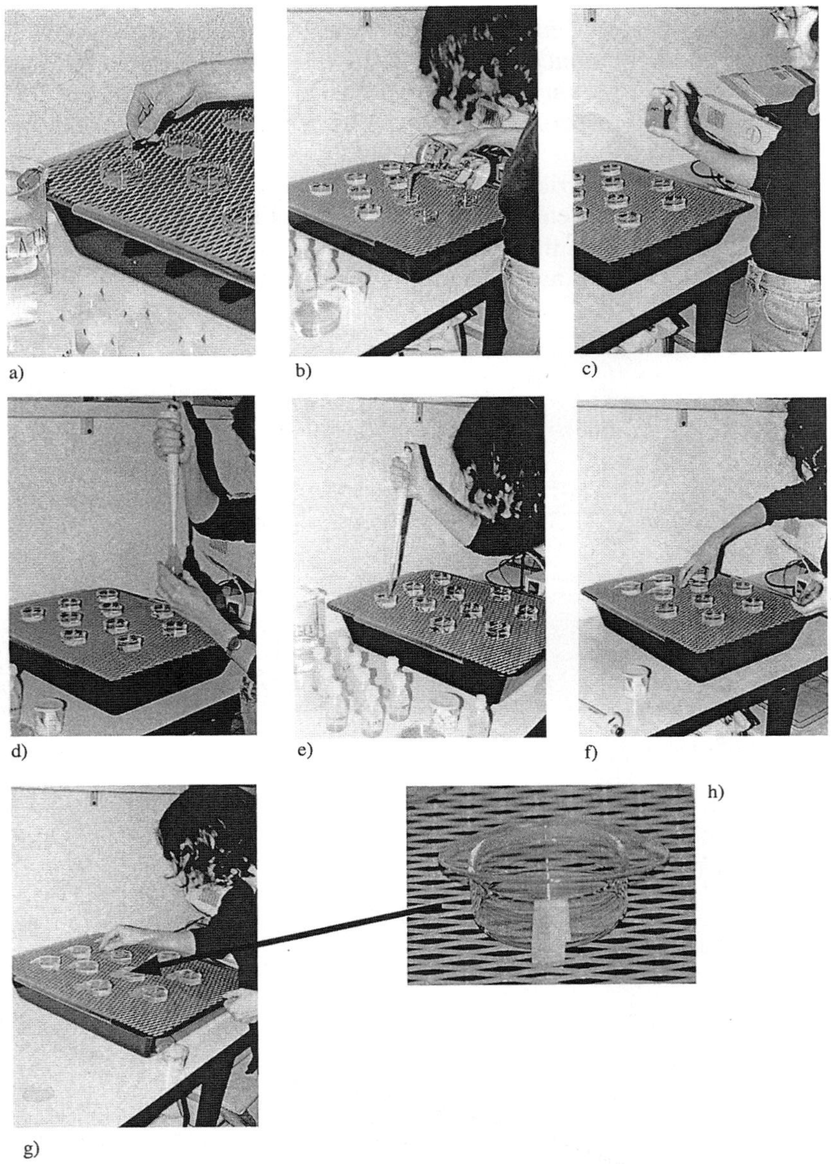

Fig. 4 a-h: The preparation steps when a sedimentation stairway/lattice is used

a) insert the cover glasses, b) fill the Petri dishes with gelatine solution , c) prepare the suspension (shaking), d) withdraw the suspension from the PE-bottle with a pipette, e) pipette the suspension and distribute eveny in the gelatine solution (pumping), then re-arrange the cover glasses, f)gravitational settling process, wait 2 hours, g) insert the paper strip, h) detail of Petri dish with paper strip cover glass bearing the spicules is removed from the Petri dish with tweezers and the sample-covered side is glued to the face coated with Canada balsam. This procedure is repeated if the slide holder is twice the size of the cover glass, and a second sample is added to the other remaining half. The sample slide is then stored horizontally under the hood for 24 hours to cool and for the mounting glue to harden. The complete preparation process using a sedimentation lattice is shown in Figures 4 a-h.

Qualitative and quantitative sample analysis

The spicules on the glass slides with are counted under a light microscope at 20 x magnification. All spicules on the cover glass are counted. The different taxa are determined after Rezvoi (1936), Masuda (1997), Efremova (unpublished) and Veinberg (unpublished). The concentration of the spicules and the concentration of each species per sample is determined by using the following formula (1):

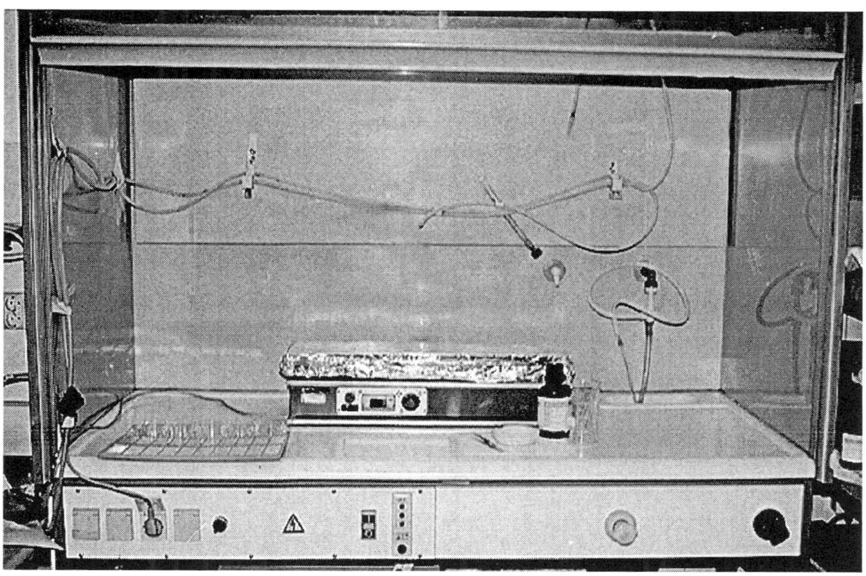

Fig. 5: Material used for preparation under the hood (hot plate, coverglasses, glass slides, mounting media, tweezers)

$$\text{Nspicules/g} = \frac{0.3925\,(n_1+n_2)\,V_{H2O}\,d^2}{V_{al}\,l^2\,M_s} \qquad (1)$$

N	-	Number of spicules per gram of freeze-dried sediment
n_1+n_2	-	Number of spicules counted on the prepared slide holder
V_{H2O}	-	Volume of distillate water added to the sample
V_{al}	-	Volume of the sample aliquot
l	-	Length of the cover glass
M_s	-	Weight of the freeze-dried sample used
d	-	Diameter of the Petri dish

The coefficient of quantity, in this case 0.3925, has to be determined each time, as it depends on the diameter of the Petri dish used. The calculation is performed by applying the following formula (2):

$$N = \frac{0.5(n_1+n_2)\,V_{H2O}\,(d/2)^2\,\pi\,d^2}{V_{al}\,l^2\,M_s} \qquad (2)$$

Concluding remarks

The fact, that only isolated spicules from sediments of Lake Baikal are available makes determination of the sponges more doubtful. This difficulty can be overcome, at least in part, by a careful study of the spicules, as identification of fossil sponges is also based on the shape of the spicules and small morphological variations in them (e.g. spines, pits, knobs, etc.). Therefore it is important that the samples are nearly clay-free. This method of preparation yields slides of high quality (Fig. 6a, b), which allow identification of individual sponge species. Hence, we are confident that by using this method of preparation we are producing sample slides that will provide reliable taxonomic results and can later be used for paleoecological and biostratigraphical interpretations. The sponge analyses using the glass slides of the samples prepared from the gravity core STX3GC represent a first attempt. We were able to document changes in sponge spicule assemblages consisting of 4 genera of the family of Lubomirskiidae and a total of 9 species (Veinberg *et al.*, 1999).

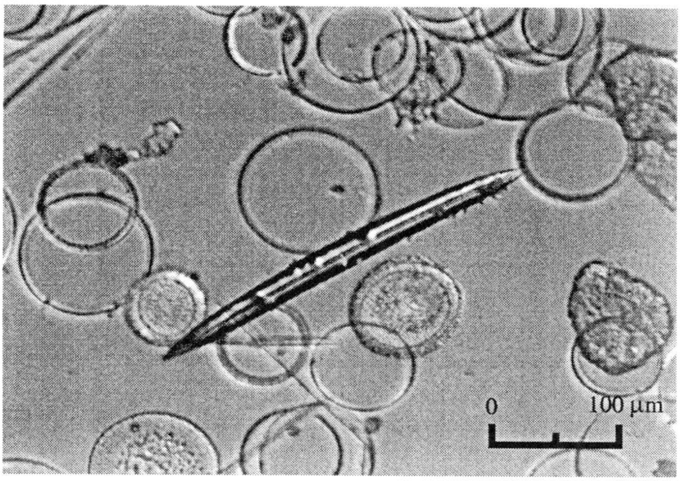

a)

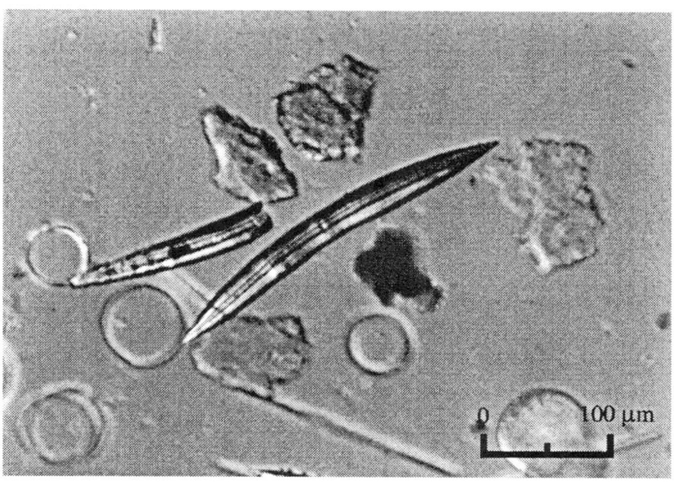

b)

Fig. 6: Slides strewn by the method described in this paper a) spicula of *Lubomirskia baicalensis.*, b) spicula of *Lubomirskia n. sp.1*

Acknowledgements
We are grateful to Dr. Mikhail A. Grachev, whose efforts have enabled international studies at Lake Baikal. We wish him all the best and a fast recovery.
C. Eckert has been supported by the Deutsche Forschungsgemeinschaft (DFG Grant: HU 378/6-1), and E. Veinberg by the Russian Foundation of Fundamental Research (Grant 95-04-13614b).

References

Martinson, G.G. (1936a): Izkopaemaya spongiofauna tretichnych otloshenii Pribaikalya (The fossil spongiofauna in tertiarnary sediments of Pribaikalian region), Doklady Akademii nauk SSSR, tom XXI, No.4, Paleontologiya, p. 212-214

Martinson, G.G. (1936b): Razpredeleniye spikul gubok v skvashine glubokovo bureniya u s. Posolska na Baikale (The distribution of sponge spicules in a sediment core from the village Posolska by Lake Baikal), Doklady Akademii nauk SSSR, tom IV (XIII), No. 6 (110), Geologiya, p. 261-264

Martinson, G.G. (1955): Iskopaemaya presnovodnaya fauna i eye znacheniye dlya stratigrafii (The fossil spongiofauna and his importance for stratigraphy), Vestnik Akademii nauk SSSR No. 12, p. 32-35

Masuda, Y., Iskovich, V., Veinberg, E.V., EfremovaA, S.M. (1997): Studies on taxonomy and distribution of freshwater sponges in Lake Baikal, in: Animal Community; Environment and Phylogeny in Lake Baikal, editor: Miyazaki N., publ.: Otsuchi Marine Research Center, Ocean Institute, University of Tokyo, p. 21-41

Rezvoi, P.D. (1936): Fauna SSSR. Presnovodnye gubki (Sem. Spongillidae i Libomirskiidae) (Fauna of USSR. fresh water sponges), Zoologicheskii institut Akademii nauk SSSR, novaya seriya No.3, izd. Akademii nauk SSSR, Moskva-Leningrad, 125 p.

Veinberg, E.V., Khlystov, O.M., Vorobyova, S.S., Kornakova, E.G., Levina, O.V., Efremova, S.M., Grachev, M.A. (1997): Distribution of sponge spicules in sediments of underwarter Akademichesky Ridge of Lake Baikal, in: Berliner Geowissenschaftliche Abhandlungen, p. 141-145

Veinberg E.V., Müller J., Eckert C. , Mehl D., Masuda Y., Efremova S.M. (1999): Extant and fossil sponge fauna of underwater Academician Ridge on Lake Baikal (SE-Sibiria), Proceedings of 5th International Sponge Symposium in Bisbane, Queensland Museum (Australia), 5 p.

Zielinski, U. (1993): Quantitative Bestimmung von Paläoumweltparametern des Antarktischen Oberflächenwassers im Spätquartiär anhand von Transferfunktionen mit Diatomeen (Quantitative estimation of paleoenvironmental parameters of the Antarctic Surface Water in the Late

Quarternary using transfer functions with diatoms), (in German), Berichte zur Polarforschung, No. 126, Bremerhaven , 148 p.

Lake Baikal
K. Minoura (editor)
2000 Elsevier Science B.V.

Evolution of freshwater centric diatoms within the Baikal rift zone during the late Cenozoic

Khursevich, G. K.[1,2*], Karabanov, E. B.[2,3], Williams, D. F.[2], Kuzmin, M. I.[3], and Prokopenko, A. A.[2,4]
[1] Institute of Geological Sciences, NAS of Belarus, Minsk 220141, Belarus, e-mail: khurs@ns.igs.ac.by
[2] Baikal Drilling Project, Department of Geological Sciences, University of South Carolina, Columbia SC, 29208, USA
[3] Institute of Geochemistry, Russian Academy of Sciences, Irkutsk, 664033, Russia
[4] United Institute of Geology, Geophysics and Mineralogy, Russian Academy of Sciences, Novosibirsk, 630090, Russia

Abstract

The evolution of freshwater centric diatoms during the Late Cenozoic was associated with processes of extinction and renewal and reflects global climatic changes. The Pliocene and Pleistocene history of centric diatoms has been documented most completely in Lake Baikal sediments penetrated by two drill holes BDP-96-1 (200 m deep) and BDP-96-2 (100 m deep) at the top of the underwater Academician Ridge. A detailed Pliocene-Pleistocene record of diatom species and diatom biostatigraphy of a 200 m sedimentary section are presented in our article. The record shows that intense speciation of diatoms corresponds to the time at 1.5 Ma (early Pleistocene) and appears to be related to the beginning of Siberian glaciations.

The Baikal Rift Zone (BRZ) extends for over 2000 km, from the Khubsugul Lake (Mongolia) in the SW, to the Oljokma River in the NE, to the Chara depression of Northern Siberia, and ranges in width from 100 to 150 km to over 3000 km (Mats, 1993). The BRZ is surrounded by the ranges and adjoins the Mid-Siberian plateau as well as the Western Trans-Baikal area, which includes the Vitim plateau and Selenga lowland.

The evolution of freshwater centric diatoms during the Late Cenozoic was associated with the processes of extinction and renewal and reflects global climatic changes. The Middle and Late Miocene diatom community within the Baikal Rift Zone extends from the Tunka depression, located 60 km west of the southern edge of Lake Baikal (Cheremissinova, 1973; Popova *et al.*, 1989). Re-examination of this diatom assemblage by SEM allowed more precise definition of the morphological features of certain centric diatoms and identification of new extinct taxa *Lobodiscus sibericus* (Lupikina, Khursevich, 1991), *Actinocyclus tuncaensis*, *Alveolophora tscheremissinovae* (Khursevich, 1994), *Aulacoseira praegranulata var. tuncaica, Cyclotella tuncaica, and Stephanodiscus tuncaensus* (Likhoshway *et al.*, 1997) to be revealed.

The Pliocene and Pleistocene history of centric diatoms has been documented most completely in Lake Baikal sediments penetrated by two drill holes, BDP-96-1 (200 m) and BDP-96-2 (100 m deep), at the top of the underwater Academician Ridge (Kuzmin *et al.*, 1997; Williams *et al.*, 1997).

The monodominant *Cyclotella* diatom flora, represented by *C. iris* with varieties, including *C. iris* var. *insueta* var. nov., has been found in the lower part of BDP 96-1 (~ 197-178 m), which corresponds to 4.9-4.5 Ma (Fig. 1). The relatively low frequency of diatoms and the monodominant character of the diatom flora in deposits of this part are indicative of unfavourable paleoecological conditions in ancient Baikal, most likely related to cooling and drying (high degree of continentality of the climate). This appears to be related to the rising of the Tibetan Plateau in the late Miocene - early Pliocene (Zhongli *et al.*, 1993), a time interval also distinguished by disintergration of the Turgai flora (Belova, 1975).

About 4.5 Ma (~ 178 m in BDP 96-1 borehole), significant reogranization of the lacustrine biota occurred in Paleo-Baikal. Dominance shifted to *Aulacoseira* aff. *islandica* and the newly appeared genus *Stephanopsis* gen. nov., which form the new distinct biostratigraphic zone at a depth of 178-142 m (ca. 4.5 - 3.5 Ma) (Fig. 1). The amelioration of paleoecological conditions in Paleo-Baikal apparently corresponds to the Pliocene thermal optimum characterized by a subtropical climate (with seasonal winter precipitation). This conclusion is confirmed by the formation of black and dark brown soils in the Lake Baikal region during the optimum, as well as by the composition of large mammalia (rhinoceroses, deer, etc.) requiring abundant vegetal food (Vorobyova *et al.*, 1995). According to Borzenkova (1992), precipitation may have increased 1.5-2 times in that part of Inner Asia.

In the second half of the Pliocene (~ 3.6-2.8 Ma, 142-112 m in BDP 96-1) a community dominated by representatives of the new genus *Stephanopsis* formed in ancient Baikal (Fig. 1). The loss of the *Aulacoseira* species from the diatom composition testifies to a less warm and wet climate by the end of Pliocene.

Above 2.8 Ma (above 112 m in BDP 96-1) the abundance of *Stephanopsis* species drastically decreases, and at about 110 m they are replaced by the genus Tertiarius (in particular, *T. baicalensis* sp.nov.) found in the sedimentary sequence of the BDP 96-1 core only in a narrow

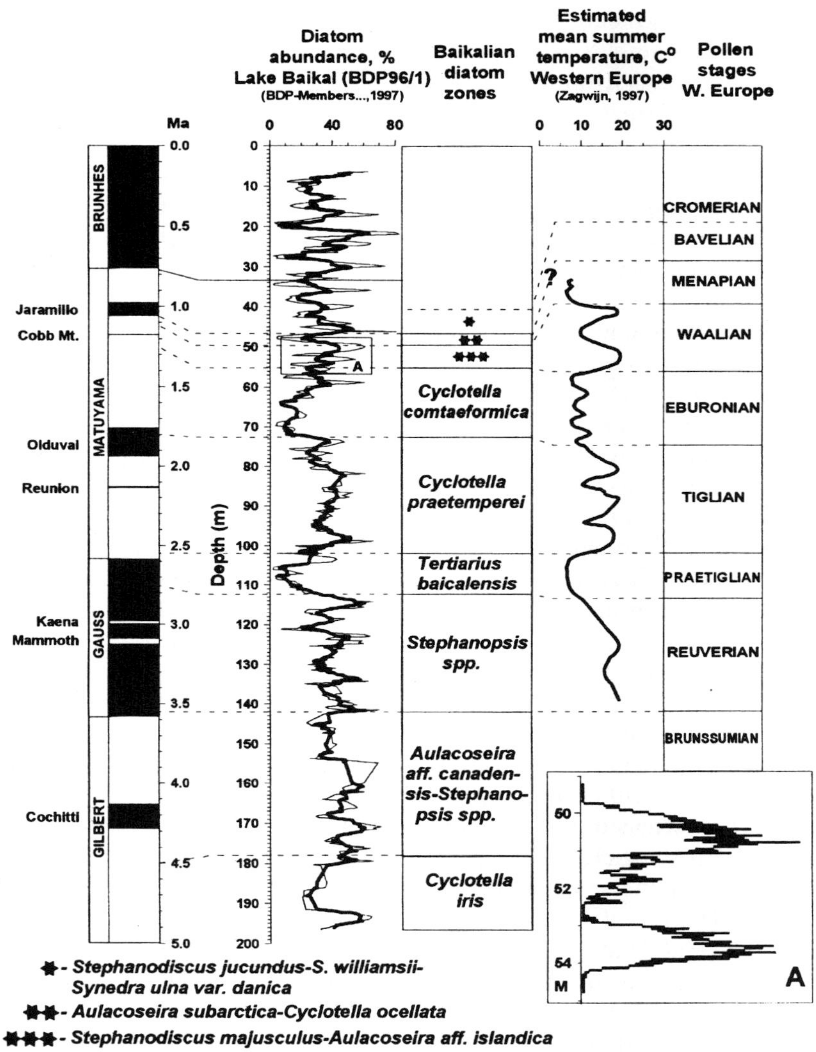

Figure 1. Plio-Pleistocene diatom biostratigraphy of Lake Baikal sediments in BDP-96-1 and BDP-96-2 cores and comparison with the West-European climato-stratigraphic scheme. The magnetostratigraphy is according to W. Berggren *et al.* (1995); the diatom abundance record was provided by the BDP-96 core description team, the smear slide observations by D. Weiel and M. Schwab, BDP Members., (1997); the biostratigraphic boundaries are shown according to the depths of the last and first appearance data of corresponding genera and species; the estimated mean summer temperatures and European pollen stages are from W. Zagwijn (1997). Inset A shows the detailed structure of the warm interglacial stages correlated with the Waalian pollen stage in Western Europe.

depth interval (~ 110-103 m) (Fig. 1). Chrisophyte cysts were abundant as well.

Above 103-102 m (ca 2.5 Ma), both genera are extinct, as is evident from the sedimentary record of Lake Baikal. It is the sharpest stratigraphic boundary corresponding to a drastic cooling event (2.6-2.5 Ma) recorded anywhere in the world. The significance of this boundary in the Lake Baikal record and the biostratigraphic (diatom) changes associated with cooling strongly suggest that the Plio-Pleistocene boundary, according to the Lake Baikal sedimentary record, should be lowered to 2.5 Ma (marine oxygen isotopic stage 100) as proposed by Van Couvering (1997) and other investigators. The biostratigraphic changes at the 2.5 Ma level can be associated with the Praetiglian/Tiglian transition (Fig. 1).

The following 2.5-1.25 Ma range (100-55 m in BDP 96-2) is characterized by the development of the *Cyclotella* diatom flora in the ancient paleobasin (Fig. 1). Their frequency repeatedly varied, testifying to the unstable character of the climate (in terms of alternation of phases of climate warmings and coolings) during the period specified above. The new extinct species *Cyclotella* praetemperei sp. nov. with varieties was typical of the period between 2.5-1.77 Ma (marine isotopic stages 99-61) according to the age model of BDP 96-2 core (BDP Members, 1997; Williams *et al.*, 1997).

At the age level of 1.8-1.7 Ma (74-72.5 m in BDP 96-2) *C. praetemperei* is replaced by the new extinct species *C. comtaeformica* sp. nov. with varieties (Fig. 1). However, this climatic and biostratigraphic boundary reflected in the change in dominant species of the same genus in ancient Lake Baikal was less dramatic in comparison with the previous boundary at 2.6-2.5 Ma, when two genera disappeared in Lake Baikal sediment. *C. comtaeformica* corresponds to the time span between 1.82-1.25 Ma (MIS 62-31) and the Eburonian period in western Europe (Fig. 1) (Zagwijn and Doppert, 1978).

A short period ca. 1.25-1.12 Ma is marked by the development of the new diatom assemblage Stephanodiscus majusculus sp. nov. - *Aulacoseira* aff. *Islandica* (Fig. 1). The replacement of the diatom flora from monodominant (*Cyclotella*) to several dominant genera is indicative of amelioration of ecological conditions in ancient Lake Baikal as compared with the previous period. This was apparently connected with obvious warming (the Waalian warm period in the West-European stratigraphic scale).

The dramatic reorganization of the Baikalian diatom flora occurred above the 49.6 m level, within the 49.6-47.4 m range in BDP 96-2, and corresponding to 1.12-1.07 Ma below the boundary of the Jaramillo subchron (Fig. 1). At this depth a new community of diatoms represented by the cold-water species *Aulacoseira subarctica* and *Cyclotella ocellata,* are very poorly preservated (corroded). These data indicate a new strong cooling episode in the Baikal region that was correlated with the Menapian cold period in western Europe.

The subsequent period, between 1.07 Ma and 0.8 Ma (47.4-33.6 m in BDP 96-2), was marked by an intense speciation of the genus *Stephanodiscus* Ehr. that was caused by frequent changes in climatic conditions (repeated alternations of warming and cooling phases) (Fig. 2). A number of new extinct and rare unique species correspond to this interval (e. g., *Stephanodiscus williamsii* sp. nov., *S.* cf. *yukonensis* and other species). The *Stephanodiscus* species dominante in communities with representatives of *Synedra* during the phases of warming, corresponding to marine isotope-oxygen stages 25, 21 and others.

The Matuyama/Brunhes boundary exhibits an immediately preceding strong monodominant peak of the cold-water species *Aulacoseira subarctica* at a depth of 34.23 - 33.6 m at the very beginning of the marine isotope stage 19 (Fig. 2). After that extreme peak, no A. *subarctica* can be found in the sedimentary record of Lake Baikal.

The evolution of diatoms during the Brunhes epoch is also associated with processes of extinction and renewal, and it clearly reflects the glacial - interglacial cycles of a paleoclimate. Distinct diatom assemblages correspond to individual isotopic stages and even substages (Fig. 2), and as a result the following five major periods can be distinguished in the development of the Baikalian diatom flora during the last 730 ka and they are correlated with isotopic events after Bassinof *et al.* (1994):

1) 730-628 ka (the marine isotope-oxygen substages 18.3-16.2) when *Cyclotella praeminuta* along with several species of *Stephanodiscus* was developing in the ancient paleobasin;

2) 594-434 ka period (substages from 15.3 to 12.2) characterized by the appearance, blooming, and extinction of specific taxon *Stephanodiscus flabellatus* var. *excentricoides* var. nov. reaching the peak in its development during the interglacial interval synchronous to stage 13;

3) 406-328 ka, when the intense development of small species of *Stephanodiscus* (extinct taxon of *S. Exiguus* included) was typical for time intervals corresponding to substages 11.3 and 9.3, whereas *Cyclotella minuta* and *Stephanodiscus flabellatus* var. *distinctus* mostly developed during the time period synchronous to substages from 11.2 to 10.2;

4) 315-79 ka (substages 9.2-5.1) characterized by the appearance, blooming, and extinction of ancient extinct species of *Stephanodiscus grandis, S. carconeiformis, S. formosus* (maximum development of these taxa corresponds to substage 5.5);

5) the last period 52 ka - recent time (marine isotopic substages 3.3-1) distinguished by the dominance of *Aulacoseira baicalensis* and *Cyclotella*

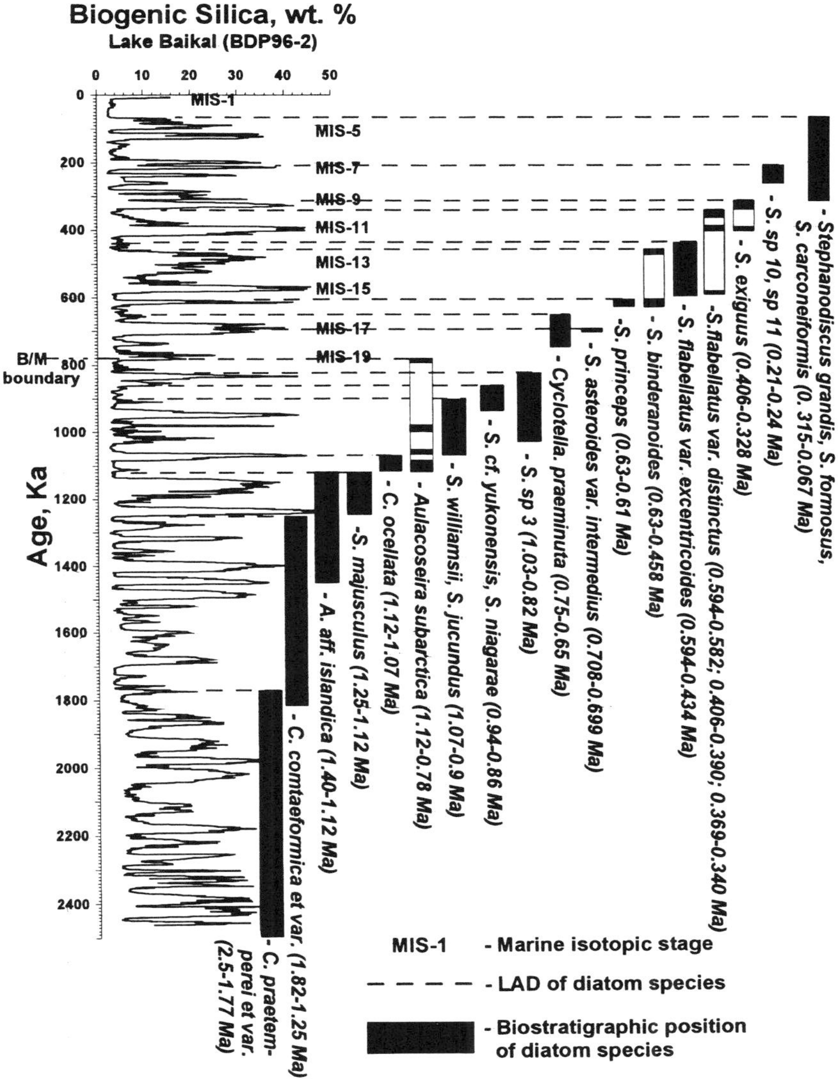

Figure 2. Biostratigraphic distribution and last appearance datum (LAD) boundaries of diatom species in Lake Baikal sediments (deep drilling core BDP-96-2) during 2.5 Ma. Biogenic silica data are plotted against age according to the correlation with ODP site 677 (Williams *et al.*, 1997). The intense speciation of diatoms corresponds to the time at 1.5 Ma (early Pleistocene) and appears to be related to the beginning of the Siberian glaciations.

baicalensis

Comparative analysis of the composition of diatom assemblages corresponding to various cold and warm intervals of Brunhes chromium showed the following:

a) cold intervals synchronous with marine isotopic stages 14 and 8 did not differ in inclement climate as compared with other cold intervals corresponding to isotopic stages 18.2, 16.2, 12.2, 10.2, 6.2, 4 and 2;

b) more or less stable paleoecological conditions existed constantly in the ancient Baikal during the warm period correlated with the isotopic stage 13;

c) similar paleoecological environment in Pre-Baikal existed during the warm intervals synchronous to the isotopic substages 9.2 - 9.1 and 7.3 - 7.1 when the diatom community of *Stephanodiscus grandis* and *S. formosus* was available, as well as during isotopic substages 15.2 - 15.1 and substage 13, when *Stephanodiscus flabellatus* war. *excentricoides* was developing;

d) excessively short dramatic coolings served as a signal for the extinction of certain species in ancient Baikal or for essential reorganization of the lacustrine biota belonging to such isotopic substages as 15.4, 7.4 and 5.4.

Acknowledgements

This work was supported by NSF grants EAR-9119537 and EAR-93-1720401, and by the Siberian Branch of the Russian Academy of Sciences. This work was implemented as part of the Baikal Drilling Project supported by NSF, the Russian Academy of Sciences, the Russian Ministry of Geology, and the Science and Technology Agency (STA) of Japan. We thank M.I. Kuzmin, M.A. Grachev and B.N. Khakhaev for help with organization of the BDP program in Russia. We also thank D. Lykov and S. Kochikov and the entire drilling team of the Nedra Drilling Enterprise and A.N Gvozdkov of the Institute of Geochemistry for technical assistance with the cores and smear-slide preparation.

References

Bassinot, F.S. Labeyrie,L.D. Vinsent, E. Quidelleur,X. Shacleton, N.J. Lancelot, Y. (1994) The astronomical theory of climate and the age of Brunches-Matuyama magnetic reversal, Earth and Planetary Science letters, V. 126, 91-108.

BDP-Members (1997) Continuous paleoclimate record of last 5 MA from Lake Baikal, Siberia, EOS American Geophyiscal Union, Transactions, V. 78, 597-604.

Belova, V.A. (1975) Vegetation and climate of the Late Cenozoic of the south of East Siberia, Nauka Press, Novosibirsk, 1985, 160 pp.

Berggren, W. A. Kent, D. V. Swisher III, C. C. Aubry, M.-P. (1995) A revised Cenozoic geochronology and chronostratigraphy. In: Geochronology, time scales and global stratigraphic correlation, W. A. Berggren, D. V. Kent, C. C. Swisher III, and J. Hardenbol, Eds., SEPM, Tulsa, Oklahoma, 129-212.
Borzenkova, I.I. (1992) Climate Changes in Cenozoic, Gidrometeoizdat Press, Sankt-Peterburg. 192 pp. (In Russian)
Cheremissinova, Ye.A (1973) Diatom flora of the Neogene deposits of Pribaikalja (the Tunka depression), Nauka Press, Novosibirsk, 68 pp. (In Russian)
Khursevich, G.K. (1994) Morphology and Taxonomy of some centric diatom species from the Miocene sediments of the Dzhilinda and Tunka depressions. In: Proceedings of the 11th International Diatom Symposium. J.P.Kociolek, ed., California Academy of Sciences, San Francisco, 269-280.
Kuzmin, M.I. Grachev, M.A. Williams, D. Kawai, T. Horie, S. Oberchansli, H. (1997) The continuous record of paleoclimates of lastt 4.5 Ma from Lake Baikal: Russian Journal of Geology and Geophysics, V. 38, 1021-1023. (In Russian)
Likhoshway, E.V. Pomazkina, G.V. Nikiteeva, T.A. (1977) Centric diatoms from the Miocene deposits in the Baikal Rift Zone (Tunka basin). Russian Journal of Geology and Geophysics, Vol. 38, N 9, 1445-1452. (In Russian)
Lupikina, Ye.G. Khursevich, G.K. (1991) Lobodiscus (Tscher.) Lupik. Et Churs. - a new genus of the class Centrophyceae (Bacillariophyta). Algologia, Vol. 1, N 3, 67-70. (In Russian)
Mats, V.D. (1993) The structure and development of the Baikal rift depression, Earth Science Reviews, V. 34, 81-118.
Popova, S.M. Mats, V.D. Chernayeva, G.P. *et al.*, (1989) Paleolimnological reconstructions (the Baikal Rift Zone), Nauka Press, Novosibirsk, 111 pp. (In Russian)
Van Couvering, J.A. (1997) Preface: the new Pleistocene. In: The Pleistocene Boundary and the Beginning of the Quaternary, J.A. Van Couvering, ed., World and regional geology, Cambridge University Press, Cambridge, xi-xix.
Vorobyova, G.A. Mats, V.D. Shimaraeva, M.K. (1995) Late-Cenozoic paleoclimates in the Baikal region. Russian Journal of Geology and Geophysics, Vol. 36, 82-96. (In Russian)
Williams, D.F. Peck, J. Karabanov, E.B. Prokopenko, A.A. Kravchinsky, V. King, J. Kuzmin, M.I. (1997) Lake Baikal record of continental climate response to orbital insolation during the past 5 million years, Science, V. 278, 1114-1117.
Zagwijn, W.H. (1997) The Neogene-Quaternary boundary in The

Netherlands. In: The Pleistocene Boundary and the Beginning of the Quaternary, J. A. van Couvering, Ed., World and regional geology, Cambridge University Press, Cambridge, 185-190.
Zagwijn, W.H., Doppert, J.W.C. (1978). Upper Cenozoic of the southern North Sea Basin: palaeoclimatic and paleogeographic evolution. In: Geol. En Mijnb., Vol. 57, 577-588.
Zhongli, D. Rutter N. Tungsheng, L. (1993) Pedostratigraphy of Chaneseloess deposits and climatic cycles in the last 2.5 myr., Catena, V. 20, 73-91.

Lake Baikal
K. Minoura (editor)
2000 Elsevier Science B.V.

Elemental composition of short sediment cores and ferromanganese concretions from Lake Baikal

Takamatsu, T.[1], Kawai, T.[2], and Nishikawa, M.[3]
[1]Soil and Water Environment Division, [2]Environmental Chemistry Division, and [3]Regional Environment Division, National Institute for Environmental Studies, 16-2 Onogawa, Tsukuba, Ibaraki 305-0053, Japan
(*Author to whom all correspondence should be addressed; Fax: +81-298-50-2576; E-mail: takamatu@nies. go.jp)

Abstract

Short sediment cores and ferromanganese concretions from Lake Baikal were analyzed for 20 elements, and the elements detected were classified into 3 groups based on their depth profiles in the sediment: Fe, Mn, As, and P (accumulated in the surface, oxidized layers), S (accumulated in the lower, reduced layers), and others (not sensitive to diagenesis). Some bio-phobic elements (Al, Ti, and V) and/or their ratios (e.g. Al/Ti) were found to be possible indicators for estimating the origin of terrigenous fractions in the sediment. The ferromanganese concretions had ordinary levels of Fe and P, but were relatively poor in Mn, As, and heavy metals compared with ferromanganese concretions from other lakes. The values calculated from a linear polynomial function that included the concentrations of 10 elements in the sediment as variables, showed a good correlation with water depth at the sampling sites and could be applied to reconstruction of paleo-water depth from the elemental composition of ancient sediments.

Introduction

When sedimentation occurs, elements sometimes redissolve in the pore water of the sediments and migrate in the sediment column due to early diagenesis, resulting in regression of elements into lake water and/or their accumulation in specific layers of the sediment (Takamatsu, 1985; Takamatsu et al., 1985a; 1985b). Thus, the elemental composition of sediment changes significantly during early diagenesis depending on several environmental factors, including redox potential, organic matter content, and sedimentation rate. An understanding of this phenomenon is essential to reconstructing the paleo-environment from the elemental composition of ancient sediments, and the concentrations of certain elements in sediment sometimes show good correlations with the water depth at which the sediment was retrieved, providing a promising means of estimating paleo-water depth (Takamatsu, 1985; Koyama et al., 1985). In view of this, the elemental composition of surface sediments and ferromanganese concretions from Lake Baikal were analyzed and assessed in relation to the environment at the sampling sites.

Materials and Methods

Samples

The specimens analyzed were 17 short cores of sediment (3 cm i.d. x ca. 25 cm) and 2 samples of ferromanganese concretions. Samples of the former were obtained by pushing small plastic pipes into box cores that had been retrieved from various sites in the lake during the summer (Aug. 30-Sep. 8) of 1996 (Fig. 1), and samples of the latter were retrieved from the surface of sediment collected offshore at Turka (near site 9; water depth: 300 m) during the same period.

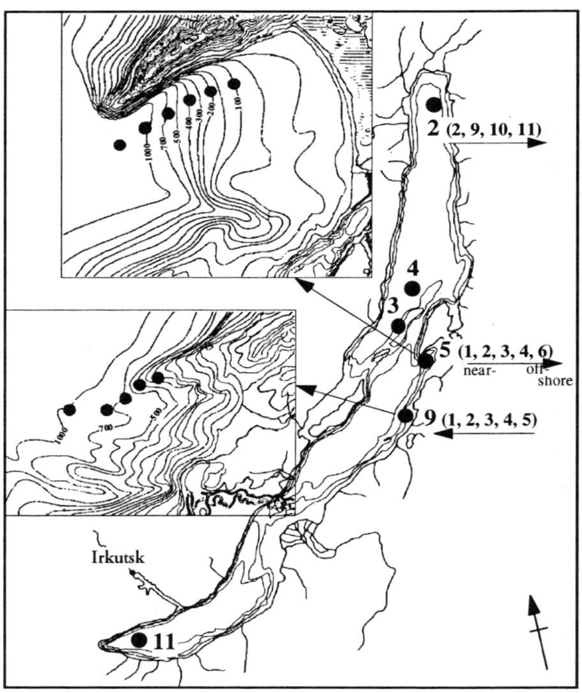

Fig. 1: Map showing the short sediment core sampling sites.
Dates: Aug. 30 to Sept. 8, 1996.

Pretreatment and analysis of the specimens

The short sediment cores and ferromanganese concretions were cut into sections 1-2 cm and 5-mm thick, respectively, and freeze-dried. The dried samples (10-20 mg) were digested with a mixture of acids ($HClO_4$, 1 ml/HNO_3, 2 ml/HF, 1 ml) in a pressurized digestion bomb and subjected to ICP-AES analysis. In the analysis, matrix effects due to high concentrations of Al, Fe, Ca, Mg, and Ti were corrected by the k-factor method. The elements analyzed were Al, Ca, Co, Cr, Cu, Fe, Mg, Mn, Ni, Zn, As, Ti, V, P, S, Sr, Ba, Pb, Sc, and Y.

Table 1. Average concentrations of elements in short sediment cores.

Element	Units	Lake Baikal		Lake Biwa##	
		Surface layers*	Lower layers**	Surface layers*	Lower layers#
WD	*m*	563 (100-1250)	563 (100-1250)	66 (5-97)	66 (5-97)
Al	%	7.28 (3.86-10.2)	8.61 (5.38-11.9)		
Ca		1.71 (0.94-3.06)	1.93 (1.26-3.76)	0.37 (0.21-0.74)	0.34 (0.24-0.63)
Fe		4.66 (1.58-7.38)	4.33 (1.39-7.66)	4.53 (3.33-5.18)	4.57 (2.36-5.11)
Mg		1.15 (0.47-1.63)	1.34 (0.50-1.97)	0.97 (0.74-1.20)	0.96 (0.38-1.26)
Ti		0.32 (0.15-0.45)	0.39 (0.20-0.55)	0.45 (0.35-0.60)	0.50 (0.23-0.64)
Co	*ppm*	14 (1-21)	17 (2-32)	18 (13-22)	18 (14-21)
Cr		55 (18-99)	57 (14-106)	67 (46-84)	70 (60-75)
Cu		47 (11-77)	52 (7-95)	76 (42-99)	52 (13-88)
Mn		3130 (240-20350)	1570 (250-14090)	3760 (690-13900)	1860 (690-3180)
Ni		40 (9-64)	46 (11-82)	33 (19-49)	30 (6-42)
Zn		96 (28-150)	110 (41-160)	200 (110-350)	140 (84-190)
As		15 (<1-58)	9 (<1-43)	55 (7-130)	23 (6-41)
V		97 (28-150)	110 (27-150)		
P		2160 (770-3620)	1410 (700-5120)	1150 (560-1710)	830 (320-1180)
S		1070 (170-7000)	2910 (47-8790)	2060 (1440-2650)	480 (380-690)
Sr		390 (160-850)	460 (250-990)	72 (56-110)	76 (62-110)
Ba		770 (420-1140)	930 (670-1430)	690 (500-820)	680 (570-740)
Pb		18 (5-34)	18 (11-25)	51 (19-76)	32 (16-52)
Sc		10 (5-16)	12 (5-19)	13 (11-15)	15 (12-16)
Y		24 (17-33)	27 (18-33)		

Values: average (minimum-maximum). WD: water depth. *: upper 2-cm layers of the sediment cores. **: below 6-cm depth: site 5(1); below 8-cm depth: sites 2(2), 2(9), and 5(6); below 10-cm depth: other sites. #: below 10-cm depth. ##: T. Takamatsu (ed): Res. Rept. NIES, Japan, No. 75, 1985; T. Takamatsu et al., Jpn. J. Limnol., 46, 115 (1985).

Results and Discussion

Elemental composition of the short sediment cores

Table 1 shows the average concentrations of the elements in the cores. Average concentrations in the Lake Biwa (Japan) sediment are listed in the table for comparison (Takamatsu, 1985; Takamatsu et al., 1985b). The elemental concentrations in the Lake Baikal sediment were generally the same as in the Lake Biwa sediment, and the same was also true of the crust. However, the Lake Baikal sediment had higher concentrations of alkaline earth metals, including Mg, Ca, Sr, and Ba than the sediments from Lake Biwa and other freshwater environments, and somewhat lower concentrations of Zn, As, and Pb (Hakanson and Jansson, 1983). The former may be the result of differences in the geological properties of the basins, while the latter are probably attributable to the water quality of Lake Baikal, which is less affected by human activity.

Fig. 2 shows the depth profiles of elements in the short sediment core sampled at site 9(1) (water depth: 1000 m). Although the profiles differed slightly from site to site, individual elements showed essentially the same patterns in all cores. By calculating the enrichment factors of elements into the upper 2-cm layers (vs. lower layers) of the sediment, elements could be classified into the following three groups: elements that are sensitive to early diagenesis and accumulate in the oxidized surface layers of the sediment {Fe (enrichment factor: 1.1), Mn (3.0), As (9.3), and P (2.0)}, elements that are sensitive to diagenesis but also accumulate in the reduced lower layers of the sediment {S (0.3)}, and elements which are not sensitive to diagenesis {Al, Ca, Co, Cr, Cu, Mg, Ni, Zn, Ti, V, Sr, Ba, Pb, Sc, and Y (0.9-1.0 for all elements)}. Elements in the last group were further classified into several subgroups by principal component analysis: 1) Al, Ti, and Mg, 2) Ca and Sr, and 3) Co, Cr, Ni, Zn, V, Pb, Sc, and Y. Ba and Cu could not be classified into any of the subgroups. Elements in the respective subgroups may have been contained in different terrigenous fractions in the sediment.

Elements were also classified into the following two groups based on their regional distribution, elements that possessed regional peculiarities in their concentrations: Al, Ti, Ca, Sr, V, etc., and elements that did not: Co, Ni, Sc, Cr, Mg, Zn, etc. It may be possible to use the elements in the former group, especially Al, Ti, and V (because these are highly biophobic) as indicators to identify the source of terrigenous fractions in the sediment. Fig. 3(a) shows the relationship between the concentrations of Al and Ti in the short sediment cores, and Fig. 3(b) shows the depth profiles of the Al/Ti ratios. The offshore sediments were considerably more homogeneous and had almost the same Al/Ti ratios, whereas several nearshore sediments

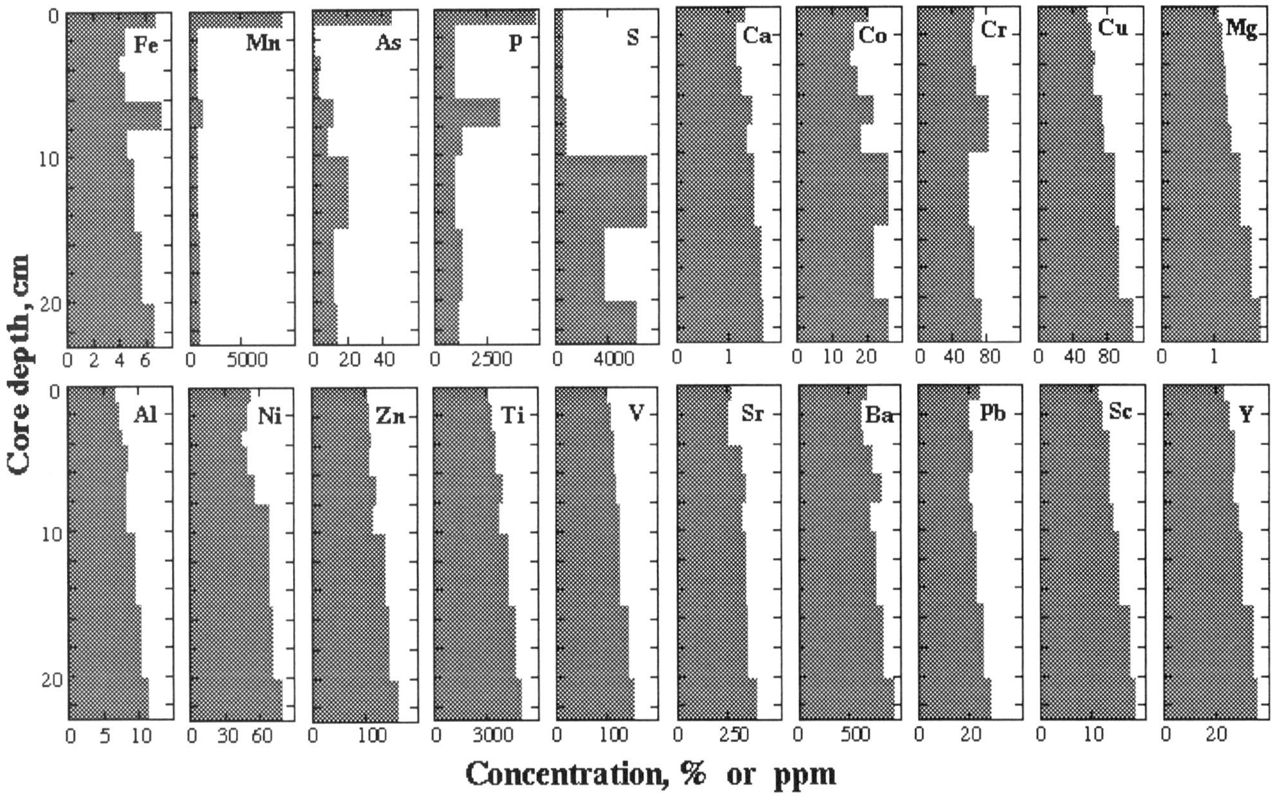

Fig. 2: Typical depth profiles of elements in the short sediment cores.
Concentration units: % for Fe, Ca, Mg, and Al; ppm for other elements. Site: 9(1). Water depth: 1000 m.

{sites 2(2), 2(9), 11, and sometimes 5} exhibited regional peculiarities in the ratios. In addition, the Al/Ti ratios at sites 2(2) and 2(9) decreased with lower core depth, while those at site 11 increased, suggesting a recent decrease in direct contribution to the sediment component of influx from the watershed.

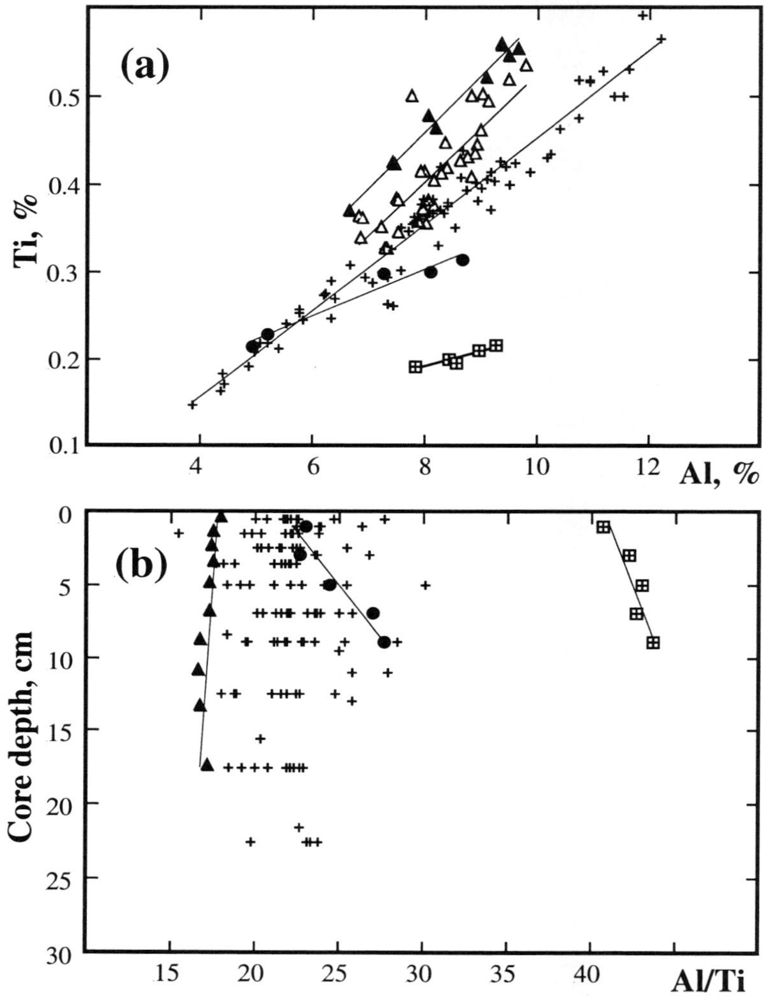

Fig. 3: Relationship between concentrations of Al and Ti (a) and depth profiles of Al/Ti ratios (b) in the short sediment cores.
Sites: (●) 2(9); (▲) 11; (□) 2(2); (△) 5(2, 3, 4, 6); (+) others.

Elemental composition of the ferromanganese concretions

Table 2 shows the concentration ranges of elements in the ferromanganese concretions from Lake Baikal and other freshwater lakes (Takamatsu, 1985; Takamatsu et al., 1985a; 1985b; 1993). The Fe and P concentrations in the Lake Baikal concretions were high as well as in the other lakes, but the Lake Baikal concretions had relatively low concentra-

Table 2. Concentration ranges of elements in freshwater ferromanganese concretions.

Element	Units	L. Baikal*	L. Biwa**	Other lakes#
Al	%	7.1-11	2.4-4.9	
Ca		1.6-2.1	0.36-0.85	1.3-3.1
Fe		7.5-15	3.1-33	2.5-40
Mg		1.0-1.8	0.12-1.0	0.9-2.2
Ti		0.29-0.52	0.05-0.29	0.3
Co	*ppm*	15-60	22-87	12-640
Cr		44-74	16-37	8.0-53
Cu		33-50	24-99	1.2-1300
Mn		840-1.4 %	7500-17 %	1.0 %-51 %
Ni		45-71	9-340	14-2400
Zn		100-140	73-220	30-2000
As		<1-75	530-1300	26-350
V		120-170	110-270	130
P		1700-8900	2700-1.1 %	2100
S		99-520		
Sr		310-430	33-100	<100-360
Ba		720-1600	560-3800	310-6.6 %
Pb		11-18	15-26	24-66
Sc		10-17	1.9-7.5	0.2-27
Y		22-35		

*: Fe-concretions on the surface of the sediment at site 9. **: Ref.: T. Takamatsu et al., Jpn. J. Limnol., 46, 115 (1985); 54, 281 (1993). #: includes data from Lake Michigan, Lake Ontario, Lake Oneida, Grand Lake, Ship Harbor Lake, Mosque Lake, Lake George, Lake Champlain, and several Canadian lakes; Ref.: D.S. Cronan and R.L. Thomas, Can. J. Earth Sci., 7, 1346 (1970); D.S. Cronan and R.L. Thomas, Geol. Soc. Am. Bull., 83, 1493 (1972); W.E. Dean and S.K. Ghosh, J. Res. U.S. Geol. Surv., 6, 231 (1978); D.E. Edgington and E. Callender, Earth Planet. Sci. Lett., 8, 97 (1970); W.S. Moore, et al., Earth Planet Sci. Lett., 46, 191 (1980).

tions of Mn, As, and heavy metals. This may have resulted from the rapid generation rates of the concretions (Granina et al., 1998) and the redox conditions at the sampling sites. The distributions of As (correlation coefficient: 0.98), P (0.97), Ni (0.91), Mn (0.85), Ba (0.71), and Co (0.69) in the Lake Baikal concretions were closely correlated with the distribution of Fe, whereas the distributions of Mg (0.98), Sc (0.98), Ti (0.97), Zn (0.95), Cu (0.91), Pb (0.86), and Cr (0.74) were closely correlated with that of Al, indicating that the former elements had been occluded in the hydrous oxides of Fe and/or Mn, and the latter in terrigenous fractions of the concretions.

Elemental composition of the sediments as an indicator of water depth

The concentration of certain elements in sediment usually changes with the depth of the overlying water. For example, in Lake Biwa, the concentrations of Mn and As in the sediment rise with increasing water depth, whereas that of Hf decreases significantly (Takamatsu, 1985; Takamatsu et al., 1985b), and this phenomenon was used as a basis for reconstructing paleo-water depth from the concentrations of these elements in ancient sediments (Takamatsu, 1985; Koyama et al., 1985). Although the concentra-

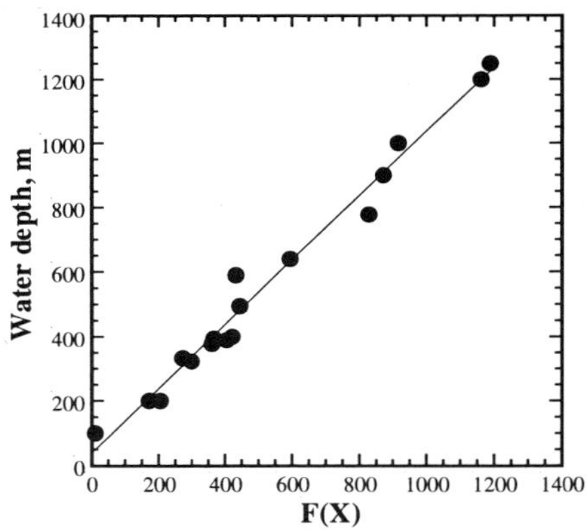

Fig. 4: Relationship between values calculated from a linear polynomial function that included elemental concentrations in the short sediment cores as variables, and water depth at sites where the core specimens were collected.

Water depth (m) = 36.7 + 1.00 x F(X), r = 0.991; F(X) = 499 + 0.00142[Al]+ 0.0990[Ca] + 150[Co] + 16.3[Cu] + 0.0320[Fe] - 0.269[Mg] - 0.151[Mn] - 41.2[Ni] - 0.0214[S] - 1.97[Sr];

[X]: average concentrations (ppm) of elements in lower layers {below 6 cm depth for site 5(1), below 8 cm depth for sites 2(2), 2(9), and 5(6), and below 10 cm depth for other sites} of the short sediment cores.

tions of Co, Cr, Cu, Fe, and Ni in the sediment from Lake Baikal rose with increasing water depth, and those of Ca, Sr, and Ba decreased, the correlations between the elemental concentrations and water depth were fairly weak (highest correlation coefficient: 0.61 for Cu), and thus the concentration of any single element may be useless as an indicator of water depth. However, as shown in Fig. 4, the values calculated from a linear polynomial function that included the concentrations of the ten elements in the sediments as variables showed a good correlation with water depth at the sites where the sediments were retrieved. Although the underlying mechanism is unclear, the chemical state of the elements that originated in rivers (e.g. dissolved or particulate) and the susceptibility of elements to early diagenesis may have been responsible for the changes in the elemental composition of the sediment with water depth, i.e, particulate elements deposit mainly in coastal shallow regions, whereas dissolved elements are transported to deep, offshore regions before being deposited, because of their relatively long residence time. Elements that are susceptible to early-diagenesis also accumulate in the sediment of deep basins due to the cycles of deposition-dissolution (Takamatsu, 1985). In any event, it may be possible to apply the above correlation to estimating paleo-water depth from the elemental composition of the ancient sediment in Lake Baikal.

References

Granina L., B. Wehrli and B. Mueller, 1998, Dynamics of Fe and Mn accumulation in the bottom sediments of Lake Baikal: implication for paleolimnology. Abst. of BICER, BDP and DIWPA Joint Inter. Symp. Lake Baikal, Yokohama, 22pp.

Hakanson L. and M. Jansson, 1983, Lake Sedimentology. Springer-Verlag, Berlin Heidelberg, 316 pp.

Koyama M., M. Kawashima, T. Takamatsu and S. Horie, 1985, Vertical distribution profiles of Mn, As, Na and Hf as possible indicators for paleo-water depth in the 1400 meter long core drilled from Lake Biwa. Proc. Japan Acad., 61(B), 407-410.

Takamatsu T., eds., 1985, Limnology and Environmental Studies of Elements in the Sediment of Lake Biwa. Res. Rept. NIES, Japan, 75, 129 pp.

Takamatsu T., M. Kawashima and M. Koyama, 1985a, The role of Mn^{2+}-rich hydrous manganese oxide in the accumulation of arsenic in lake sediments. Water Res., 19, 1029-1032.

Takamatsu T., M. Kawashima, R. Matsushita and M. Koyama, 1985b, General distribution profiles of thirty-six elements in sediments and manganese concretions of Lake Biwa. Jpn. J. Limnol., 46, 115-127.

Takamatsu T., M. Kawashima, J. Takada and R. Matsushita, 1993,

Characteristics in elemental composition of ferromanganese concretions from Lake Biwa. Jpn. J. Limnol., 54, 281-291.

Lake Baikal
K. Minoura (editor)
2000 Elsevier Science B.V.

Mercury distribution in the bottom and stream sediments of Lake Baikal,water reservoirs of the Angara river cascade, and adjacent drainage basins

Koval, P. V.*, Kalmychkov, G. V., Geletyi, V. F., and Andrulaitis, L. D.
Vinogradov Institute of Geochemistry, P.O.Box 4019, Irkutsk-33, 664033, Russia.
Fax: 7 3952 464050; E-mail: koval@igc.irk.ru
(*corresponding author)

Abstract

This paper summarizes data on Hg distribution in the bottom and stream sediment of Lake Baikal and adjacent areas of Siberia that are strongly endangered in terms of technogenic Hg emission into the environment. Cold vapour atomic absorption spectrometry was used to make the measurements (detection limit 2: ppb).
The average natural concentrations (ppb) of mercury were: bedrock 11; soils 17-29; recent overbank sediment 40.
The average Hg concentration (ppb) in bottom sediment was as follows: Lake Baikal, 15-66; Irkutsk hydropower water reservoir, 40 (30, with the bays included); the Bratsk reservoir, which drains the main industrial and agricultural areas of region, 960 -1,050; the Ust'-Ilimsk reservoir, 547-766. The silt fractions with particles smaller than 0.04 mm were the main Hg concentrators in sediments and suspended matter in waters. In polluted waters, their Hg concentration was one order higher (up to 3,500 - 4,500,000 ppb) than in the sand-aleurite fractions. The relatively weakly bound portion of Hg in polluted sediments and suspended matter was much greater than in natural sediments and suspensions.
In general, the region can be considered a low Hg pollution area, except for the basins downstream of the Irkutsk dam, where the Hg pollution problem is very real.

Introduction

The anthropogenic mercury input into the global environment is comparable in scale with the natural flows of the pollutant in recent biosphere wheels, and the primary concern in this regard are the atmospheric fluxes and pollution of aquatic systems by industrial sources (gold mining, chlorine production, etc.). The pollution of aquatic systems is especially harmful to people and wildlife because of the greater methyl mercury accumulation (Rudd, 1998).

The chain consisting of Lake Khubsuguil (in Mongolian: Hovsgol Nuur) - Selenga River - Lake Baikal (in Russian: Ozero Baykal)- the Angara River cascade of hydropower water reservoirs (Irkutsk - Bratsk - Ust'-Ilimsk) - Enisey River (Fig. 1) is a unique object for comparative study of natural and anthropogenic sources in recent continental aquatic systems because of the high diversity of technogenic pollution in particular parts of it. Technogenic loads vary from minimal (Lakes Baikal and Khubsugul) to very high (Bratsk water reservoir).

The region under discussion is one of the most highly Hg-polluted regions in Siberia, and accounts for approximately 24% of the technogenic Hg emission into the Siberian environment (Yagolnizer et al., 1996). The Hg sources are: the chemical industry, which uses Hg in technology (Usolie-Sibirskoe, Sayansk), other enterprises and settlements, agricultural pesticides, and, possibly, gold amalgamation. The main technogenic anom-

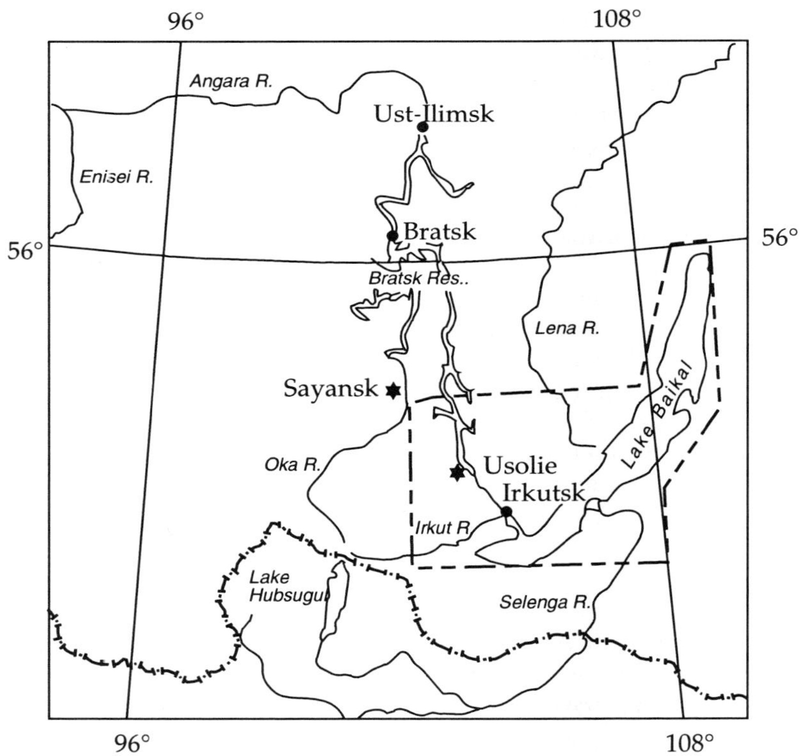

Figure 1. Sketch map of Lake Baikal and adjacent territories.
The Baikal geoecological polygon is outlined by the broken line. The mercury electrolysis sites are indicated by stars.

alies, which are related to the "Usoliekhimprom" and "Sayanskkhimprom" chemical plants, appear to be similar to man-made "deposits". Their total metallic mercury waste during the last 27 years has exceeded 1,000 tons. In spite of the decision to stop mercury electrolysis in the region and abandoning the process in "Usoliekhimprom", these sites could remain pollution sources for a long time after shutting down Hg-using devices. The Baikalsk pulp and paper mill and Selenga River are among the main sources of technogenic contamination of Lake Baikal.

In addition, some natural low-contrast Hg anomalies related to bedrock and active faults have been identified in the region (Belogolova, Koval, 1995; Koval, et al., 1999). Lake Baikal and man-made water reservoirs (notably the Bratsk reservoir) located on the Angara river flowing from Baikal can be considered as accumulators of contaminants from the areas drained.

This contribution summarises the data on the Hg distribution in the bottom sediments of Lake Baikal and stream sediments in adjacent areas mainly obtained from investigations related to the Baikal Drilling Project (BDP), the programs "Multipurpose Geochemical Mapping and Geoecology of Russia", "Global changes of environment and climate", Irkutsk Regional Ecological Fund, and grants RFBR 97-05-96397 and 99-05-64162.

Materials and Methods

Stream sediment was sampled in the Baikal geoecological polygon (BGEP), with 1 sampling site per 100 km^2 (<0.18 mm fraction), and from an area of 3500 km^2 adjacent to Irkutsk (ISZ, 1 sampling site/13-16 km^2, <0.25 mm fraction). More details on sampling areas and procedures are given in Koval et al., 1993 and 1995 and in Belogolova et al., 1995.
Some features of the aquatic systems under discussion are shown in Table 1.

The Irkutsk water reservoir is a less contaminated natural-technogenic system than the Bratsk and Ust'-Ilimsk reservoirs. The main anthropogenic Hg sources, including two producers of chlorine by electrolysis on Hg cathodes, are concentrated in the basin of the man-made Bratsk reservoir.
Drill core BDP96 was collected at a site located on Academic ridge, between the northern and central basins of Lake Baikal (Members of BDP, 1998). The bottom sediments of the Irkutsk, Bratsk, and Ust-Ilimsk water reservoirs were sampled with a GOIN-1 system bottom tube sampler at all sites and, in some cases, by dredger. Granulometric analysis was performed on the samples of Bratsk reservoir sediment. The grain size of all of them was less than 0.18 mm. In the most cases, the <0.05 mm fractions

Table 1. Some characteristics of Lake Baikal and the reservoirs of the Angara River cascade

Parameters	Baikal	Reservoirs Irkutsk	Bratsk	Ust'-Ilimsk
Nature	Natural lake	Man-made	Man-made	Man-made
Age (years)	~30 000 000	43	38	25
Length (km)	636	55	570	292
Max. breadth(km)	79.4	7	25	12
Area (km2)	31 500	154	5 470	1 892
Max. depth (m)	1 630	35	150	90
Volume (km3)	23 000	2.1	169.3	58

accounted for 95-99%, with the < 0.001 mm fraction representing about 60-85%. Bottom sediment thickness differed as follows (cm): 4-77 (Irkutsk reservoir), 1-46 (Bratsk), <0.1-18 (Ust'-Ilimsk). The suspended matter (<0.45 μm pore filter) was taken from Bratsk reservoir water (one sampling site) and the drainage ditch of the "Usoliekhimprom" plant.

The cold vapour atomic absorption spectrometry method (detection limit: 2 ppb) was used for Hg analysis. The bulk samples of bottom sediment were commonly analysed. The results of the reference analyses in different laboratories are presented in Koval et al. (1999).

Results

The analytical data available for the material under discussion are shown in Table 2.

The average Hg content of bedrock, soils, and stream sediment was lower or comparable to the lower means for the average concentration (30-90 ppb) in the Earth's crust (Vinogradov, 1962; Gevis and Ferguson, 1972). It is remarkable that the maximum concentrations in the upper horizon of alluvial soil and overflood sediment were very high and that there was some increase in the average content. Very small difference between 1 mm fraction of BC soil horizon and <0.18 mm fraction of stream sediment is the case.

The data obtained for the bottom sediment of Lake Baikal by the different research teams were mostly of the same order. A higher concentration range was reported in Kuznetsova et al., 1991 for the area adjacent to the Baikalsk pulp and paper mill, but it conflicted with later obtained data by V.F.Vetrov and A.I.Kuznetsova, (1997). There were no large differences between the mercury content of sediment having different lithological characteristics, but the averages increased slightly from sand to diatoma-

Table 2. Mercury content (ppb) in the different media in the Baikal area and adjacent region

Sample type	n	Cmin	Cmax	Average
Bedrock, BGEP (Koval et al., 1999)	514	< 2	140	11
Slope soil, BC horizon (< 1 mm), BGEP (Koval et al., 1999)	786	< 2	120	17
Slope soil, A horizon (< 1 mm), BGEP (Koval et al., 1999)	784	< 2	230	22
Stream sediment (< 0.18 mm), BGEP (Koval et al., 1999)	658	< 2	200	21
Alluvial soil, A horizon (< 1 mm), BGEP (Koval et al., 1999)	712	< 2	2400	29
Overflood sediment (< 0.25 mm), ISZ, Belogolova, et al.,	-	< 5	1500	40
Recent bottom sediment, Lake Baikal (Pampura et al.,1993): Baikalsk pulp and paper mill (Kuznetsova, et al., 1991)	-	40	140	-
Recent bottom sediment, Lake Baikal (Pampura et al.,1993):				
average	154	-	-	66
sand	9	-	-	40
aleurite silt	91	-	-	50
pelitic silt	270	-	-	70
diatom silt	84	-	-	80
Fe-Mn crusts	19	-	-	70
Recent bottom sediment, Lake Baikal (Vetrov, Kuznetsov, 1997): silt, Selenga shallow area	3	< 2	180	~60
silt, area adjacent to Baikalsk pulp and paper mill	8	7	14	10
Bottom sediment, Lake Baikal(Leemakers et al.,1996)	21	5	72	40
Bottom sediment, Lake Baikal BDP-96 (this paper)	32	5	50	15
Bottom sediment, Irkutsk reservoir (this paper)	33	12	65	30
Bottom sediment, Irkutsk reservoir, central part without bays (this paper)	12	10	60	40
Bottom sediment, Bratsk reservoir (this paper)				
top (≦25mm) layer	317	2.5	5000	1050
average for cores*	317	10	4600	960
bays	22	40	3000	700
Bottom sediment, Ust'-Ilimsk reservoir (this paper)				
top (≦25mm)layer	18	500	2800	766
average for cores *	34	250	1700	547

n - number of samplers, *-number of sites sampled, average was calculated as a weighted mean on sediment thickness for samples of each sediment core. A bulk sample was also used for some sites.

ceous silt (Tab. 2). The latter was also mentioned in Leermakers et al., 1996, to explain the Hg concentration in sediment with depth. No marked Hg changes with age were found in pelitic or diatomaceous silt until ~4.5 Ma in the BDP-96 core, but there were some strong fluctuations (Fig. 2). The highest concentrations were reported in the central basin of the lake, the Selenga shallow area, and the area adjacent to the town of Slyudyanka (670-840 ppb) (Kuznetsova et al., 1991; Leermakers et al., 1996).
The Hg content in the bottom sediment of the Irkutsk reservoir is very close to or slightly lower than in Lake Baikal (Table 2). The deep central channel of the reservoir is characterised by a higher average (40 ppb) than the shallower parts and bays (24 ppb).

In contrast to Irkutsk, the Bratsk reservoir had an approximately 30-fold higher Hg concentration in the bottom sediment, corresponding to the highly polluted aquatic systems (Sukhenko, 1995). The lateral and vertical Hg distribution in the bottom sediment is very changeable as in the reservoir as a whole, as well as in particular sites (Fig. 3), and it is strongly dependent on the lithological characteristics of the sediment (Fig. 4). The most polluted area is associated with wedging out of affluent and located 55-175 km downstream of the main pollution source (Figs. 1 and 3).

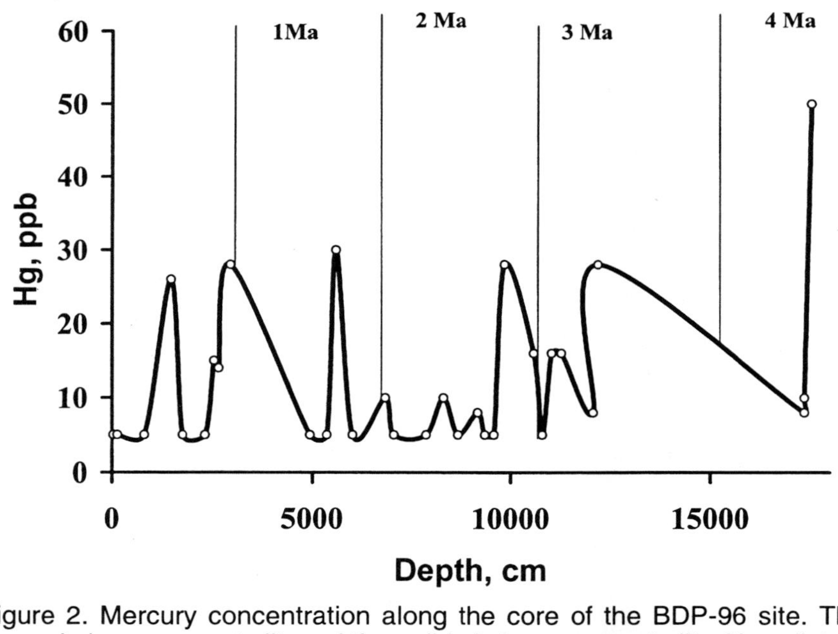

Figure 2. Mercury concentration along the core of the BDP-96 site. The open circles represent silt, and the solid circles represent silt with a diatom content ≥50%.

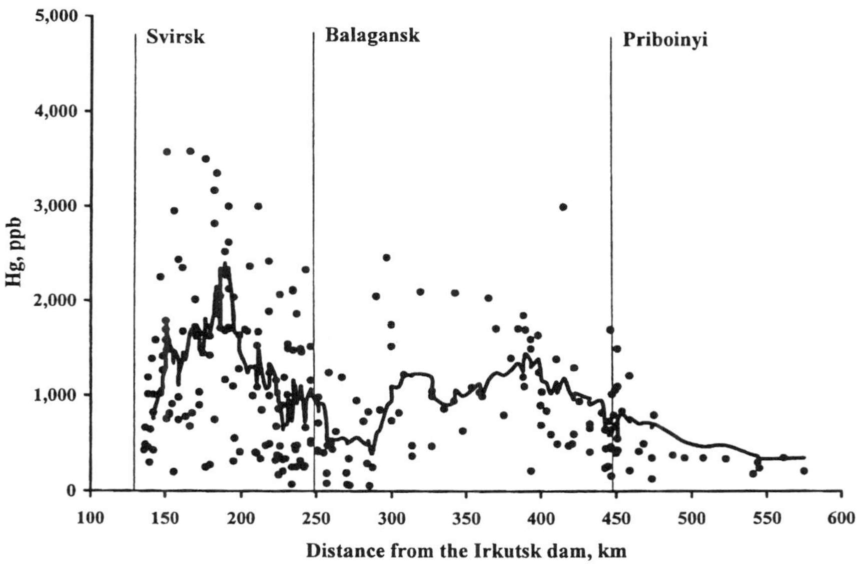

Figure 3. Hg distribution in the surface layer of the most polluted bottom sediment, the Bratsk water reservoir, the portion downstream of the "Usoliekhimprom" chemical plant (the Svirsk - Priboinyi section). The solid curve is the plot smoothed with a 10-km window.

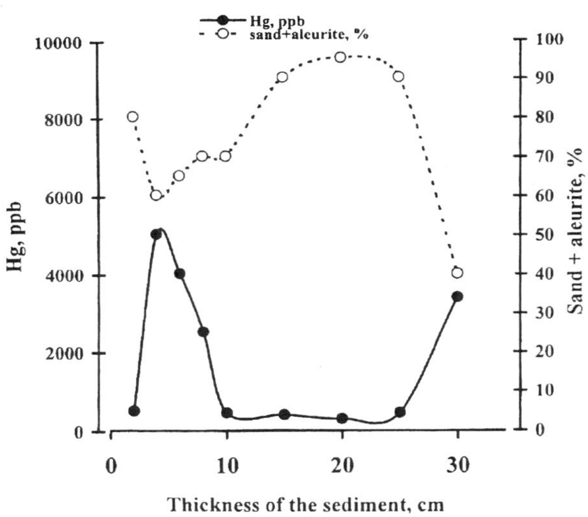

Figure 4. Correlation between Hg concentrations and sand-aleurite fraction content in the bottom sediment core, Bratsk reservoir, Site B-105.

The main Hg concentrators are in silt fractions smaller than 0.04 mm, which predominate (90-99%) in the typical sediment of the central part of water reservoir channel (Table 3). The Hg concentration in filtered suspension at site B-315 (3750 ppb) appeared to be comparable to the Hg content in the bottom sediment of most polluted zones. Suspended fine fractions appear to also be the main Hg concentrator in the water of the drainage ditch of the "Usoliekhimprom" chemical plant (Table 4).

Hg concentrations indicative of a polluted water system were also obtained in bottom sediment of the Ust'-Ilimsk reservoir (Tab. 2). Higher Hg concentrations were also found in the bottom sediment of the main channel and in the top 25 mm portion of the sediment cores.

Discussion and conclusions

The regional natural Hg baselines for the lithochemical media of Lake Baikal and adjacent Angara River and tributary basin areas (11-70 ppb) can be considered to correspond to the lower portion of the average concentration range in the Earth's crust. Although mercury anomalies have been reported in soil in relation to faults (Belogolova and Koval, 1995), there

Table 3. Mercury concentration in different fractions of the bottom sediment of Bratsk water reservoir sampling site B-326

Fraction, mm	Hg, ppb
0.2 - 0.1	120
0.1- 0.4	220
< 0.04	3,000

Table 4 Hg concentration in the water and suspended matter of the drainage ditch of the "Usoliekhimprom" chemical plant , October 1995

Site No.	Unfiltered water, ppb	Filtered water, ppb	Amount of suspended matter, mg/l	Hg in suspended matter, ppb
3	30	0.6	26.45	2000000
4	40	0.5	24.35	2100000
5	30	0.8	24.85	2400000
6	40	1.0	15.40	2100000
7	40	0.8	19.10	2400000
8	30	1.0	18.00	2300000
9	50	1.6	17.15	2600000
10	50	4.4	11.25	4500000

has never been any evidence of a strong influence of active fault "breathing" in the data available thus far. The higher average Hg content in the bottom sediment of Lake Baikal and deep parts of the Irkutsk reservoir, than in the shallow parts of the reservoir and the stream sediment can be attributed to more effective sedimentation of fine fractions in the deep areas of these basins. In general, Lake Baikal, adjacent territory, and the Irkutsk reservoir basin can be considered a region of low Hg pollution. The Irkutsk water reservoir can be regarded as a background man-made reservoir in comparison with the other water reservoirs of the Angara River cascade.

In contrast, the downstream reservoirs of the Angara River cascade, especially the man-made Bratsk Sea, which drains the main industrial and agricultural areas of the region, are known to be polluted (Ust'-Ilimsk) and highly polluted (Bratsk). This is supported by the data on Hg content in the biota of the Bratsk reservoir (Koval et al., 1998 and 1999). The main bulk of pollution was found downstream of the main pollution sources in the zone of wedging out of affluent, which thus can be considered a sedimentary geochemical bar. The Hg contamination penetrates through the Bratsk hydropower dam downstream to the Ust'-Ilimsk water reservoir system. On the other hand, the many-fold increases in Hg content in the upper horizon of alluvial soil and overflood sediment relative to the regional baseline is a sign of technogenic pollution in the corresponding basin.

Silt fractions less than 0.04 mm are the main Hg concentrators in sediments, and, as suspended matter, in waters (Tabs. 3 and 4). This is especially remarkable for polluted sediment with very large differences in Hg content (over one order) between the sand-aleuritic and silt fractions (Fig. 4, Tab. 3) as compared with natural sediments, where the distinction is not so high (Tab. 2). This provides reliable evidence that the relatively weakly bound (sorbed) portion of Hg in polluted sediments and suspended matter is strongly increased, as opposed to natural sediments and suspensions. The data on the correlation between Hg concentrations and the sedimentation rates (Koval et al., 1999) favour this view. A pyrolysis study of mercury speciations also found a predominance of the relatively weakly bound sorbed mode of occurrence in contaminated sediments of the Bratsk reservoir and the drainage ditch of the "Usoliekhimprom" chemical plant (Tauson et al., 1996).

The available data highlighted the problem of Hg pollution in the regional environment and provide key topics for the further scientific investigations: the mode of Hg occurrence in the main natural and man-made reservoirs and routes of migration, Hg balance in particular geochemical cycles (industrial, local related to the main industrial anomalies, regional in the basin of the Bratsk water reservoir and Baikal region, etc.), and approaches to conserve and rehabilitate extremely polluted sites.

Acknowledgements
The authors are grateful for the support and cooperation of their colleagues in programs mentioned. The authors are indebted to the organisers of BICER, BDP, and DIWPA Joint International Symposium on Lake Baikal (November 5-8, 1998, Yokohama) which stimulated this presentation. Assistance in editing the English version by T. Bounaeva is appreciated.

References

Belogolova G.A. and P.V. Koval, 1995, Environmental geochemical mapping and assessment of anthropogenic chemical changes in the Irkutsk-Shelekhov region, southern Siberia, Russia. J. Geochem. Expl., 55, 193-201.

Belogolova G.A., P.V. Koval and V.D. Pampura, 1995, Rapid geochemical assessment of agricultural-industrial pollution from stream sediment dispersion flows, South Siberia (Irkutsk Region). Sci. Total Environ., 162, 1-11.

Gevis J. and J.F. Ferguson, 1972, The cycling of mercury through the environment. Water Res., 6, 989-1008.

Koval P.V., G.A. Belogolova, E.K. Burenkov and V.D. Pampura, 1993, The Baikal polygon: international and national projects of geochemical mapping and monitoring of the environment. Russ. Geol. and Geophys., 34, No10-11, 207-220.

Koval P.V., E.K. Burenkov and A.A. Golovin, 1995, Introduction to the program "Multipurpose Geochemical Mapping of Russia". J. Geochem. Expl., 55, 115-123.

Koval P.V., G.V. Kalmychkov, V.F. Geletyi and L.D. Andrulaitis, 1998, Mercury distribution in bottom and steram sediments of the Baikal Lake, Bratsk and Ikuitsk man-made water reservoirs and the adjasent drainage basins, In: BICER, BDP and DIWPA Joint International Symposium on Lake Baikal, Abstracts, Yokohama, November, 5-8, 51 pp.

Koval P.V., G.V. Kalmychkov, V.F. Gelety, G.A. Leonova, V.I. Medvedev and L.D. Andrulaitis, 1999, Correlation of natural and technogenic mercury sources in the Baikal polygon, Russia. J. Geochem. Expl., 66(1-2): 277-290.

Kuznetsova N.A., E.I. Grosheva and Z.A. Klimashevskaya, 1991, Trace elements in bottom sediment of the Lake Baikal, In: Monitoring sostoyaniya ozera Baikal, Gidrometeoizdat, Leningrad, pp.94-97 (in Russian).

Members of the Baikal Drilling Project, 1998, A continuous record of climate changes of last 5 million years stored in the bottom sediments of Lake Baikal. Geologiya i geofizika, 39, No2, 139-156 (in Russian).

Leermakers M., C. Meuleman and W. Baeyens, 1996, Mercury Distribution and Fluxes in Lake Baikal, In: Global and Regional Mercury Cycles: Sources, Fluxes and Mass Balances, eds. W. Baeyens, R. Ebinghaus and O. Vasiliev, Kluwer Academic Publishers, Dordrecht. 563 pp.

Pampura V.D., M.I. Kuzmin, A.N. Gvozdkov, V.S. Antipin, I.S. Lomonosov and A.P. Khaustov, 1993, Geochemistry of recent sedementation in Baikal Lake. Russ. Geol. Geophys., 34(10/11): 41-57.

Rudd J.W.M., eds., 1998, Forth International Conference Mercury as Global Pollutant. Special Issue. Biogeochemistry, 40, 374 pp.

Sukhenko S.A., 1995, Mercury in reservoirs: a new feature of man-made environmental pollution. Analiticheskiy obzor, Novosibirsk, 59 pp. (in Russian).

Tauson V.L., V.F. Gelety and V.I. Men'shikov, 1996, Mercury Speciation in Mineral Matter as an Indicator of Sources of Contamination, In: Global and Regional Mercury Cycles: Sources, Fluxes and Mass Balances, eds. W. Baeyens, R. Ebinghaus and O. Vasiliev, Kluwer Academic Publishers, Dordrecht. 563 pp.

Vetrov V.A. and A.I. Kyznetsova, 1997, Trace elements in the natural media of the Lake Baikal region. SB RAS, SPC UIGGM, Novosibirsk, 234 pp (in Russian).

Vinogradov A.P., 1962, Average contents of the elements in the main types of eruptive rocks of the earth. Geokhimiya, 7, 555-571 (in Russian).

Yagolnitser M.A., V.M. Sokolov, A.D. Ryabtsev, A.A. Obolensky, N.A. Ozerova, S.Ya. Dvurechenskaya and S.A. Sukhenko, 1996, Industrial Mercury Sources in Siberia, In: Global and Regional Mercury Cycles: Sources, Fluxes and Mass Balances, eds. W. Baeyens, R. Ebinghaus and O. Vasiliev, Kluwer Academic Publishers, Dordrecht, 563 pp.

Lake Baikal
K. Minoura (editor)
2000 Elsevier Science B.V.

Correlations between geochemical features of recent bottom and stream sediments in the Baikal geoecological polygon

Koval, P. V.*, Gvozdkov, A. N., and Romanov, V. A.
Vinogradov Institute of Geochemistry, P.O.Box 4019, Irkutsk-33, 664033, Russia
Fax: (7) 3952 464050, e-mail: koval@igc.irk.ru
(*corresponding author)

Abstract

The paper presents correlations of data on major elements (Si, Ti, Al, Fe, Mn, Mg, Ca, Na, K, P, and S) and trace elements (B, V, Cr, Co, Ni, Zn, Sr, Ba, Zr, Mo, Pb, and U) in recent stream and bottom sediment obtained from regional geochemical mapping and other projects in the Baikal geoecological polygon.

The composition of stream and bottom sediments in general reveals the more basic composition of the upper crust of the region as opposed to the upper continental crust. The main difference between stream and bottom sediments consist in the enrichment of bottom sediments in Fe, Mn, P, Cr, V, Ni, Co, Si, U, B and Mo, and lower Ca, Mg, and Sr levels.

The Baikal bottom sediment area is perfectly outlined by Mn-P-Fe associations on the geochemical association map of major elements. The trace element associations are mainly common to stream and bottom sediments and related adjacent drained areas. The strong geochemical difference between the western and eastern side of the Baikal region is evident on maps of geochemical associations of major and trace elements.The sedimentation in Lake Baikal is somewhat similar to the sedimentation in oceanic basins, except that it has very low mineralized water and less Ca and Mg.

Introduction

An understanding of the natural geochemical baselines is essential as a foundation for assessing chemical changes in the environment of the Lake Baikal area. Regional geochemical mapping, as a universal approach to the study of the chemical changeability of the environment, is applicable to the basic problems of human beings: fundamental geochemistry, mineral exploration, pollution assessment, agriculture, medicine, land use, etc. (Darnley et al., 1995). In this paper, the authors have tried to correlate regional geochemical mapping data and the results of other projects with the chemistry of stream and bottom sediments obtained in the Baikal geoecological polygon.

The Baikal geoecological polygon (an area of 110,000 km^2 that excludes the aquatic area of Lake Baikal, see Fig.1 in Koval et al., this volume) is one of six testing areas selected for implementation of the methodological stage of the programs "Geochemical Map of Russia" and "Geoecology of Russia" (1991-2005) aimed at producing a set of maps containing basic information on lithogeochemistry, mineral exploration, environmental assessment, agrogeochemistry, and land-use planning (Koval et al, 1993, Koval et al, 1995). This area is a part of the larger Baikal polygon. In a general scientific sense the latter is considered to be the physical-geographical region (about 850,000 km^2) that occupies the watershed of Lake Baikal and the adjacent areas of the Angara and Lena river basin and it united by the processes of water and air mass transfer (Koval et al., 1993). It is marked by a variety of geological features, landscapes, and mineralization. Lake Baikal, which contains about 20% of the world˥s fresh water, and the Irkutsk and, in part, the Bratsk water reservoirs are located within the Baikal geoecological polygon. The natural ecosystems of the region are very sensitive to anthropogenic influences. The polygon is also characterized by broad diversity of anthropogenic transformation of landscapes: from pure practically non-polluted areas to places under highly industrial pressure (chemical, aluminium, pulp and paper, heavy machinery, coal mining, cement producing, light industry enterprises, etc.).

Data and methods

The analytical data selected for discussion are given in Table 1. They represent data obtained by two programs, "Multipurpose Geochemical Mapping of Russia - Geoecology of Russia" (MPGM), polygon stage, 1991-1995 (Koval et al., 1993, Koval et al., 1995) and the Baikal Drilling Project with related research (LBP), 1989-1996 (Kuzmin et al, 1998), as well as data from some sources in the literature. The first program includes sampling stream sediment, soils, river water, snow, vegetation on 800 - 900 sampling sites (approximately 1 site per 100 sq. km), with more comprehensive analytical work at 62 basic sites (stations). All solid materials were analysed by semi-quantitative optical emission spectrometry (SQOES) for 49 elements, and by gamma ray spectrometry for U, Th, K, and ^{137}Cs. X-ray fluorescent emission spectrometry (XRF) and quantitative arc optical emission spectrometry (OES) were also used for the basic station samples. No more than 10-15% of the samples from which the data under discussion were obtained had element content below the detection limit of the analytical methods used. The LBP made the bottom (~450)

Table 1. Chemical component of stream and bottom sediments in the Baikal geoecoogical polygon

	Stream sediment			Bottom sediments of Lake Baikal						
Object	MPGM	MPGM basic Stations	Tributaries of Lake Baikal	MGPM	LBP	LBP, * - Pampura et al., 1993				
Fractions	< 0.18 mm	< 0.18 mm	< 0.5mm	< 0.18mm	< 0.5mm	< 0.5 mm	> 0.1 mm	0.1-0.01 mm	< 0.01 mm	> 30% of diatom SiO2
No. Samples	854	62	251	101	450		54	259	79	35
Si (%)	-	27.99	25.89	-	28.77	28.41	29.98	28.11	29.69	33.89
Ti (%)	0.47	0.42	0.42	0.39	0.55	0.40	0.45	0.58	0.50	0.34
Al (%)	6.13	5.89	7.01	5.52	9.03	6.58	8.32	9.21	8.91	5.98
Fe (%)	2.99	3.23	4.66	4.29	6.76	4.81	4.28	7.20	6.79	5.58
Mn (%)	0.11	0.09	0.09	0.50	0.37	0.43	0.13	0.37	0.54	0.69
Mg (%)	2.36	1.60	1.68	1.53	1.58	1.13	1.33	1.68	1.44	1.12
Ca (%)	2.74	3.48	2.52	1.39	2.08	1.41	2.69	2.15	1.65	1.61
Na (%)	1.42	1.42	1.66	1.01	1.94	1.41	2.72	2.01	1.43	1.39
K (ppm)	1.66	1.56	1.83	1.14	2.19	1.65	2.18	2.23	2.09	1.38
P (ppm)	0.05	0.14	0.10	0.09	0.14	0.17	0.08	0.15	0.16	0.17
S (ppm)	-	-	0.12	-	0.08	0.08	0.06	0.09	0.07	0.06
Ba (ppm)	715	-	750	685	622	750	605	695	645	540
Zr(ppm)	435	-	260	357	206	150	440	186	154	120
V (ppm)	120	139	77	116	143	154	93	141	171	116
Cr (ppm)	114	82	45	80	104	107	76	110	103	84
Ni (ppm)	45	47	14	50	51	50	26	54	56	43
Co (ppm)	18	9	10	22	17	16	11	17	18	16
Pb (ppm)	14	-	9	11	18	21	15	18	21	15
Cu (ppm)	39	-	47	62	46	45	26	47	51	44
Zn (ppm)	91	-	58	87	86	80	61	86	96	71
Mo (ppm)	4.2	-	2.7	5.0	2.9	3.2	1.2	2.8	4.0	5.0
Sr (ppm)	328	-	400	254	292	150	200*	138*	152*	44*
B (ppm)	27	-	13	26	26	24	22*	23 *	25*	17*
U (ppm)	2.7	-	-	-	11	10.8	-	-	-	-

sediment of the lake and stream sediments in its tributaries (~250) available. Samples were analysed by quantitative OES, XRF and atomic absorbtion spectrometry (AAS). More details on sampling and analysis are given in Pampura et al. (1993).

Multidimensional Geochemical Field Analysis software (Evdokimova, 1978) was used for compiling the multielement geochemical maps (maps of geochemical associations). Taking into account the principal differences in major and trace element distribution and the specificity of the method, two maps were prepared, one for major elements and the other for trace elements (Figs. 1 and 2).

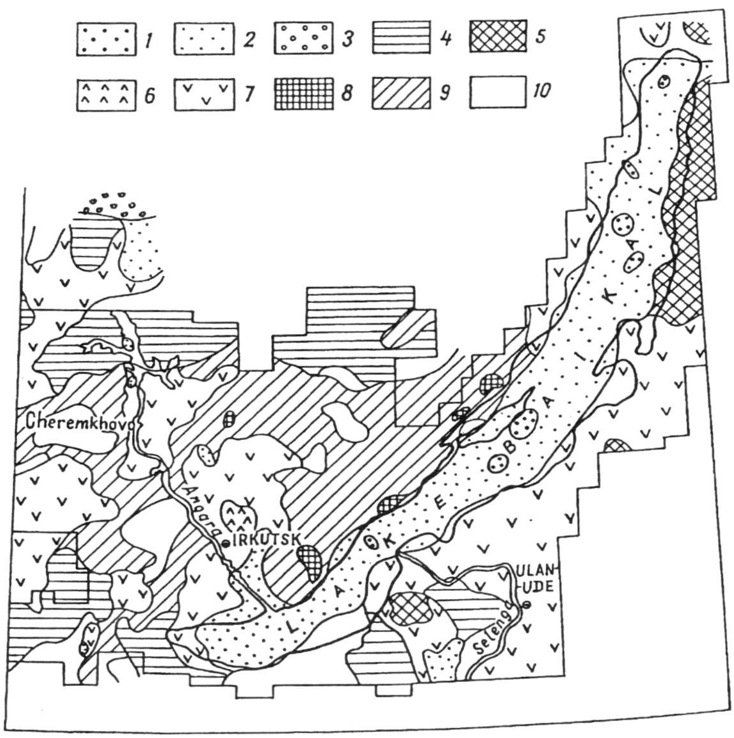

Figure 1. Map of geochemical associations of major elements in the Baikal geoecological polygon. The characteristics of the associations are given in Table 2. The numbers of the associations in Fig. 1 correspond to those in Table 2.

Table 2. Geochemical associations of major elements in stream and bottom sediments in the Baikal geoecological polygon, MPGM (see Fig. 1)

Group of Associations	Association No. on Fig, 1	Number of sites	Ti	Fe	Mn	Mg	Ca	Na	K	P
Mn - P - Fe	1	19	0.4	**1.5**	**11.5**	0.4	0.4	0.5	0.2	**2.2**
	2	90	1	**1.5**	**3.7**	1	0.7	1	0.7	**2**
Mn - Fe	3	10	1	**1.2**	**2.0**	0.9	0.6	0.5	0.4	0.3
Ca	4	260	0.9	0.9	0.9	1.1	**1.5**	1.2	0.9	0.4
P	5	37	1.2	1.1	1.2	1.7	2.2	1.7	1.1	**2.8**
Na	6	7	0.7	0.6	0.8	0.3	0.4	**1.9**	**1.4**	0.5
	7	267	1.1	1.2	1.2	1.1	1.1	1.5	1	1.5
Fe	8	3	1.1	**1.6**	0.9	1.5	0.3	0.9	**1.5**	1.3
Mg	9	220	1	1.1	1	**1.4**	1	1.1	0.7	0.8
	10	16	Sites which not included in associations							
Background, %			0.45	2.7	0.087	1.8	2.0	1.0	1.8	0.045

Note : Digits in the element columns in Table 2 and 3 are coefficents of contrast, which are concentrations normalized to background. The coefficents of contrast of the indicator elements for particular associations are shown in bold print.

Correlation between the chemical composition of recent stream and bottom sediments

Major elements

The average chemical composition of stream and bottom sediments was generally rather similar, except for Mn and P, which were much higher in bottom sediment (Tab. 1, Fig. 3). Both stream and bottom sediments were essentially more basic (enriched in Fe, Mg, Ti, and Mn) than average continental crust and lower in alkali metals (Taylor and McLennan, 1985). This appears to reflect the more basic composition of the upper lithosphere in the region around Lake Baikal.

Assessment of chemical trends in sedimentation and the relation between sediment and source chemistry is greatly hampered by lithological differences between sediments. Nevertheless, Baikal bottom sediment is believed to be enriched, in Fe and Si, in addition to Mn and P, and to contain reduced levels of Ca and Mg, and probably S. This may be related to the chemical and biochemical features of the sedimentation process in the huge volume of low-salt-containing Lake Baikal water. Maximum Si was found in diatom silt.

We should also report the stable level of average Ti content in all materials under discussion: stream sediment (MPGM), tributary sediment (LBP), and bottom sediment (LBP). This can be used as a normalising factor in comparative studies of sediments having different lithological characteristics. The remarkable distinction (Ti = 0.94 %, Pampura et al., 1993)

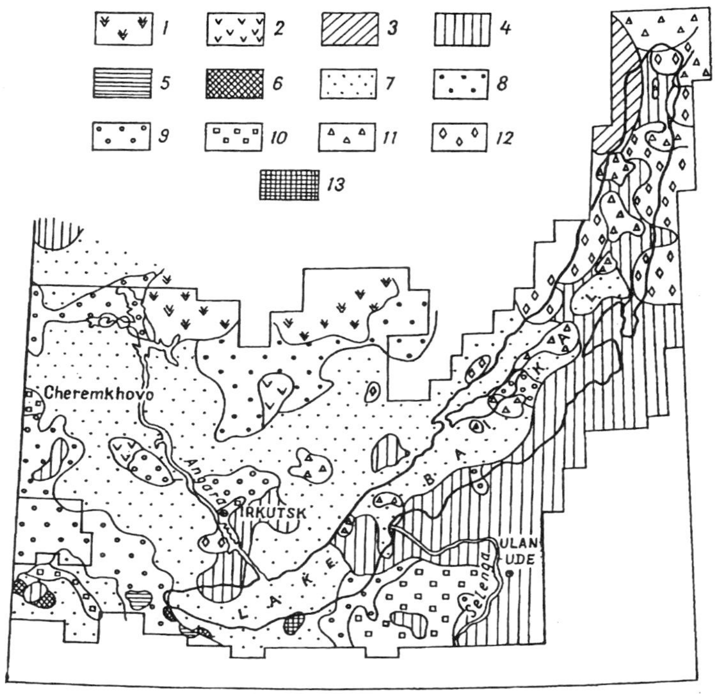

Figure 2. Map of geochemical associations of trace elements in the Baikal geoecological polygon. The characteristics of the associations are given in Table 3. The numbers of the associations in Fig. 2 correspond to those in Table 3.

is the characteristic of sand enriched in heavy minerals, especially in the beach zone.

Trace elements

As might be expected from their major element content, compared with the average upper crust the stream and bottom sediments of the region are enriched in certain sidero-chalcophile elements (V, Cr, Cu), as well as Ba and Mo (Tab. 1, Fig. 4). Not surprisingly Baikal sediment is notably enriched in iron-group metals (Cr, V, Ni, Co), as well as U, and probably B, Mo, and Pb, compared to tributary sediment. Accumulation of uranium is especially essential. It should be noted that an increasing content in bottom sediment is generally characteristic of certain elements (Mn, Ni, V, Mo) that have been reported to be increased in tributary water relative to

Table 3. Geochemical associations of trace elements in stream and bottom sediments in the Baikal geoecological polygon, MPGM(see Fig. 2)

G	No.	n	Ba	Be	Zr	V	Cr	Ni	Co	Pb	Cu	Zn	Mo	Li	Yb	La	Nb	Sr	B
Cr-Li-Cu	1	20	0.7	0.3	0.8	1	2.5	1	0.9	0.9	1.2	1	0.7	1.5	0.8	0.9	0.7	0.9	0.2
Cr	2	7	1	0.5	1.1	1.1	2.4	0.9	0.7	0.7	1	0.5	0.6	0.6	1	0.9	0.6	0.9	1
Yb-Nb-Pb	3	2	3.2	1.7	10	0.9	0.9	1.3	1.3	4.4	1.1	1.9	4.1	1.3	10.7	6	10	0.9	0.9
Zr-Nb-Pb	4	116	1.5	1.5	2	1.4	0.9	1	1.1	1.6	1	1.1	1.3	1.1	1.4	1.5	1.8	1.4	0.9
La-Yb-Zr	5	5	0.9	0.7	2.2	1.2	0.8	1	0.8	1.6	0.9	1.1	0.6	1.2	2.3	3.8	1.6	0.8	0.3
B	6	2	0.8	0.7	0.5	0.6	0.3	0.2	0.6	1.5	0.6	0.7	0.7	1	1.4	1.1	0.7	1.3	3.6
Li-V	7	288	1.1	0.9	1.2	1.4	1.2	1.2	1	1.1	1.1	1	1	1.8	1.2	1.2	1.2	1.1	1.1
Li-Cr	8	42	0.8	0.3	0.7	1	1.4	1	0.7	0.8	1.3	0.9	0.7	1.8	0.9	0.9	0.6	0.7	1.1
Li-Pb	9	264	1.1	0.8	1.1	1.2	1	1	0.9	1.2	0.7	0.9	0.9	1.3	1	1.2	1.2	1	0.5
Pb-Sr	10	86	1.5	1.5	0.9	0.7	0.4	0.4	0.6	2.7	0.3	0.7	1.1	1.1	1	1.2	1.4	1.7	0.5
Mo-V-Cr	11	20	0.9	0.9	1.8	4.1	3	2.3	2.8	1.1	2.2	1.3	6.9	0.5	2.5	1.1	1.7	1.2	1.3
Nb-Li	12	33	1.9	1.8	2	2.4	1.6	1.7	1.3	1.8	1.4	1.3	1.6	2.7	1.9	2.2	2.7	1.2	1
Sr-Li	13	5	1.2	1.3	0.7	0.9	0.9	0.9	1.1	1.1	0.8	1.2	0.5	1.4	1.2	0.8	0.6	1.8	0.3
Background. $N \cdot 10^{-3}\%$			60	0.37	30	9	9	4	1.8	0.9	4	9	0.37	1	0.35	2.5	1	27	2.8

Note: G – geochemical association, No. – number of the association in Fig. 2 n – number of sites.

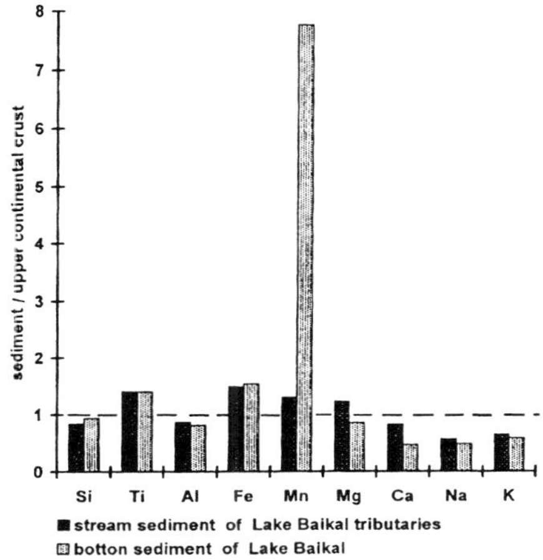

Figure 3. Comparison of the major element content of stream and bottom sediments in the Lake Baikal area. The data are normalized to the upper continental crust (Taylor and McLennan, 1985).

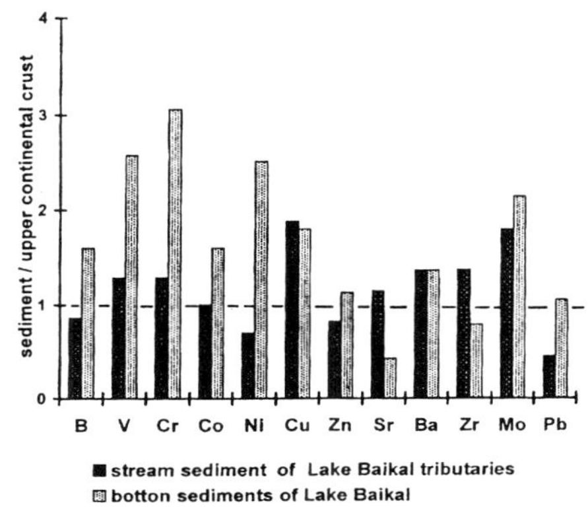

Figure 4. Comparison of the trace element content of stream and bottom sediments in the Lake Baikal area. The data are normalized to the upper continental crust (Taylor and McLennan, 1985).

Lake Baikal (Pampura et al., 1993).

The exceptions are such good water migrants as Cu and Sr.

Geochemical associations

The most striking feature of the map of geochemical associations of major elements is the clear distinction outlined by Mn-P-Fe associations in the Baikal area (Fig. 1). Even taking the low sampling density and SQOES analysis used into account, the conformity must be admitted to be very good. This is in keeping with the above-described chemical characteristics of the bottom sediment. There is no doubt that this picture manifests the peculiar features of the sedimentation process in Lake Baikal. The next feature is the chemical asymmetry of the region, with different associations on the western side, mainly formed by ancient, more basic, formations than on the eastern side, where granitoids predominate.

In contrast, the trace element associations convey quite a different image (Fig. 2). The Baikal area has differentiated into several associations, which are generally common to the adjoining drained areas. The "spots" of Mo-V-Cr association (No. 11, Fig. 2) seem to be the only signs related to specific Baikal sedimentation proper. This is supported by the presence of a Mo-V-Cr association field in the area of Academician Ridge, where the possible influence of mud flows (turbidity) has been completely ruled out. However, the above-described asymmetry of the western and eastern sides of the lake is real. There is no doubt, that more sophisticated analytical procedures and denser sampling would provide significant advantages to this approach for geochemical correlation of bottom sediments in relation to changing conditions of erosion and sedimentation in the Lake Baikal area.

It follows from the data presented that the geochemical trend of sedimentation in Lake Baikal, in particular the increase in Fe, Mn, P, Cr, V, Ni, and Co, has a certain similarity to the sedimentation in deep seas and oceans (Ivanov, 1994 and 1996).

Conclusions

Lake Baikal, as a strong sedimentary geochemical bar, is the natural accumulator of the lithochemical and hydrochemical drainage of a huge territory. The composition of stream and bottom sediments in general reveals the more basic, as opposed to the upper continental crust, composition of upper crust of region discussed. Although there is a correlation between the total composition of stream and bottom sediment, as compared with the upper continental crust, Baikal sediments do not fully reflect the average composition of the area drained. The main differences constitute the enrichment in Fe, Mn, P, Cr, V, Ni, Co, Si, U, B, and Mo, and lower levels of Ca, Mg and Sr. However, the Baikal bottom sediment area is perfectly outlined by Mn-P-Fe associations on the geochemical association map of the major elements. This process is somewhat similar to sedimentation in oceanic basins, but differs in terms of the low mineralization of fresh water and lower levels of Ca and Mg (Ivanov, 1994 and 1996).

Trace element associations are mainly common to stream and bottom sediments and related adjacent drainage areas. The strong geochemical asymmetry of the western and eastern sides of the Baikal region is evident on both the major and trace element maps of geochemical associations. The geochemical association approach is a very promising method of searching for regularities in sedimentation in relation to changing climate and the geological condition of erosion in the areas drained.

Acknowledgements

The authors are grateful to A.E. Gapon for the Baikal tributaries sediment samples and to their colleagues on the Baikal polygon project for their support. This work was supported by RFBR grants 97-0596397 and 98-07-90069. Assistance in editing the English version by T. Bounaeva is appreciated.

References

Darnley A.G., A. Bjorklund, B. Bolviken, N. Gustavsson, P.V. Koval, J.A. Plant, A., Tauchid, M. Steenfelt and Xie Xuejing, 1995, A Global Geochemical Database for Environmental and Resource Management. UNESCO Publishing, Earth Sci. Rep.19, 122 pp.

Evdokimova V.N., 1978, Automated processing of geologo-geochemical data multidimemional fields, In: Geochemical Methods of Prospecting for Ore Deposits in Siberia and Far East, eds. V.V. Polikarpochkin, Nauka, Novosibirsk, pp. 3-26 (in Russian).

Ivanov V.V., 1994, Environmental Geochemistry of Elements: Reference Book, Vol.2, Ekologiya, Moscow, 303 pp. (in Russian).

Ivanov V.V., 1996, Environmental Geochemistry of Elements: Reference Book, Vol.4, Ekologiya, Moscow, 416 pp. (in Russian).

Koval P.V., G.A. Belogolova, E.K. Burenkov and V.D. Pampura, 1993, The Baikal polygon: international and national projects of geochemical mapping and monitoring of the environment. Russ. Geol. and Geoph., 34, No10-11: 207-220.

Koval P.V., E.K. Burenkov and A.A. Golovin, 1995, Introduction to the program "Multipurpose Geochemical Mapping of Russia". J. Geochem. Expl., 55: 115-123.

Koval P.V., G.V. Kalmychkov, V.F. Geletyi and L.D. Andrulaitis, Mercury distribution in bottom and stream sediments of the Baikal Lake, water reservoirs of the Angara river cascade and the adjacent drainage basins (this volume).

Kuzmin M.I., E.M. Karabanov, V.F. Gelety, V.S. Antipin, A.V. Goreglyad, G.V. Kalmychkov, A.N. Gvozdkov, V.A. Kravchinskiy and A.A. Prokopenko, 1998, A continuous record of climate changes of last 5 million years stored in the bottom sediments of Lake Baikal, In: Global Change of Environment, N.L. Dobretsov and V.I. Kovalenko, eds., Pulished by SB RAS, SPC UIGGM, Novosibirsk, pp. 58-72 (in Russian).

Pampura V.D., M.I. Kuzmin, A.N. Gvozdkov, V.S. Antipin, I.S. Lomonosov and A.P. Khaustov, 1993, Geochemistry of recent sedimentation in Baikal Lake. Russ. Geol. Geophys., 34(10/11): 41-57.

Taylor S.R. and S.M. McLennan, 1985, The continental crust: its composition and evolution. Blackwell Scientific Publications, Oxford a.o, 312 pp.

Lake Baikal
K. Minoura (editor)
2000 Elsevier Science B.V.

Remote sensing methods in studies of Lake Baikal environment

Semovski, S. V.
Limnological Institute SB RAS, Irkutsk, Russia
Fax: +7-3952-466933, E-mail: Semovsky@lin.irk.ru

Abstruct

Satellite imagery applications to investigations of Lake Baikal environment are reviewed. Examples of different data use for the Lake investigations are presented, namely: high-resolution imagery, data bases of NOAA AVHRR images, SAR images analysis, multispectral remote sensing applications, satellite altimetry and other methods.

Introduction

Modern satellite remote sensing is a powerful tool for natural environment studies. Regular, low-cost surveys with high spatial and spectral resolution is an only way to describe processes on the Earth surface in their space and time dynamics. For the highly variable water objects there are still many problems, that are limit systematic use of satellite data. Whereas for open parts of the ocean during past two decades several space platforms have been launched and data was collected and used for description of ocean circulation, productivity and mesoscale variability, the studies of highly variable, the most productive and important for mankind coastal and lacustrine environment still run into difficulties (see Jaquet, 1989 for the review on problems arising in application of remote sensing for lake studies).

Fortunately, for big lakes such as Great Lakes of North America, Caspian Sea or Baikal many of methods can be used, that were successfully used in marine studies. At the first time, seemingly, perspectives of satellite data using for studies of lake Baikal have been discussed in (Galazij *et al.*, 1980). Many physical and biological processes in the Baikal have marine analogies, so that the lake can be used as natural laboratory for calibration and validation of satellite data. However, it should be noted that use of remote sensing for studies of Baikalian environment is on the very early stage due to low availability of data sources, structure of previous scientific programs, etc.

Coastal studies

Imagery produced by satellite-based platforms designed for thematic mapping of the Earth surface, spy satellites data and photos made by astronauts during manned space missions have the best spatial resolution to give representation of coastline details, relief of the shore zone, etc. Data of Landsat-5 TM with 30 m resolution and Resurs MK-4 with 8 m resolution was used for classification of land use types in Institute of Geography SB RAS, Irkutsk (for Selenga delta, see Hern *et al.*, 1998; and for Olkhon island, unpublished). Landsat data can be useful also for paleolimnological studies as it was shown in (Zou, 1987), for bottom topography in coastal zone, etc..

One of the limited factors for systematic use of high resolution imagery for Baikal studies is its high cost and irregularity. As a rule, only high quality images are available on the market and for low-populated areas they are usually unique for every region. Nowadays many commercial satellit–based platforms of new generation with very high spatial and spectral resolution are working on the orbit or are under development (announced Russian "Resurs" series with 8 m resolution and 8 visual spectral bands, advanced USA and Russian satellites with 1-2 m resolution, etc.) Probably, new sources of data can would be useful in Baikal studies. These intensive data flow would need new computer platforms and software.

As an example of high-resolution imagery of Lake Baikal we present here photo of Angara source area made by astronauts during manned "Shuttle" mission (colour photo is converted here in black and white), see Fig.1.

Analysis of hydrophysical fields of Baikal by satellite methods

Representative data bases of satellite imagery should be developed to realise statistical analysis of variable hydrophysical fields of the lake Baikal and for justified conclusions of climate, annual and small-scale variability. Moderate space and spectral resolution data of AVHRR (Advanced Very High resolution Radiometer) instrument based on NOAA meteorological satellites can be suitable source of data for these studies, because NOAA satellites on polar orbits produces daily several image of every region on the Earth surface. Pioneer paper of Bolgrien *et al.* (1995) was the first attempt to use AVHRR data for detection of Baikal hydrophysical features. Main object of studies have been pointed out in this work, namely, temperature fronts, river plumes, processes of freezing and break-up, etc.

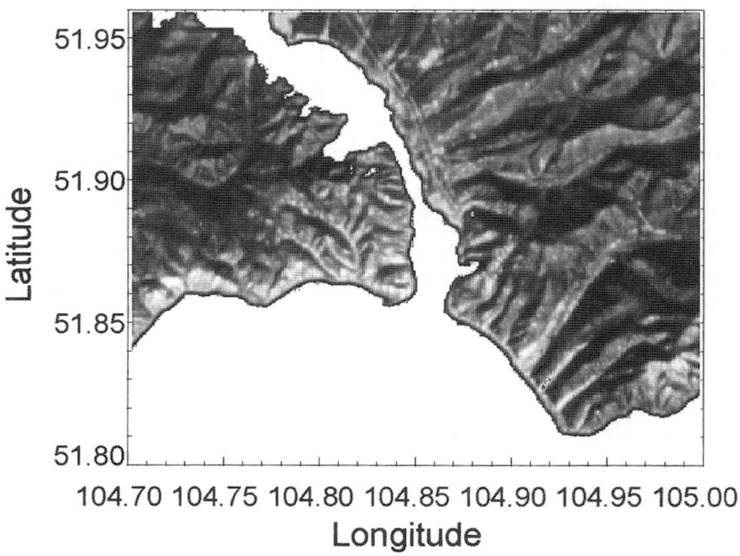

Fig.1. Angara source image (cut from a South Baikal photo taken by astronauts during a manned "Shuttle" mission, courtesy NASA, USA).

Regular application of AVHRR data for Baikal studies became possible after NOAA satellites receiving station installation in the Irkutsk Institute of Solar-Terrestrial Physics SB RAS. This station, based in the Centre of Space Monitoring, was designed mainly for forest fires detection in Irkutsk region. However, with participation of Limnological Institute, data bases for lake Baikal was collected on CD ROM's. XV software package (Zacharov *et al.*, 1995) is used for data pre-processing and georeferensing.

It is well-known that accuracy of surface temperature determination for water objects by passive remote sensing from space in the infrared band is small enough, usually not less then 1 deg.C (see, e.g., McClain *et al.*, 1985). However, zones of high gradients on the water surface can be detected with good precision. This fact is a basis for studies of temperature fronts variability by remote sensing methods. For the lake Baikal thermal fronts are very important zones. Spring thermobar is a zone of deep mixing, river plume areas are regions of pollutant spreading and deep "cascading" of more dense river water. In the paper (Semovski *et al.*, 1998a) Baikal thermal fronts were studied using AVHRR data base. The velocity of spring thermobar movement was estimated for shallow areas, evolution of temperature gradients was derived for Selenga delta area, some calcula-

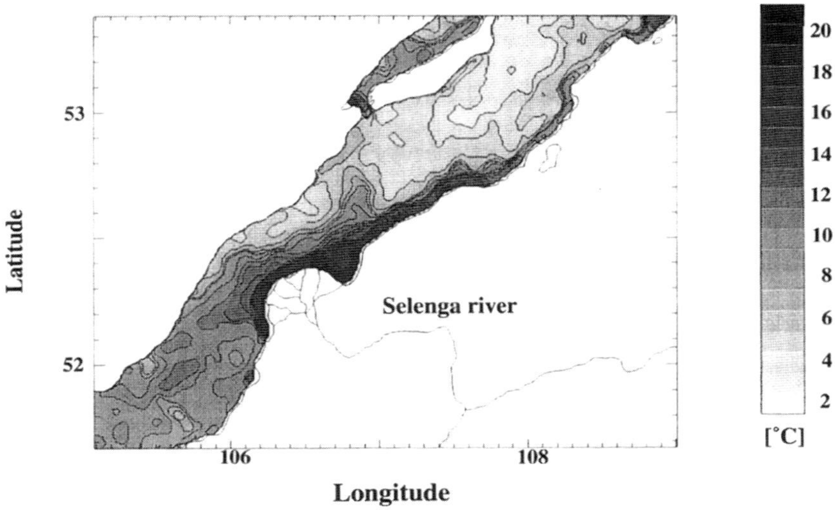

Fig.2. AVHRR-derived surface temperature map (4.07.1996). NOAA data courtesy to Space Monitoring Centre ISTP SB RAS.

tions were made for surface velocity field using sequences of satellite temperature maps. Fig.2 presents late spring - early summer surface temperature distribution for Selenga delta area. Coastal transport of warmer river waters can be traced clearly on this map in the north-east direction, isolated patterns of river water penetrate the south basin, thermal fronts position can be seen in the Maloe More straight (north-west corner of the map).

Lake Baikal is covered by ice about half of the year. Processes of freezing and break-up are well-defined indicators of climatic variability. Optical properties of ice and snow cover play an important role in the formation of under-ice phytoplankton bloom conditions. Between proposed explanations of well-known interannual variability in spring phytoplankton concentration there is hypothesis on crucial meaning of ice properties for development of spring bloom. According to this, presence of peculiar Baikalian very clean and transparent ice patterns is very significant for suitable light conditions formation and intensive under-ice density convection. Multispectral satellite images can be used for investigations of different ice classes distribution during winter and for studies of their interannual changes. In the paper (Sitnikova *et al.*, 1984) air survey data were coupled with images of Soviet meteorological satellite for some preliminary studies of Baikal ice variability. Methods of multivariate statistical analysis were used in (Semovski *et al.*, 1998b) to derive classification of main Baikal ice and snow cover types during winters 1995-1998. Distribution of main ice classes for March 16, 1997 is presented in Fig.3. Different classes are correspond to ice cover types from clear transparent ice to dry white snow cover.

It is well known that the best source of information for ice studies is the data of SAR (Synthetic Aperture Radar) instruments, that are independent of atmospheric conditions, have high spatial resolution (for ERS satellites it is about 25 m) and highly sensitive to differences in ice cover properties (see, e.g. Jeffries *et al.*, 1994). However, availability of SAR images is depending on receiving station position. During autumn 1997 and summer 1998 ESA (European Space Agency) mobile receiving station of ERS satellites were installed in Ulan-Bator, Mongolia. In the framework of ESA announcement of opportunity about 80 images were placed in our disposal (with collaboration of Prof.W.Alpers and Dr.C.Schrum, Institute fuer Meerskunde, Univ.of Hamburg, Germany). Surface roughness field, detected by SAR, contains information on wind field patterns, thermal fronts

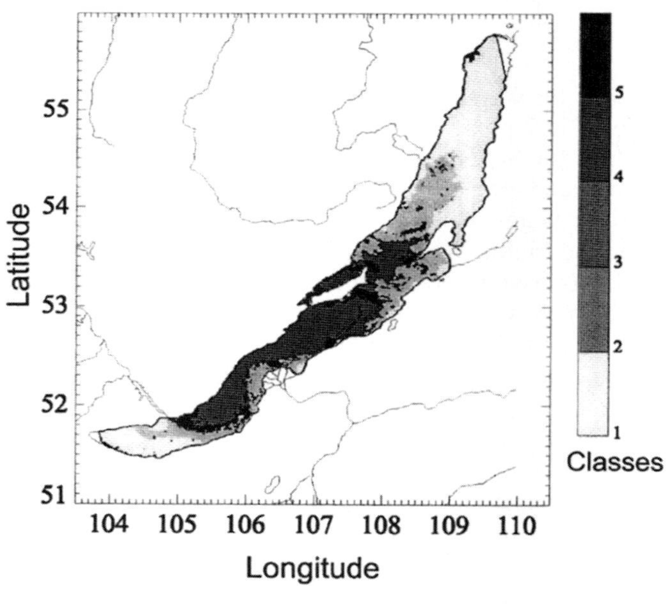

Fig.3. Distribution of main classes of ice and snow cover (16.03.1997) derived from AVHRR multispectral image. Classes 4 and 5 correspond to clean transparent ice; classes 1 and 2 corre-spond to white snow cover.

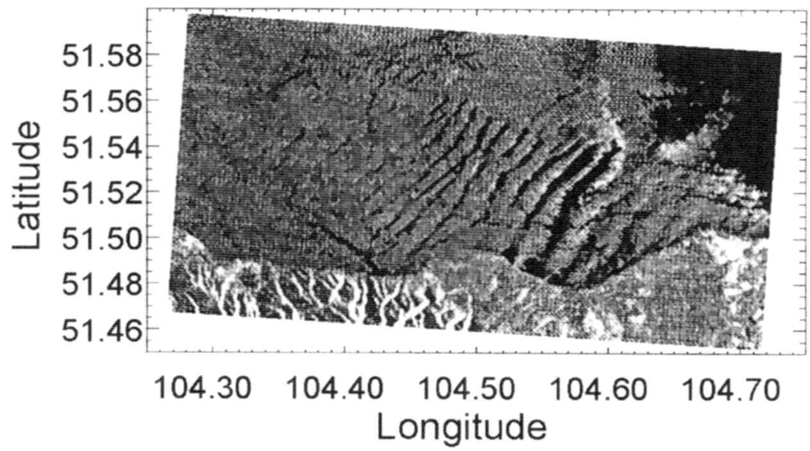

Fig.4. South part of South Baikal, train of internal wave transformation on the Murino Bank, ERS SAR image (27.08.1998), data courtesy of ESA.

position, currents, internal waves manifestation, some biological processes. Preliminary analysis suggests that some new features were firstly detected on the lake Baikal surface, such as numerous internal waves manifestations, fronts, films of biological origin and other structures (see examples in Figs.4, 9a).

Satellite altimetry is the active method, which makes possible precise measurement of satellite position relatively to Earth surface. For the ocean studies due to satellite–based altimeters using we produced detailed maps of wave heights, gravitational anomalies, etc. For the lake Baikal only one study of Birkett (1975) were published, where some measurements of lake level were presented using Topex/Poseidon data (see Fig.5). More advanced application of this data in Baikal studies is a subject of future investigations.

Remote sensing methods in the Lake Baikal ecosystem studies

Physical processes such as vertical and horizontal exchange, currents, temperature variability, etc., play an important role in the lake ecology. Detailed description of these processes by satellite remote sensing produces new knowledge of Baikal ecosystem. As an example we can mention spreading of fish larvae with rivers water, that can be traced using infrared imagery. However, some data sources can be used for detection of biologi-

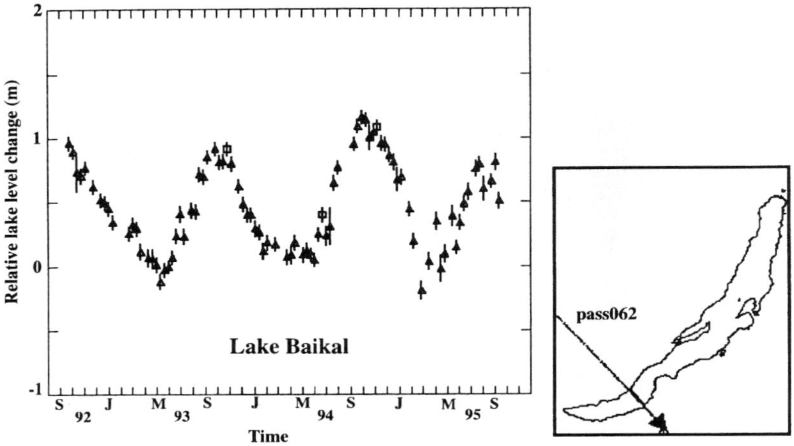

Fig. 5. Mean level variations on the transect shown on the right ploat, Topex/Poseidon mission data (Birkett, 1995; with permission of the author).

cal features on the lake surface. The most important data for this is multi-spectral imagery in visual band, produced by satellite-based scanners specially designed for studies of water environment, such as CZCS, MOS-IRS, MOS-Priroda, OCTS, SeaWiFS and future MERIS. Development of these systems gives a possibility to use their data not only for clean ocean waters (CZCS) but for coastal and lake environment as well. Special algorithms are developed for analysis of this information in many world centres. Data of universal multispectral systems (Landsat, Resurs, SPOT, *et al*) can be also used for studies of chlorophyll, suspended matter, other optically active substances distribution and dynamics in lakes, however, peculiar algorithms should be developed for every water object. Results of three-year experiment of Landsat data analysis for Chicon lake, Arkansas, USA was presented by Schiebe and Harrington (1992).

Ability to use modern satellite images for Baikal study is limited by availability of images in suitable resolution. SeaWiFS imagery, for example, is distributed free of charge between registered users for scientific and educational purposes, but for the areas without receiving stations this data has low (4 km) resolution. Figure 6 present Baikal SeaWiFS image for August 1998 (chlorophyll concentration).

AVHRR imagery collection can be also used for Baikal ecosystem studies. As it is well-known, main bloom of diatom algae (endemic *Aulocaseira baicalensis*) is occurs under the ice during March-May. As

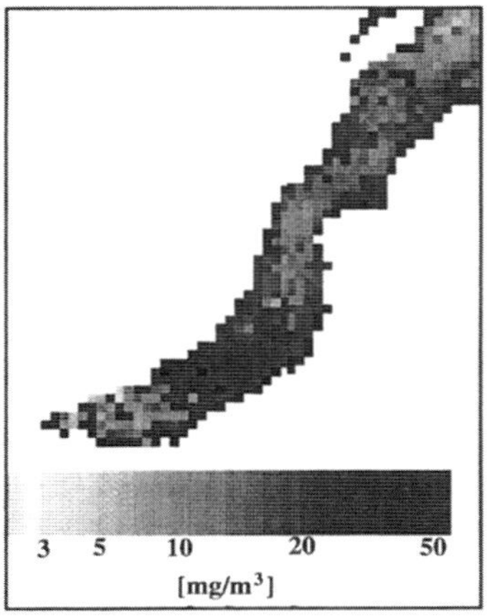

Fig.6. Baikal SeaWiFS-derived surface chlorophyll concentrations (26.08.1997).

was noted above, multispectral data was used for main ice and snow cover classes delineation (Fig.3). If bio-optical ecodynamic model (Semovski, 1999) is applied for estimation of total production and vertical structure of phytoplankton, then measurements of the ice optical properties should be incorporated for the underwater irradiance field simulation. The most influenced on the lake productivity are patterns of peculiar Baikal transparent ice (light transmission up to 80%, Sherstyankin, 1998).

It is well known that AVHRR imagery can not be used effectively for chlorophyll pigments concentration estimation. Some estimations of total suspended matter can be made, however, during spring, when the water has high transparency and phytoplankton is the main suspension, see Fig.7 and Semovski *et al.* (1998c). For future application of modern SeaWiFS and MERIS multispectral data the specific Baikal bio-optical model should be developed, based on peculiar optical properties. One of actual problems for the Baikal studies is the discrimination between main phytoplankton taxa (diatoms and picophytoplankton) based on multispectral remote sensing. The background of this procedure is the difference in specific attenuation and scattering of pigments in living cells.

Nutrient limitation can occur in surface layer during summer. Spreading of nutrient-rich river waters is detectable by their higher temperature and corresponds to productive areas. Figs. 2, 8 illustrate warmer Selenga waters distribution, which are visible in surface temperature band and in the AVHRR channel 1 albedo field due to contained suspended matter. Part of this suspensions should have biological origin. Summer bloom development on the Southern basin can be also seen on this image. Circulation patterns in the area of river influence produce response in phytoplankton spatial structure.

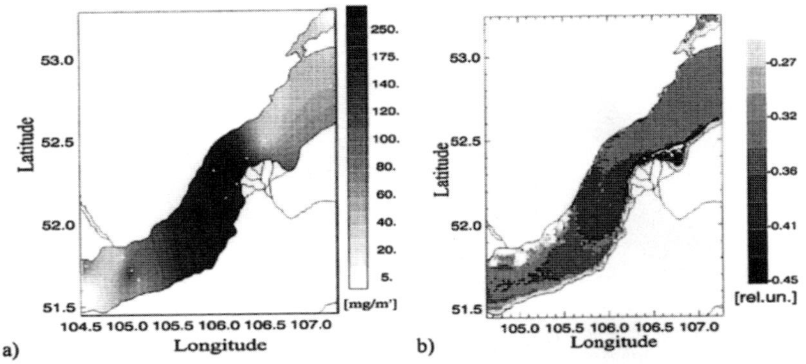

Fig.7. (a) Column phytoplankton concentrations, end of May 1995, Central Baikal. White asterisks indicate station positions (data courtesy to N.A. Bondarenko, see Semovski et al., 1998). (b) Normalised AVHRR vegetation index for 28 May 1995, Central Baikal.

Cold areas of coastal upwellings usually correspond to higher nutrients supply from bottom layers. Warmer regions in shallow areas without significant external inflow also can correspond to the higher phytoplankton production areas due to picophytoplankton dependence on the temperature. Thermal fronts can border productive coastal zones with their specific ecosystems. This feature can be clearly seen on Figs.2, 8 in Maloe Morye (Small Sea) straight. Surface temperature patterns there are in correspon-

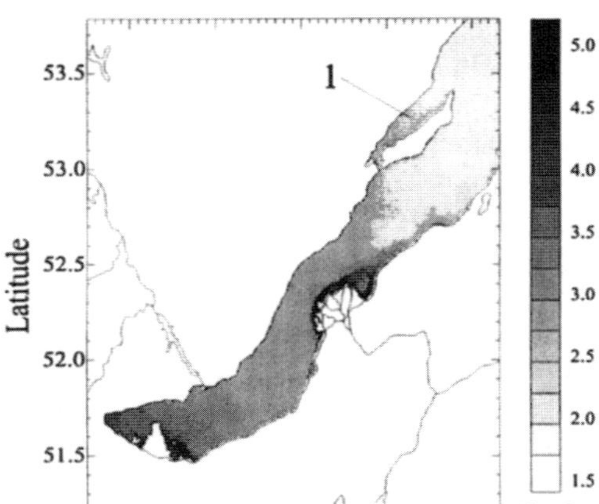

Fig.8. AVHRR Channel 1 albedo, 4 July 1995, Southern and Central ba-sins. 1 – Maloe Morye Strait.

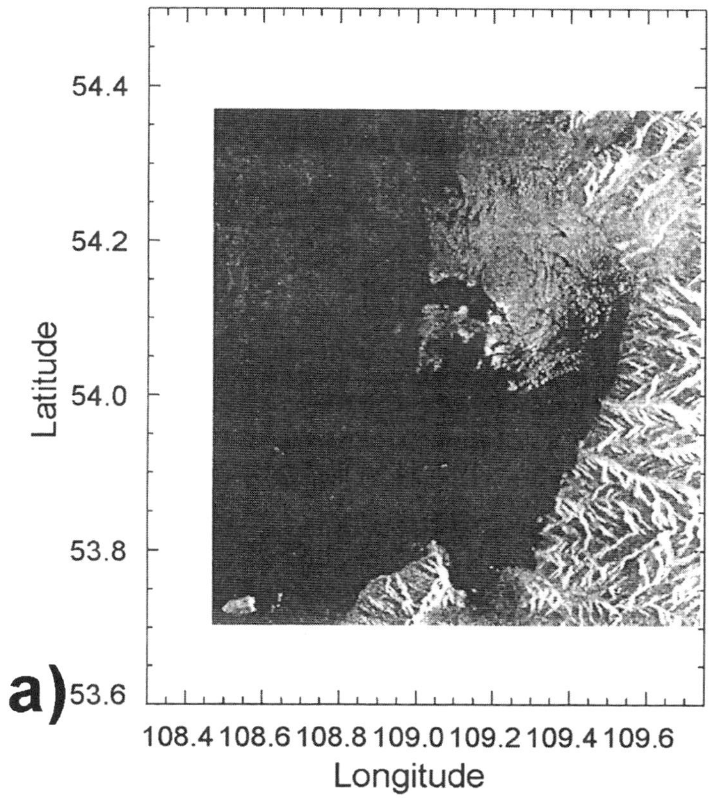

Fig.9. (a) ERS SAR image of North Baikal (27.09.1998). Films of bi-ological origin can be seen at the upper right.

dence with features in the suspended matter field. These suspensions are of biological origin, because Maloe Morye has no significant tributaries. Well known is, for example, local fish (Corregonius migratorus autumn.) population, for which the food base seems to be supported by local hydrodynamics conditions.

High-resolution SAR imagery can be very useful in Baikal ecosystem studies. For the autumn 1998 ERS SAR image (see upper) shows films of biological origin on the North Baikal (Fig.9a). Probably, intensive bloom of green-blue Anabena lemmermanii was detected there. This *Cyanophita* forms patterns of high concentration (5-10 g/m^3) on the lake surface during low wind conditions (Popovskaya, 1988). Normalised AVHRR channel 1 albedo shows high concentration of suspended matter in the same area,

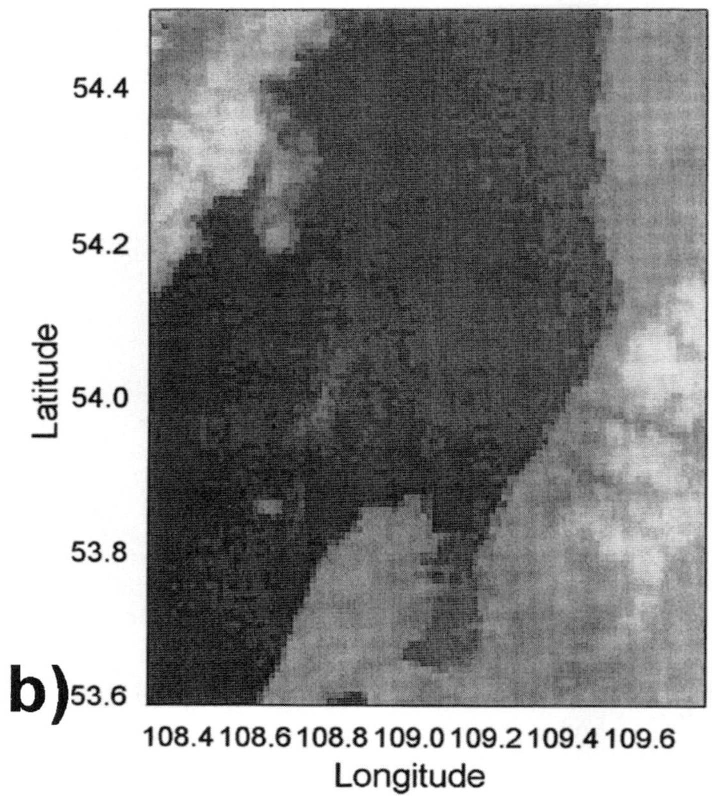

Fig.9. (b) NOAA AVHRR normalised albedo for Channel 1 (27.09. 1998). The lighter grey colours correspond to higher albedo values.

which has no significant river influence and other sources of suspensions (Fig.9b).

Conclusion

Application of satellite remote sensing in lake studies is still limited due to limited availability of low cost, high spatial and spectral resolution imagery. However, development of space-based instruments can gives an opportunities for new efforts in studies of inland water objects.

Now possibilities are under discussion to design low-cost specialised instruments with spectral and spatial resolution good enough for inland water basins study. Results of last discussion (November-December 1998) in Internet mailing group, joining specialists in remote sensing of inland water objects, show that general requirements should be formulated for satellite–based scanner of new generations, specially constructed for investigations of freshwater basins with extended SeaWiFS spectral bands and spatial resolution of 50-300 m. This Internet discussion was organised in the framework of International Alliance of Marine Remote Sensing for the preparation of proposal for the NASA small satellites program. Another discussion on possible requirements to source of free or low cost AVHRR data with 50 m resolution will be organised in the framework of the program supported by European Union.

Concerning Lake Baikal studies, new possibilities can be realised with SeaWiFS receiving station installation in Eastern Siberia region. Planned on the Spring 2000 launch of European Envisat platform with recording facilities and special "host" transmitting satellite can give high quality data flow for Baikal research.

From the point of view of the lake physics and ecology investigations using remote sensing methods, the following problems can be formulate:

– investigations of annual and interannual variability of surface temperature, lake area delineation on the types of seasonal cycle;
– thermal fronts studies, analysis of interannual variability of thermal bar formation and movement;
– detailed studies of interannual variability of ice cover conditions, their correlation with changes in air temperature, precipitation and global variability (ENSO, NAO, etc.).
– integrated analysis of SAR data, investigations of internal waves occurrence and variability, thermal fronts manifestation, films of biological origin and other features detection on SAR imagery using simultaneous AVHRR imagery;
– investigations of annual cycle of suspended matter distribution for different parts of the lake, field studies of correlation between suspensions and phytoplankton concentration;
– multispectral satellite imagery incorporation for studies of chlorophyll distribution and estimations of other optically active components concentration, validation and calibration using *in situ* bio-optical measurements.

Acknowledgement
It gives me great pleasure to thank Prof.M.N.Shimaraev and Prof.G.I.Popovskaya for advice in Lake Baikal physics, climate and phytoplankton ecology. Fruitful collaboration and discussions with Prof. P.P.Sherstyankin, Dr. N.P.Minko, N.Yu.Mogilev, Dr. N.A.Bondarenko, Dr. C.Schrum. Prof. W.Alpers, Dr. C.Birkett, Prof. J.A.Maslanik, Prof. V.S.Micheev and Dr. A.D.Kitov were essential for fruitful development of

satellite methods applications in Baikal research. This study was supported by Russian Foundation for Basic Sciences grant 99-05-64814.

References

Birkett C.M. ,1995, The contribution of TOPEX/POSEIDON to the global monitoring of climatically sensitive lakes. J.of Geophys.Res., 100(C12), 25, 179-204.

Bolgrien D.W., N.G. Granin and L.A. Levin, 1995, Surface temperature dynamics of Lake Baikal observed from AVHRR images. Photogrammetric Engineering and Remote Sensing, 61, 211-216.

Hern P.P., D.R Plamli., S.G. Shapchaev, A.D. Kitov, V.S. Micheev and A.K. Cherkashin, 1998, GIS of the river Selenga area: content and structure, In: Proceedings of Int. Conference "GIS for nature use optimisation for the sustainable development of territories", Barnaul, 1-4.07., 502-509.

Galazij G.I., G.I. Ladejshchikov, P.P. Sherstyankin and M.N. Shimaraev, 1980, Perspectives of satellite information use in limnological studies. Issledovanija Zemli iz Kosmosa (Earth Studies from Space). 6, 103-106, (in Russian).

Jaquet J.M., 1989, Limnology and remote sensing: Present situation and future developments. Rev.Sci.Eau, 2(4), 457-481.

Jeffries M.O., K. Morris, W.F. Weeks and H. Wakabayashi, 1994, Structural and stratigraphic features and ERS 1 synthetic aperture radar backscatter characteristics of ice growing on shallow lakes in NW Alaska, winter 1991-1992. J.of Geophys.Res., 99(C11), 22,459-22,471.

McClain E. P., W. G. Pichel and C. C. Walton., 1985, Comparative performance of AVHRR based multichannel sea surface temperatures. J.of Geophys. Research, 90, 11587-11601.

Popovskaya G.I., 1987, Phytoplankton of the deepest lake in the World, In: Marine and freshwater plankton, Proceedings of the Zoological Institute 182, 107-115, (in Russian).

Schiebe F.R. and J.A. Harrington, 1992, Remote sensing of suspended sediments: the Lake Chicot, Arcansas project. Int.J.of Remote Sensing, 13(8), 1487-1509.

Semovski S.V., M.N. Shimaraev, N.P. Minko and R.Yu. Gnatovsky, 1998, Satellite observations using for the lake Baikal thermal fronts studies. Issledovanija Zemli iz Kosmosa (Earth Studies from Space), 1998, 5, 65-75, (in Russian).

Semovski S.V., N.Yu Mogilev. and N.P. Minko, 1998, Multiparameter AVHRR-based description of the Lake Baikal ice cover variability, In: Hyperspectral Remote Sensing and Application, eds. R.O. Green and Q. Tong, Proc.SPIE, 3502, 270-277.

Semovski S.V., N.A. Bondarenko, P.P. Sherstyankin, N.P. Minko and N.Yu.

Mogilov, 1998, Lake Baikal phytoplankton annual cycle studies using AVHRR imagery collection, In: Proceedings of 27th Int.Symposium on Remote Sensing of Environment, 7-12 June 1998, Tromso, Norway, 320-323.
Semovski S.V., 1999, The Baltic Sea and Lake Baikal underwater bio-optical fields simulation using ecodynamical model. Ecological Modelling, 116(2/3), 149-163.
Sitnikova G.V., M.S. Furman and N.N. Yanter, 1984, Remote and ground-based methods for researching the dynamics of ice processes on Lake Baikal, In: Remote and groundbased studies of the dynamics of natural processes in Siberia, Institute of Geography, USSR Academy of Science (Siberian Branch), Irkutsk, 72-81. (in Russian)
Sherstyankin P.P., 1998, Optical properties, In: Lake Baikal, Evolution and Biodiversity, eds. O.M. Kozhova and L.R. Izmest'eva, Backhuys Publ., Leiden, pp.44-55.
Zakharov M.Yu., E.A. Loupian, R.R. Nazirov, A.A. Mazurov and E.V. Flitman, 1995, Experimental system for satellite data acquisition and processing. Space Bull., 2(2), 22-24.
Zou.S., 1987, Study of modern vicissitudes of the Jianghan Lake group by using remote sensing techniques. Oceanol.Limnol.Sin., 18(5), 469-476.

Lake Baikal
K. Minoura (editor)
2000 Elsevier Science B.V.

Environmental impact on the dynamics of Lake Baikal phytoplankton taxonomic groups: modelling attempt

Semovski, S. V.
Limnological Institute SB RAS, Irkutsk, Russia
Fax: +7-3952-466933, E-mail: Semovsky@lin.irk.ru

Abstract

Lake Baikal phytoplankton community dynamics is simulated for modern conditions and for few hypothetical paleoclimate scenarios. A one-dimensional bio-optical model of pelagic ecosystems is used. The aim of the investigation is to simulate the influence of a few factors that have been proposed as possible reasons for full extinction of diatoms during cold climatic conditions. The effects of the silicon influx change, duration of the winter period, variability of the optical properties of ice, and glacier-generated suspended matter impact are studied. Diatom production is shown to be highly sensitive to the optical properties of ice and of the surface layer. The most probable reason for diatom extinction during cold climatic conditions is the optical impact of glacier-generated suspended matter and the variability of the optical properties of the ice cover.

Introduction

Because of Lake Baikal's size (maximal depth 1,642 m, surface area 31,470 km^2), many phenomena are similar to those that occur in a marine environment. The methods and models being developed for coastal sea and open ocean studies can be used to investigate Baikal water column hydrodynamics, deep mixing processes, and current and eddy structure, as well as for ecodynamic studies and modelling. A bio-optical vertically-resolved phytoplankton dynamics model is presented (see Semovski *et al.*, 1996; Semovski, 1999a,b) for assessment of Baikal phytoplankton community variability under possible paleoclimate conditions. The model includes a spectral bio-optical block, primary production algorithm based on quantum yield of photosynthesis, and simple model of water column hydrodynamics.

Information about the climatic changes and geological history of Asia during the past 25-30 million years is "encoded" in the Baikal sediments (up to 10 km thick), and the sedimentary record has revealed multiple dramatic changes in the phytoplankton communities (see Grachev *et al.*, 1997). Full extinction of planktonic diatoms was observed in the lake ecosystem during periods corresponding to global cold climates (oceanic Stages 2 and 4), but their concentrations were high during the periods of warmer climates - Holocene, oceanic Stages 3 and 5. Genetic studies, however, demonstrate the continuous development of the whole Baikal pelagic community from top to bottom throughout the history of Lake Baikal.

According to current concepts (Nagata *et al.*, 1994; Bondarenko *et al.*, 1996), in the modern period, primary production of picophytoplankton (the most abundant being endemic *Synechocystis limnetica Popovsk.*) comprises an average of about 60% of the total production in Lake Baikal. Against this background, the diatom community (prevailing species: *Aulacoseira baicalensis*) exhibits great interannual variability in the intensity of its under-ice spring bloom (from less then $2x10^5$ to more then $4x10^8$ cells per litre).

However the factors influencing the marked variability in the diatom community due to global climatic changes are still unknown, the following hypothetical scenarios have been suggested:

(a) silicon is assimilated by diatom frustules, and competition between diatoms and picoplankton due to non-linear population dynamics can cause changes in the dissolved silicon supply during cold climates that favor the complete dominance of picoplankton;

(b) diatom production is sensitive to longer winters and shorter ice-free periods;

(c) the under-ice diatoms bloom takes place only when part of the lake ice has extreme transparency (up to 80%), and this feature is now regularly observed in Lake Baikal (Sherstyankin, 1998);

(d) glaciers, which were abundant in the Baikal area during cold climatic periods, produced suspended matter, and their optical impact in the surface layer limits the light for diatom production because of their higher sinking rate.

The mechanisms controlling the interannual variability of diatoms are still not clearly understood, and scenarios similar to (a)-(c) have been proposed by different authors to explain it.

Recent analysis of modern sediments (Mackay *et al.*, 1998) has demonstrated that *A.baicalensis* frustules have prevailed in the Baikal diatom community for only the past 150 yrs. Prior to that time autumnal *Ciclotella minuta* dominated the diatom community. This change seems to correlate with the end of the "Little Ice Age" and corresponding changes in the Eastern Siberian climate. We do not take up the problem of explaining this variability, because a detailed study would require information on precipitation in the Baikal area, nutrient influx, etc.

Model

A vertically resolved model for the dynamics of several algal species was analytically assessed by Britton and Timm (1993), and they showed that stationary distribution can exist, and that in some cases vertical segregation can occur. The qualitative results cannot be directly applied to the Baikal pelagic ecosystem studies, however, the general principles of population dynamics can be used in our numerical model.

The model presented is a generalisation of the bio-optical model of the Baltic pelagic ecosystem (Semovski, 1999a,b), and it includes two photosynthesyzing taxa, diatoms and picophytoplankton. Their dynamic properties differ as follows:

– nutrient assimilation rate of picoplankton increases with rising temperature (see Nagata *et al.*, 1994);

– diatoms use silicon (*Si*) as an additional nutrient, and their production is limited by the supply of dissolved silicon;

– sinking rate of diatom frustules is highr than picoplankton because of their larger size and siliceous skeleton.

Note that increasing temperature becomes a limiting factor for A.baicalensis production, however, other diatoms can survive in warmer water. No functions that depend on temperature are used in the equations for diatom dynamics. The life cycles of diatoms (see Jewson, 1992) can play an important role in the functioning of the phytoplankton community. Some sophisticated systems of mathematical equations should be used, for example, to describe the A.baicalensis survived strategy. The model presented, however, is not suitable for reproducing the dynamics of individual species.

Photosynthetic primary production

Primary production is a basic process for the carbon and energy cycles in the marine ecosystem. We chose the Wozniak model (Wozniak *et al.*, 1995) from a number of primary production models describing its dependence on a spectral downward irradiance field (Morel, Platt *et al.*). This algorithm is based on the quantum yield of photosynthesis attempts. It was used with success in the world ocean and the Baltic studies. The main equations of the method can also be found in Semovski *et al.*(1996) and Semovski (1999 a and b).

Microbial loop and dissolved organic matter

A bacterioplankton block in the model is necessary for simulation of dissolved organic matter (DOM) turnover. Some of the DOM constituents (mostly humic substances) may be associated with optically active "yellow substance" (or Gelbshtoff) (Carder *et al.*, 1989; Karabashev *et al.*, 1989). The DOM production and destruction processes include DOM arrival with river runoff and with partial solution of dead plankton cells, and DOM assimilation by heterotrophic bacteria. In the past few years progress has been made in microbial loop studies and modelling of the marine environment. We use some modifications of the Lancelot and Billen (1990) equations for heterotrofic bacteria and DOM dynamics.

Dinoflagellates, zooplankton, higher trophic levels

It should be noted that we use the terms flagellates and zooplankton in the model for a wide class of species considered to be grazers for the two taxa of autotrophic phytoplankton. These groups can include fish species, which are grazers on a larger phytoplankton fraction. Flagellates and zooplankton state variables are not included in the optical block, and they can be considered as balance and ecological limitation terms in the bio-optical model framework.

Nutrients, silicon

The state variable corresponding to nutrients (substrate) is a usual part of ecosystem models. Its role is to describe some limited environmental factor, which can differ according to the specific geographical conditions and even for different annual cycle stages. This parameter is often associated with nitrogen, phosphorus, or ammonium. A nutrient limitation in the upper layer, which arises after the spring bloom and during the summer, is one of the reasons for subsurface phytoplankton maximum formation. Under the specific conditions of Lake Baikal, it is appropriate to use two nutrient factors. The first we call "nutrients" and corresponds to a group of components (nitrates, phosphorus, etc.) partially recycling in water columns because of the microbiological loop and arriving again in their initial dissolved form. The dissolved silicon (*Si*) is a separate nutrient, because according to modern information the main source of it in the Baikal is river influx. There have stet estimates of silicon recycling due to partial dissolution of dead diatom frustules (smaller fractions).

Biologically produced and mineral suspended matter

We use two suspended matter fractions, namely, the detritus and the biologically neutral mineral suspensions arriving with river runoff, eolithic, etc., in the model.

The physical model

The physical block of the model can be chosen from as large class as the formulation of the problem requires. The processes of vertical movement and mixing are of decisive importance for the dynamics of the biological variables. In the present study the two-layer model with different vertical mixing intensities in the top and bottom layers was used as the simplest decision (see Semovski *et al.*, 1996, 1999a,b). Monthly values for the depth of the top mixed layer and corresponding coefficients of turbulent mixing were taken as the climatic values (Shimaraev and Verbolov, 1998). We must stress that during the spring and autumn periods of homothermy, when the vertical gradient between the top and deep layers change their signs, the depth of the mixed layer can reach 200 m, and the value of the mixing coefficient increases by an order of magnitude.

Note that according to modern observations and models (Kelley, 1997; Granin *et al.*, 1998) the difference in under-ice mixing producing by density convection generated by sun irradiance penetrating transparent ice patterns can have a significant influence on diatom production. If a suitable convection model can produce estimates of the vertical mixing coefficient and depth of the mixing layer, these values can be used in our bilayer model. The qualitative model of Kelley, 1997, however, cannot be used for this purpose. The observation data collected are still not representative enough to obtain good estimates.

The optical model

Most existing models of underwater optical fields were developed only for so- called case 1 waters according to Morel terminology. Another type of waters, so-called case 2 waters, typical of coastal area and lakes, is the most productive type. Baikal waters cannot be described by models suitable for ocean waters because of the detectable concentration of so-called "yellow substance" (Karabashev *et al.*, 1989). In our previous studies of ocean and coastal seas ecosystems (Semovski *et al.*, 1996; Semovski, 1999a,b) we used the Wozniak bio-optical model (Wozniak *et al.*, 1995). This model is based on optical observations, C_a (chlorophyll a + pheophytin a) observations, and the quantum yield of the photosynthesis database.

A simple generalisation of the Semovski, (1999b) model of the two-component phytoplankton community is used in the present study. The representation of the spectral diffusive attenuation $Kd(\lambda),[m^{-1}]$ is used in the form of the sum of components corresponding to the optical active constituents of water.

$$K_d(\lambda) = K_w(\lambda)+K_{pl}(\lambda)+K_{y1}(\lambda)+ K_{y2}(\lambda)+K_{S1}(\lambda)+K_{S2}(\lambda),$$

where λ is wavelength [nm] and $K_w(\lambda)$ is diffusive attenuation of pure water (see values in Wozniak *et al.*, 1995). Diffusive attenuation by pigments contained in the diatom and picophytoplankton phytoplankton cells $K_{pl}(\lambda)$ can be calculated by corresponding formulas in Wozniak *et al.*, 1995.

The respective terms for the two kinds of "yellow substance" $K_{y1}(\lambda)$, $K_{y2}(\lambda)$ correspond to two kinds of DOM in the biological portion of the model. The specific form of the expressions can be controlled by local conditions (see Karabashev *et al.*, 1989). This is particularly true for terrigenous (allochtonous) DOM and the corresponding "yellow substance" fraction. We use some general estimations after Carder *et al.* (1989), because it is commonly believed that mineral suspended matter attenuation has a non-spectral character. Some empirically derived formula based on Baltic data is used for the detritus (Semovski, 1999b).

Simulation phytoplankton bloom under the ice cover calls for incorporation of Lake Baikal ice spectral light transmission in the photosynthetically active radiation band (400-700 nm). We used observed values taken from Sherstyankin (1998), for clean Baikal ice, which is exceptionally transparent (transmission up to 80%), and for the ice covered by snow (transmission is one order of magnitude lower). A random choice between clean and snow-covered ice was also used in the simulation.

3. Results

The model annual cycle of main variables is in general agreement with modern-day knowledge (see Semovski, 1999b). Higher picophytoplankton concentrations are observed during mid-summer (July-August). If transparent ice conditions are used, diatom bloom can occur in early spring (March-April), and higher concentrations of diatoms are observed during autumn (September-October)as well (see.Fig.3). These calculations were used as the basis for sensitivity studies of the diatom-picoplankton relationship to variations in some environmental factors in the model.

Variations in Si supply

The main source of dissolved *Si* in Baikal is river inflow. In modern period during the spring diatom bloom only "weak" limitation by *Si* usually occur. This means that the *Si* concentration in the surface layer decreases significantly during the bloom (down to order of magnitude, see Votincev, 1975), but concentrations of other nutrients (nitrogen, phosphorus) remain relatively high. The lower *Si* supply tends to be associated with decreasing concentrations of diatoms, however, the simulation results (Fig.1) show that the variations in diatom biomass due to limitation by *Si*

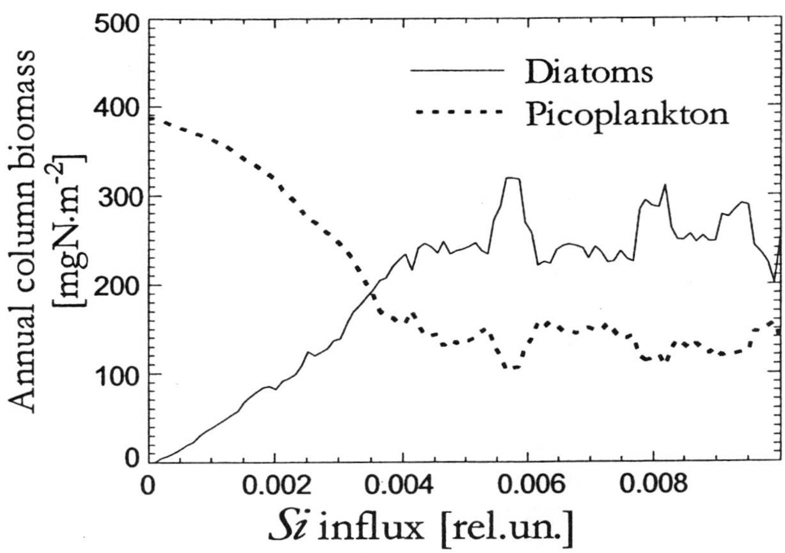

Fig. 1. Annual integrated phytoplankton biomass dependence on external Si influx. One point corresponds to a 10-year period.

are "smooth" and proportional to the *Si* influx. This probably means that *Si* supply variability is not the main reason for diatom extinction during cold climatic conditions, because there is some sediments information of *Si* presence in lake waters during these periods (Grachev *et al.*, 1997).

Variations in ice-free period duration

Simulations show that diatom populations answer on variations in winter (ice-cover) period from 3 to 10 months seems moderate (see Fig.2). According to recent observations, the main annual biomass of diatoms, is produced during the spring under-ice bloom. Picoplankton production, by contrast, is highly dependent on the temperature, and consequently declines when the duration of summer shortens.

Variations in the optical properties of Baikal ice

As noted earlier, one of the peculiar properties of Baikal ice is the regular occurrence of patterns with high transparency (transmission up to 80%). Ice covered by snow usually has low transparency (less then 10%). Results of the simulations shown in Figure 3 demonstrate that the great diatom bloom depends on the optical properties of the ice. This result, however,

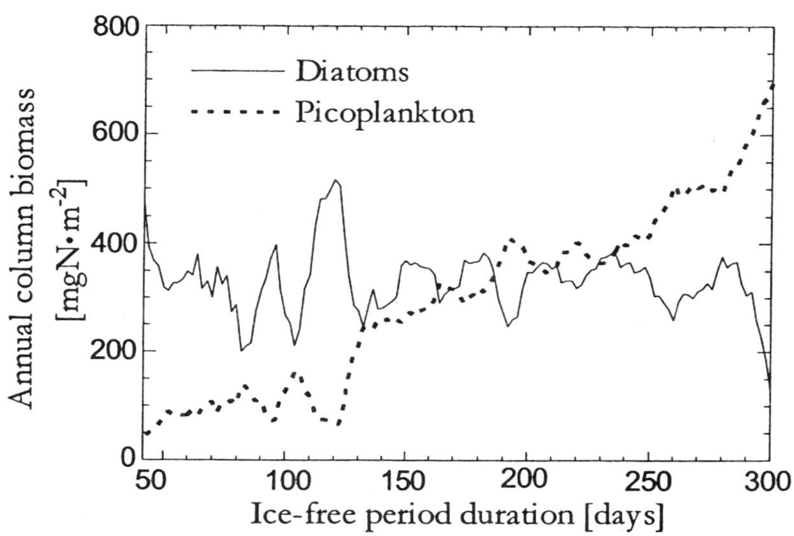

Fig. 2. Annual phytoplankton biomass dependence on the length of the ice-free period. One point corresponds to a 10-year period.

hardly have something common with paleoclimatic studies, because no data are available for the optical properties of Baikal ice during the cold periods. Modern interannual variations in diatom blooms can be "triggered" by the properties of the ice, but not enough observational data is available to support this conclusion. Use of multipsectral satellite data to study Baikal ice was proposed in Semovski *et al.* (1998). Figure 3 demonstrates that the intensity of autumn bloom is stable under different ice conditions. During "little ice ages" local precipitation and wind conditions prevent transparent ice pattern formation, and variability in ice properties may be the factor governing the predominance of *C.minuta* in the Baikal diatom community (Mackay *et al.*, 1998). However, the processes that control the variability of the Baikal snow cover during this period should be the subject of future studies. A general positive correlation between annual net phytoplankton concentration and mean winter temperature was demonstrated by Kuzevanova (1986).

Influence of the suspended matter in the surface layer

The glaciers that were abundant in the Baikal depression during the cold climate periods produced great quantities of suspended matter, and it must have a significant impact on the optical properties of the surface layer. The smallest clay particles should have a very low sinking rate and thus should be present in the water column even during winter. The simulation results in Figure 4 show significant diatom community dependence on the optical

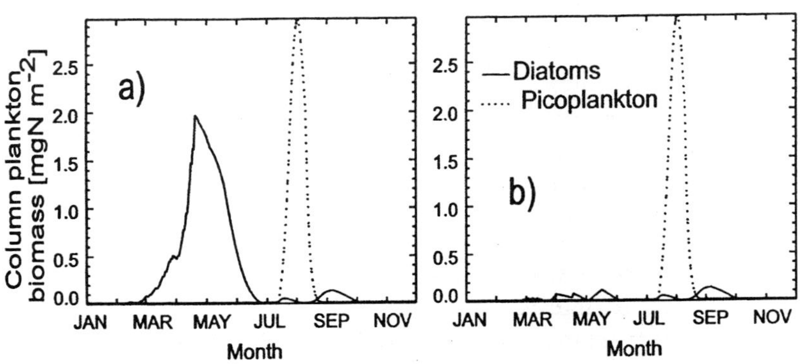

Fig. 3. Simulated annual dynamics of diatoms and picoplankton integral column biomass: (a) transparent ice during winter occurs randomly 55% of the time; (b) 5% of the time.

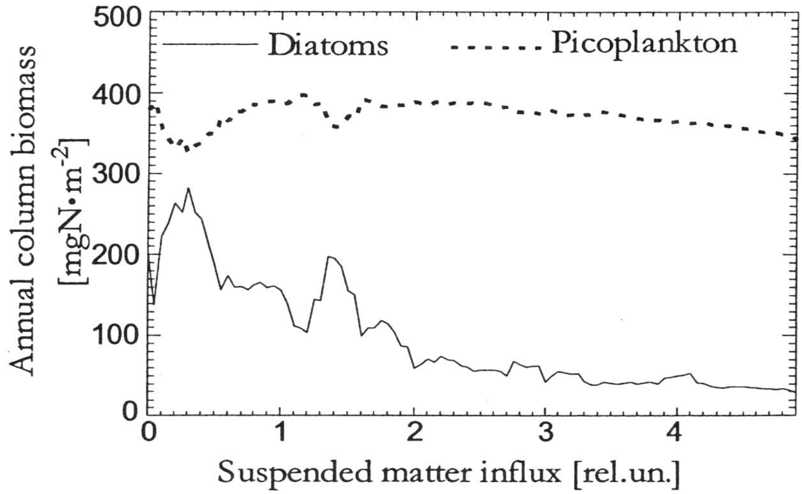

Fig. 4. Annual phytoplankton biomass dependence on the suspended matter influx. One point corresponds to a 10-year period

properties of the surface layer. The effect on the picophytoplankton is not very important because of its lower sinking rate.

Interannual variability simulation

As noted earlier, the factors that influence the interannual variability of under-ice spring diatom bloom intensity are not yet known. Among the numerous candidates are riverine silicon influx variability and interannual changes in the optical properties of ice. Silicon accumulation in the top layer over a two-three year period until it reaches a level sufficient for blooming to occur was proposed by Votintcev (1975) as an explanation for differences in bloom intensity. According to our calculations, when the model silicon influx is set at the level corresponding to "weak" limitation of spring diatom production, non-linear population dynamic features produce strong interannual oscillation in the total diatom concentration (see Fig. 5b). When silicon influx limits diatom development in the model, their concentration remains stable (Fig. 5a) It is possible that these simulated features represent actual interaction processes between the two phytoplankton groups.

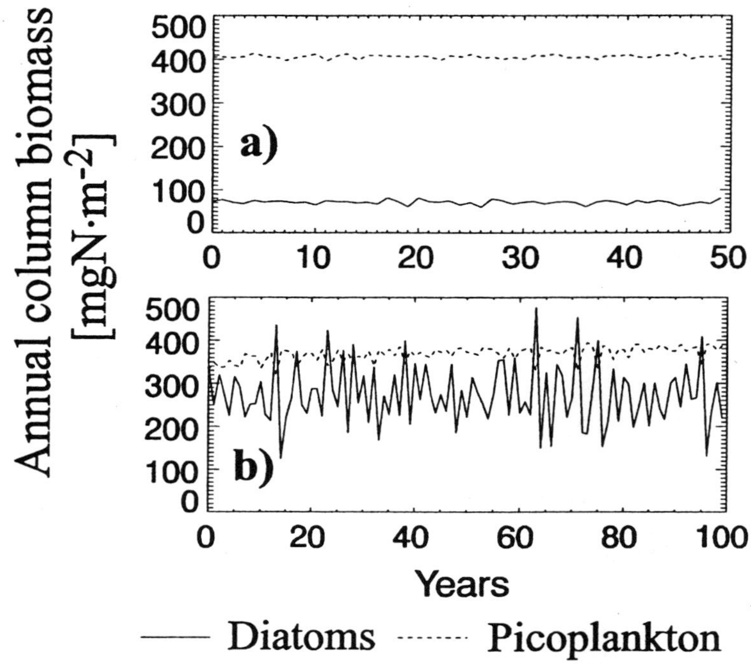

Fig. 5. Interannual phytoplankton biomass variability simulation: 100-year period for different Si external influx values: (a) 0.001 rel.un., (b) 0.003 rel.un. (coresponds to modern period).

Conclusions

The results of simulation seem to support the hypothesis that the reason for diatom extinction in Lake Baikal during cold climate periods is the optical influence of glacier-generated suspended matter. The effect of other factors on the diatom-picopankton relationship was not very important. However, the results of application of a rather simple bio-optical model can be considered only preliminary. To begin with, we have not studied the combined effects of several factors. Second, in the framework of ecodynamical modelling it is impossible to investigate the problem of evolution and species substitution within a single taxonomic group.

The dissimilarities between northern Baikal and the central and southern basins should be described in this connection. The northern Baikal basin area belongs to a different climatic zone (region under Arctic

Yakutian anticyclone influence). It has been estimated on the basis of a box budget model (Semovski *et al.*, 1997) that silicon assimilation by diatoms per unit area is more then ten times lower in the northern basin than in southern basin. This feature is most likely governed by the lower river inflow and the lower silicon supply. Another reason is the smaller diatom fraction prevailing there, as noted by several researchers (see e.g., Bondarenko et. al., 1996). Smaller diatom frustules usually are not recognised in sediments (Grachev *et al.*, 1997), and this may mean that they are completely dissolved in the water column or in the active top sediment layer. However, we do not yet have a complete description of silicon turnover in the Baikal water column or at the sediment-water interface.

Another possible method of investigating Baikal paleoecology is study of the water objects that are currently exposed to similar climatic and geochemical conditions. The glacier-influenced lakes of the Siberian Arctic, Alaska, and New Zealand can be cited as examples. Discussion of the West Spitzbergen semi-enclosed fjord environment with Prof. Jan-Martyn Weslawski (Institute of Oceanology, Poland), however, has revealed some differences. Significant under-ice diatom blooms are regularly observed in the fjord despite high summer concentrations of glacier-produced suspensions. The deciding factor in the difference may be, as first pointed out by Prof. M.N.Grachev (private conversation), a different rate of clue suspended matter particles sedimentation in fresh and saline waters, that tend to suspended matter sedimentation in fjord during winter, and, possibly, winter fjord ventilation by marine waters.

In any event, bio-optical models coupled with adequate hydrodynamics are a powerful tool for comparable studies of the Lake Baikal ecosystem during cold climate periods and can be used for its modern analogues. Suitable application of simulations can reveal the essential differences and similarities. Underwater special field surveys should be conducted to study bio-optical properties because of their important impact on ecodynamics, phytoplankton production, and taxonomic composition.

Acknowledgements

The author wishes to thank Prof. M.A. Grachev for formulation of the problem and Prof. G.I. Popovskaya, Dr. N.A. Bondarenko, Prof. M.N. Shimaraev, Prof. P.P. Sherstyankin, and Prof. J.-M. Weslawski for useful discussions. This study was supported by Russian Foundation for Basic Sciences grant 99-05-64814.

References

Bondarenko N.A., N.E. Guselnikova, N.E. Logacheva and G.V. Pomazkina, 1996, Spatial distribution of phytoplankton in Baikal, spring 1991. Freshwat.Biol., 35, 517-523.

Britton N.F. and U. Timm, 1993, Effects of competition and shading in plankton communities. J.of Math.Biology, 31, 655-673.

Carder K.L., R.G. Steward, G.R. Harvey and P.B. Ortner, 1989, Marine humic and fulvic acids. Their effect on remote sensing of ocean chlorophyll. Limnol. Oceanogr. 34(1), 68-81

Grachev M.A., E.V. Likhoshvay, S.S. Vorobieva *et al.*, 1997, Signals of paleoclimates of upper Pleistocene in the sediments of Lake Baikal. Russian Geology & Geophysics, 38, 957-980.

Granin N.G., D.H. Jewson, M.A. Grachev *et al.*, 1998, Physical processes as a factor for developing diatoms under the ice. Russian Academy of Science Doklady, in press. (in Russian).

Jewson D.H., 1992, Size reduction, reproductive strategy and the life cycle of a centric diatom. Phil. Trans. Royal Soc., Lond., B, 336, 191-213.

Karabashev G.S., A.F. Kuleshov and P.P. Sherstyankin, 1989, Spectral transparency of the Baikal waters in ultraviolet and visual spectral bands. Doklady AN SSSR, 306(5), 1091-1094, (in Russian).

Kelley D.E., 1997, Convection in ice-covered lakes: effects on algal suspension. J. Plankton Res., 19(12), 1859-1880.

Kuzevanova E.N., 1986, Peculiarities of the long-term phytoplankton and zooplankton dynamics in southern Baikal. Preprint, Irkutsk, 23 pp.

Lancelot C. and G. Billen, 1990, Joint EEC Research Project on the dynamics of Phaecystis blooms in nutrient enriched coastal zones (contract EV4V-0102-B(GDF)). 2nd annual progress report.

Mackay A.W., R.J. Flower, A.E. Kuzmina *et al.*, 1998, Diatom succesion trends in recent sediments from Lake Baikal and their relation to atmospheric pollution and to climate change. Phil.Trans.R.Soc.Lond. B, 353, 1011-1055

Nagata T., K. Takai, K. Kawanobe *et al.*, 1994, Autotrophic picoplankton in southern Lake Baikal: abundance, growth and grazing mortality during summer. J.Plankton. Res., 16, 945-959.

Semovski S.V., B. Wozniak, Hapter R and A. Staskiewicz, 1996, Gulf of Gdansk spring bloom physical, bio-optical and biological modelling. J.of Marine Systems, 7(2-4), 145-160.

Semovski S.V., M.N. Shimaraev, M.A. Grachev and V.N. Domysheva, 1997, Inverse stochastic box models for the Lake Baikal: Silica vertical balance parameters estimation, In: 7th International Conference on Lakes Conservation and Management, LACAR97, Sao Martin de los Andes, Argentina, 2, 51-54

Semovski S.V., N.Yu. Mogilev and N.P. Minko, 1998, Multiparameter AVHRR-based description of the Lake Baikal ice cover variability, In: Hyperspectral Remote Sensing and Application, eds. R.O.Green and Q.Tong, Proc.SPIE, 3502, 270-277.

Semovski S.V., 1999, Water ecosystems: from satellite observations to mathematical modelling. Inst.of Geography, Irkutsk, 200 pp.

Semovski S.V., 1999, The Baltic Sea and Lake Baikal underwater bio-optical fields simulation using ecodynamical model. Ecological Modelling, 116(2/3), 149-163.

Sherstyankin P.P., 1998, Optical properties, In: Lake Baikal, Evolution and Biodiversity, eds. O.M. Kozhova and L.R. Izmest'eva, Backhuys Publ., Leiden, 44-55.

Shimaraev M.N. and V.I. Verbolov, 1998, Water temperature and circulation. Ibid., 26-43.

Votincev K.K., A.I. Meshchryakova, G.I. Popovskaya, 1975, Organic matter turnover in the Lake Baikal. Novosibirsk, Nauka, 188 pp, (in Russian).

Wozniak B., J. Semovski, S.V. Dera *et al.*, 1995, Algorithm for estimating primary production in the Baltic by remote sensing. Studia i Materialy. Oceanologia. Marine Physics, 68, 91-123.

Lake Baikal
K. Minoura (editor)
2000 Elsevier Science B.V.

Nonlinear stability near the temperature of maximum density and thermobaric instability in Lake Baikal during summer stratification.

Granin, N.G.[1]*, Gnatovsky, R. Yu.[1], Kay, A.[2], and Galkin, L.M.[3]
[1] Limnological Institute of the Siberian Division of the Russian Academy of Sciences, P.O.Box 4199, Irkutsk 664033, Russia, Fax: 7 3952 460405, E-mail: nick@lin.irk.ru
[2] Department of Mathematical Sciences, Loughborough University, Loughborough, Leicestershire, LE11 3TU, U.K
[3] Institute of Geochemistry of the Siberian Division of the Russian Academy of Sciences, P.O.Box 4199, Irkutsk 664033, Russia
(*corresponding author)

Abstract

The stability of freshwater bodies near the temperature of maximum density T_{md} is investigated by using a particle dynamics formalism. T_{md} decreases linearly with depth, and density is an approximately quadratic function of temperature near T_{md}. Hence, the Brunt-Vaisala frequency N, which is calculated by using a linearisation of the temperature–density relationship, is not an adequate measure of stability. In particular, a water column that is at T_{md} throughout its depth is absolutely unstable even when $N = 0$. A water column slightly warmer than T_{md} and with temperature decreasing with depth (as is commonly found in Lake Baikal in summer) is stable to very small disturbances, but unstable to larger disturbances. This explains recent observations of convective overturning events in the depth range 50 – 200 metres in Lake Baikal.

Introduction

The anomalous thermal expansion of fresh water is of primary importance in determining the circulation of deep temperate lakes [*Carmack & Farmer*, 1982]. It gives rise to instabilities that are absent in a fluid with linear dependence of density on temperature. Furthermore, the approximately linear decrease of T_{md} with pressure, and hence with depth [*Eklund*, 1963], opens up the possibility of two types of thermobaric instability. First, there is a conditional instability requiring the mechanical displacement of a body of water downwards across the T_{md} profile [*Farmer & Carmack*, 1981]; this has been invoked as a mechanism for deep ventilation in Lake Baikal [*Weiss et al.*, 1991; *Shimaraev & Granin*, 1991]. Second, *Granin* [1995, 1996] has noted that a water column with temperature everywhere equal to the local value of T_{md} is unstable, and has proposed this instability to explain some recent observations of vertical mixing events in Baikal. A further consequence of the decrease of T_{md} with depth is the asymmetry between spring and autumn convection [*Shimaraev & Granin*, 1991], a feature which has been absent from most mathematical models of thermal bar circulations.

The range of possible phenomena in deep lakes with temperature profiles close to T_{md} has led to the use of increasingly complex numerical models to simulate the flow, such as the 3-dimensional simulation of deep ventilation by *Walker & Watts* [1995]. However, simpler mathematical formalisms can often provide considerable insight into the dynamic processes. In the present study, we investigated the instabilities described above by considering the unidimensional motion of a water parcel under buoyancy forces following an initial displacement from equilibrium in a water column with a given temperature profile. The initial displacement need not be infinitesimal; internal waves, for instance, may result in substantial vertical displacements, and thus it is important to consider the stability of a water column subjected to large disturbances.

Instabilities in waters at temperatures near maximum density

The thermobaric instability for that exists in situations in which cold water overlies slightly warmer water is well known [*Carmack & Weiss*, 1991; *Weiss et al.*, 1991; *Shimaraev & Granin*, 1991]. The reverse situation (a layer of warm water overlying a layer of cooler water) is not so well documented. Suppose the interface between the layers is situated below the T_{md} profile, but with both layers colder than 3.98°C (the value of T_{md} at the surface) (see Fig. 1.) While this configuration is stable, if the interface is moved upwards through the local compensation depth (defined by *Carmack & Weiss* [1991] as the depth at which the two layers have equal density), it will become unstable (Fig. 1), and convection will ensue.

Consider also the "profile of maximum density", in which the temperature is equal to T_{md} throughout the water column. Under such circumstances any vertical displacement of a water parcel will then generate a buoyancy force directed upwards (because the surrounding water is denser than the displaced parcel). Although this appears to be a restoring force in the case of a water parcel initially displaced downwards, such a parcel would "bounce" back above the equilibrium, and then accelerate

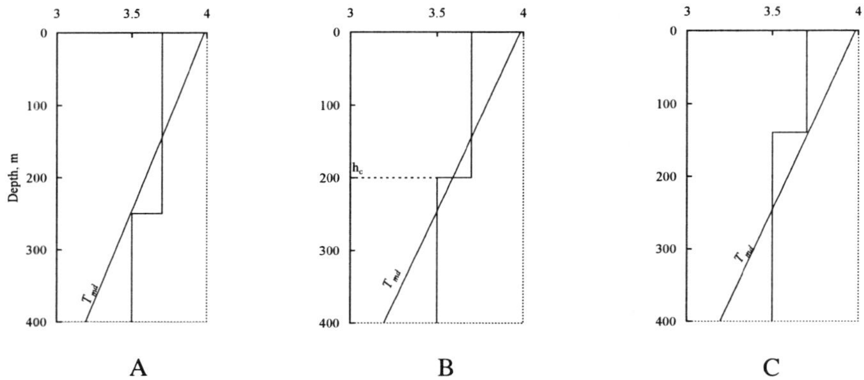

Figure 1. Schematic representation of the "reverse thermobaric instability" (adapted from *Weiss et al.* [1991]). (A) The cooler deep water is denser (closer to the T_{md} line) than the warmer overlying water, so the interface is stable. (B) If the interface is lifted to the local compensation depth h_c it becomes neutrally stable. (C) If it is lifted any further, the deep water becomes less dense than the overlying water, and convective mixing of the layer above the interface may ensue.

upwards. Thus, the water column is unstable (in the sense that any small disturbance of the equilibrium will lead to large-scale motion): even though the Brunt-Vaisala frequency is zero, the profile of maximum density is not neutrally stable. The Brunt-Vaisala frequency describes local stability, but the concept of "local" becomes inoperative in the special case of the profile of maximum density.

This is because calculations of Brunt-Vaisala frequency for a given temperature profile usually use only the coefficient of thermal expansion $d\rho/dT$, neglecting the second derivative $d^2\rho/dT^2$ [*Millard et al.*, 1990]. When calculating Vaisala frequency by the Hesselberg Swerdrup method, expanding density in a Taylor series,

$$\rho(S,T,p)=\rho(S_0,T_0,p_0)+\frac{\partial\rho}{\partial T}\Delta T+\frac{\partial\rho}{\partial S}\Delta S+O\left(\Delta T^2,\Delta S^2\right) \quad (1)$$

one usually neglects the higher order terms, $O\left(\Delta T^2,\Delta S^2\right)=0$. The resulting equation for calculation of the Vaisala frequency is as follows:

$$N^2 = g^2\left[\frac{\partial\rho}{\partial S}\frac{\partial S}{\partial p}+\frac{\partial\rho}{\partial T}\left(\frac{\partial T}{\partial p}-\left(\frac{\partial T}{\partial p}\right)_{adia}\right)\right] \qquad (2)$$

This approximation is invalid when the coefficient of thermal expansion is zero, as is the case at T_{md}. It is also impossible to neglect the second derivative, $d^2\rho/dT^2$, when dealing with significant displacement at temperatures near T_{md} .

Both the above instabilities will lead to upward mixing and are possible mechanisms for transport of nutrients up to the photic zone in summer, provided that the convection penetrates into the zone where there are significant nutrient concentration gradients. In this paper we will use a simple "particle dynamics" formalism to describe the instability of the profile of maximum density and other phenomena that occur in freshwater bodies near T_{md}. We emphasise the importance of the second derivative in the density–temperature relationship for these phenomena. Our calculations do not take into account any adiabatic heating of the water parcel as its pressure varies: *Eklund* [1965] has shown that at the temperature of maximum density adiabatic compression is isothermal, and below we show that adiabatic heating remains negligible within the temperature range of interest in our study. We do not include salinity effects, even though salinity gradients must be the dominant determinant of stability at temperatures sufficiently close to T_{md}, because the salinity gradient is very small within the range of depths at which the temperature is close to T_{md}. Nor do we attempt to represent viscous forces in our equations, even though they may be of comparable size to buoyancy forces, since the latter are small near T_{md}. Although a linear (Stokes' law) resistance could easily be included, there is no way of determining the magnitude of the proportionality constant for our hypothetical water parcel. Finally, temperature changes due to viscous dissipation of kinetic energy are ignored, despite the sensitivity of the dynamics to small temperature changes near T_{md}.

Buoyancy forces near the temperature of maximum density

The quadratic approximation to the equation of state,

$$\rho = \rho_m - \lambda(T - T_{md})^2 \qquad (3)$$

in which $\lambda = 8.25 \times 10^{-3}$ kg·m^{-6}· °C^{-2}, is sufficiently accurate to illustrate the essential features of water parcel motion near the temperature of maximum density. The variation of T_{md} with depth is linear,

$$T_{md} = T_{m0} + Mz \tag{4}$$

where $T_{m0} = 3.98°C$ and $M = 0.002°C.m^{-1}$, and the vertical coordinate z is measured upwards from the water surface (and so is negative throughout the depth). The maximum density, ρ_m, is also a function of depth (through compressibility effects), but this does not concern us since we shall only need to compare the density of a water parcel with that of its immediate environment. However, it should be noted that the coefficient of thermal expansion, α, does vary with depth at a fixed temperature, thus:

$$\alpha \equiv -\frac{1}{\rho}\frac{\partial\rho}{\partial T} = \frac{2\lambda}{\rho}(T - T_{md}) = \frac{2\lambda}{\rho}(T - T_{m0} - Mz) \tag{5}$$

The above formulation is equivalent to that of *Farmer & Carmack* [1981], but without salinity effects; it would be straightforward to include a linear dependence of density on salinity in our model.

Note that the adiabatic lapse rate is

$$\Gamma \equiv -\left(\frac{\partial T}{\partial z}\right)_{\text{isentropic}} = \frac{g\alpha T}{c_p},$$

[*Gill*, 1982] which, with α given by (5) yields

$$\Gamma = 1.1 \times 10^{-5}(T - T_{md}).$$

We shall be mainly interested in temperature profiles less than 0.2°C away from T_{md}, so Γ is three orders of magnitude smaller than M. Thus, we ignore adiabatic heating when considering motion of water parcels in an environment with a lapse rate of order M.

Consider a water parcel whose temperature and density differ by T' and ρ', respectively, from the values in the surrounding water. The buoyancy force on the water parcel is

$$F = -g\frac{\rho'}{\rho_p} \tag{6}$$

per unit mass, and the equation of state (3) then yields

$$F = -\frac{g\lambda}{\rho_p}\left\{2(T_{md} - T_e)T' - T'^2\right\} \tag{7}$$

here, the subscripts p and e refer to the parcel and to the surrounding water, respectively. Equation (7) constitutes the linear and quadratic terms in a Taylor series in T', in contrast to *Millard et al.*'s [1990] formalism which neglects the quadratic term. While the linear approximation is acceptable in the oceans or in lakes far above the temperature of maximum density, it is clear from (7) that it is not valid in a freshwater body close to T_{md}.

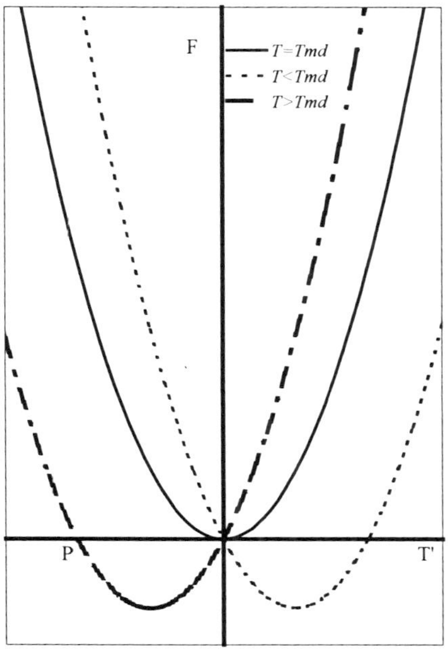

Figure 2. Buoyancy force F on a water parcel with temperature differing by T' from that of the surrounding water, for cases in which the water temperature T_e is (a) higher, (b) lower, and (c) equal to the temperature of maximum density T_{md}. Note that downward motion in the water column implies rightward motion along the curves in this figure, provided the water column temperature gradient is positive. A water parcel at P is in unstable equilibrium.

Figure 2 shows the buoyancy force (7) as a function of the temperature difference T' for three cases, $T_e < T_{md}$, $T_e = T_{md}$, and $T_e > T_{md}$. In general, the buoyancy force is downward when T' lies between zero and $2(T_{md} - T_e)$, and upward when T' is outside this range. Consider, for example, a water column in which the lapse rate dT_e/dz is equal to dT_{md}/dz , so that a single curve represents the entire water column. Upward motion of a water parcel with constant temperature is then represented by leftward motion along the curve. If $T_e > T_{md}$ a small displacement up or down from equilibrium ($T' = 0$) will induce a restoring force on the water parcel, leading to oscillatory motion (internal waves), provided that the parcel never moves above the point where $T' = 2(T_{md} - T_e)$: the water column is *locally* stable. On the other hand, if $T_e < T_{md}$ the buoyancy force will be in the same direction as the displacement, and the water parcel will accelerate away: the water column is unstable and convection will occur. Finally, if $T_e = T_{md}$, the buoyancy force is upward for any non-zero T' , so that any displacement of a water parcel will result in it eventually accelerating upward. This instability was identified by

Granin [1996] as an explanation for observations of convective mixing around the temperature of maximum density in Lake Baikal.

Motion of a water parcel for a temperature profile parallel to the profile of maximum density

We shall solve the equation of motion

$$\frac{d^2 z}{dt^2} = F \qquad (8)$$

with F given by (7), for a parcel in a water column with a temperature profile parallel to the profile of maximum density,

$$T_e = T_{md} + \Delta T \qquad (9)$$

for some fixed ΔT, so that the water column has a uniform positive temperature gradient equal to M: this is the usual temperature profile during the summer period in the depth range of 50 – 200 metres in Lake Baikal (see Fig. 3-5 and *Shimaraev & Granin* [1991]). The water parcel starts from rest at $z = z_i$, having been impacted by an initial temperature perturbation T_i' or equivalently, having been brought to z_i by a vertical displacement $-T'/M$ from equilibrium. A positive temperature gradient is locally stable

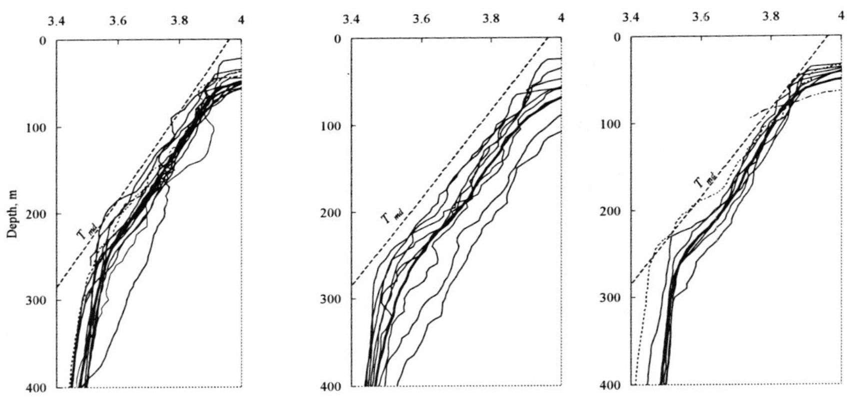

Figure 3. Temperature profiles measured in three Lake Baikal basins in September 1998; A - southern basin, B - central basin, and C - northern basin. Averaged profiles (thick lines) are approximately parallel to the T_{md} profile (below the strongly stratified surface layer).

at temperatures above T_{md}, but unstable below T_{md}, because of the change in sign of the coefficient of thermal expansion.

A water parcel at the same temperature as its surroundings is then in a stable equilibrium. However, if it is cooled to a temperature $T_{md} - \Delta T$ or raised through a distance $2\,\Delta T/M$ it will again have the same density as its surroundings, but this is an unstable equilibrium. Consider a water parcel initially released from rest at a distance z_+ $(< 2\Delta T/M)$ above the stable equilibrium. It will oscillate between this point and a point at some distance z_- below equilibrium, where it has the same potential energy as at its release point; z_- will be smaller than z_+, because the restoring force is greater at a given distance below equilibrium than at the same distance above it (see Fig. 2). In terms of dimensionless distances

$$\zeta_\pm \equiv \frac{Mz_\pm}{\Delta T} \tag{10}$$

z_- and z_+ are related by

$$\zeta_- = \frac{1}{2}(\zeta_+ - 3 + 3Z) \tag{11}$$

where

$$Z = \left(1 + \frac{2\zeta_+}{3} - \frac{\zeta_+^{\,2}}{3}\right)^{1/2} \tag{10}$$

The equation of motion may again be solved in terms of elliptic integrals to find the frequency of oscillation

$$\omega = \frac{\pi\sqrt{Z}}{F\left(\pi, (\zeta_+ + \zeta_-)/3Z\right)} N \tag{11}$$

where N is the Brunt-Vaisala frequency, which applies within the limits of small oscillations:

$$N^2 \equiv \frac{g}{\rho_p}\frac{d\rho}{dT}\frac{dT_e}{dz} = \frac{2g\lambda M}{\rho_p}\Delta T \tag{12}$$

The frequency may be approximated by

$$\omega \approx N\left(1 - \frac{5}{48}\zeta_+^{\,2}\right) \tag{13}$$

for small ζ_+. For an initial displacement close to the unstable equilibrium (i.e., for ζ_+ close to 2),

$$\omega \approx \frac{\pi}{\ln \frac{24}{2-\zeta_+}} \qquad (14)$$

A water parcel that moves beyond the unstable equilibrium point, because of having an initial displacement greater than $2\,\Delta T/M$ upward or $\Delta T/M$ downward (the latter gives it sufficient potential energy to "bounce" upward beyond the unstable equilibrium), will then accelerate upward to infinity.

Consider the case $\Delta T > 0$, in which we expect the water column to be stable. A water parcel displaced from equilibrium will execute nonlinear oscillations, but only if the initial displacement is sufficiently small. The frequency of the oscillations is equal to the Brunt-Vaisala frequency within the limits of small amplitudes, but decreases with increasing amplitude (equations (10) – (15)). The frequency becomes zero if the initial vertical displacement of the water parcel is $2\,\Delta T/M$ upward (i.e., to point *P* in Fig. 2 where it has the same density as its surroundings) or $\Delta T/M$ downward (from where it will rise to point *P* and no further). Any greater initial displacement would raise the parcel to a point at which it is less dense than its surroundings, so that it would accelerate upward to infinity. Thus convective instability is possible even at $\Delta T > 0$, if sufficiently large disturbances occur. The closer the water column profile is to the T_{md} profile, the smaller the disturbance required to trigger this instability.

A linear (i.e. local) stability analysis would predict only stable oscillations when ΔT is positive. The instability described above is a nonlinear effect arising from the quadratic term in the Taylor series for the buoyancy force (7). As ΔT decreases, the maximum size of disturbances that will preserve linear dynamics falls, until at $\Delta T = 0$ motion is purely nonlinear and we have the absolutely unstable situation of a water column at maximum density throughout. Note that even though this profile of maximum density is the borderline between the linearly stable case $\Delta T > 0$ and the unstable case $\Delta T \leq 0$, it is not neutrally stable. Using the formalism proposed, it is possible to describe a parcel motion in the cases: a) water temperature is equal to the temperature of maximum density, b) the water column is homothermal, and c) linear temperature decreases with depth.

Experimental data

It is known that during positive temperature stratification in Lake Baikal the pattern of vertical temperature distribution within the depth range from

50-100 m to 250-300 m is close to the T_{md} profile, often being parallel to the latter (Shimaraev & Granin, 1991; Shimaraev et al., 1994). At temperatures near T_{md}, vertical gradients of dissolved solids may be of great importance to stability (equation 2). In order to account for vertical fluctuations in total dissolved solids (TDS), special experiments were performed to determine the pressure and temperature dependence of electrical conductivity. Our results in regard to the pressure dependence of electrical conductivity agreed with those published by Hohmann and co-authors, but the results for temperature dependence differed substantially from those published (Hohmann et al., 1997). To recalculate electrical conductivity for fixed temperature, we used our equations based on experimental results, and to diminish errors, electrical conductivity was calculated for 3.5°C, the predominant temperature in the Baikal water column. We followed the procedure proposed by Hohmann et al. 1997 to recalculate electrical conductivity in total dissolved solids.

Measurements in 1995-1999 with CTD probe SBE-25 allowed us to obtain new data on the vertical distribution of the temperature and electrical conductivity. The new temperature profiles confirmed our previous results, whereas the electrical conductivity profiles are grenerally completely new. The new experimental data demonstrate that thermobaric instability is rather common during summer temperature stratification. We regard the temperature profile crossing the T_{md} profile and the existence of fields characterised by unstable stratification (for example, field 1 in Fig. 3 A) as evidence of thermobaric instability. This is exemplified by the data obtained in lake's three basins in September 1998. ΔT changed from 0.05°C or below to 0.3°C, corresponding to the temperature profile displacement downward from 25 m or less to 150 m T_{md} (Fig. 3). Most of the temperature profiles show indications of thermobaric instability. The vertical distribution of the temperature averages for the three basins yielded a ΔT from 0.05°C to 0.2°C (Fig.3). The average temperature profiles essentially paralleled the T_{md} profile. Minimal ΔT not exceeding 0.05°C within the depth range from 70 m to 150 m, is observed in northern Baikal. In southern Baikal ΔT is slightly below than 0.1°C, and in Central Baikal it is slightly above than 0.1°C. The differences between the average temperature profiles is of great interest. In northern Baikal the minimal ΔT occurs at a depth of 50-100 m, and ΔT increases below 150 m. In contrast to northern Baikal, the temperature profiles in central and southern Baikal parallel the T_{md} profile within a wider range of depths – from 50 to 250 m. This difference in average temperature profiles is caused by differences in the intensity of wind-wave mixing and vertical turbulent tensity of wind-wave mixing and vertical turbulent exchange because of higher wind velocity in central Baikal than in northern Baikal (Shimaraev et al., 1994).

At temperatures near T_{md}, the contribution of TDS gradients to density gradients may be relatively high because of the small coefficient of temperature expansion. When the temperature is equal to the T_{md}, the change in density in response to a temperature change of 0.1°C is the same as to a TDS change 0.1 mg/l. Within the range of depths in which we were interested, electrical conductivity changes less than 0.1 mSm/cm (Fig.4), corresponding to a TDS change of approximately 0.1 mg/l. This means that the

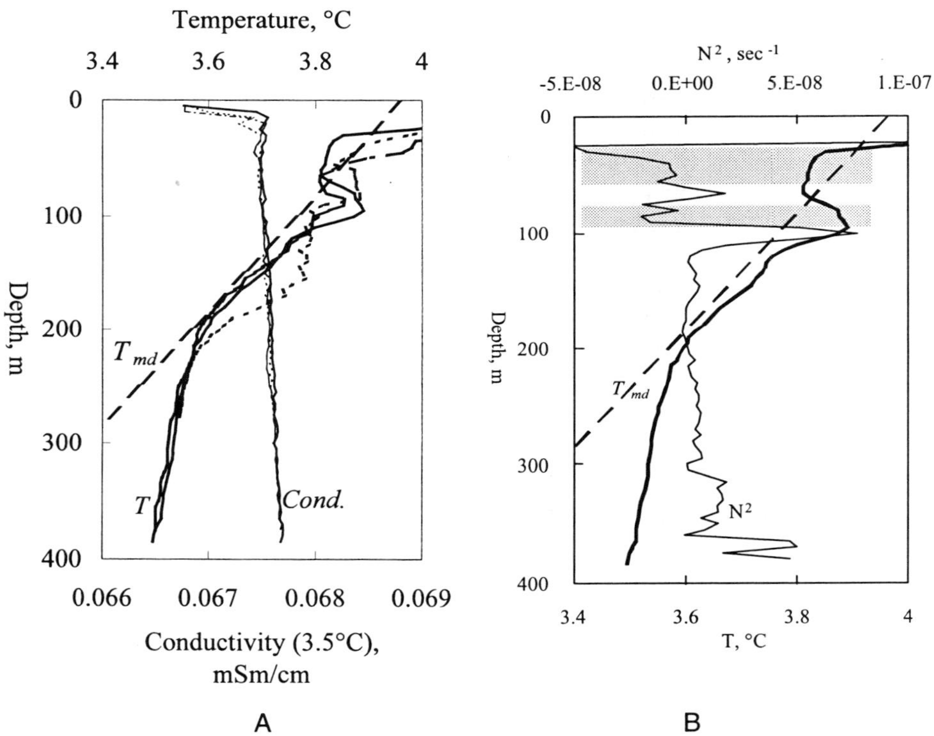

Figure 4. Temperature, electrical conductivity and the T_{md} profiles in the central part of the northern basin, September 1997 - A; profiles of temperatures and squares of the Vaisala frequency - B.

contribution of temperature to changes in density is comparable to that of

TDS. In spite of the small coefficient of temperature expansion, the contribution of TDS is not the crucial factor when T' is sufficiently high.

It is possible to determine whether the stabilising influence of the vertical TDS gradient is adequate to compensate for the destabilising influence of the vertical temperature gradient by calculating the Vaisala frequency. The profile of the square of the Vaisala frequency is highly consistent with the vertical temperature distribution pattern (Fig. 4, 5). The square of the Vaisala frequency is negative in the fields characterised by unstable tem-

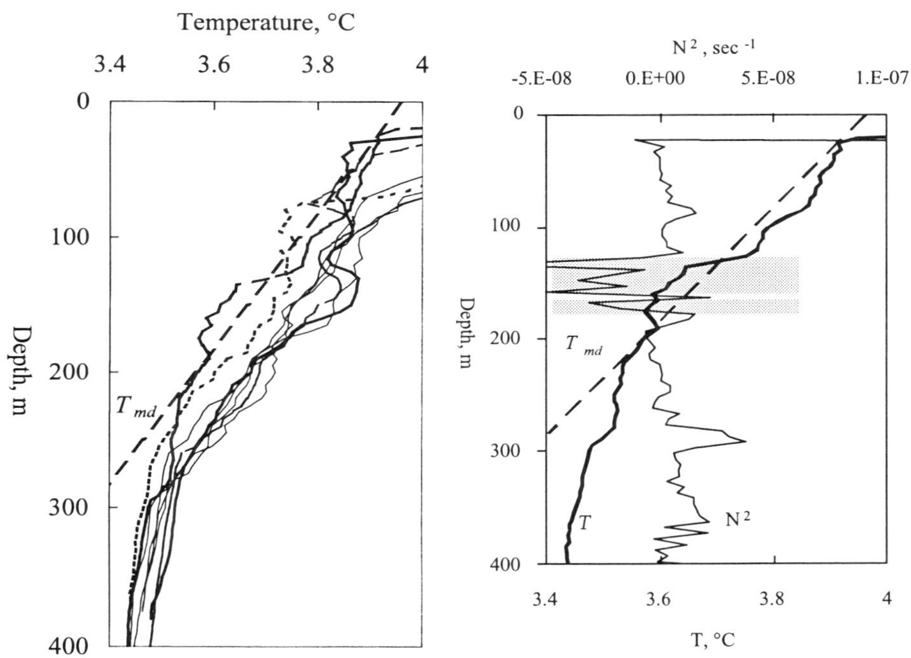

A B

Figure 5. Temperature profiles measured on the crossection Listvyanka – Tankhoy (southern Baikal), July 1998 - A; profiles of temperatures and squares of the Vaisala frequency at the station located 8 km from Tankhoy - B.

perature distribution (shaded area). This demonstrates that TDS gradients do not compensate for the destabilising action of the temperature gradient. Density stratification becomes unstable upon generation of thermobaric instability, and, consequently, a convection needs to be existing.

Some regions of the lake display thermobaric instability more often. These regions coincide with the zones of current divergence and with the centrers of cyclonic circulation. In other words, there is a greater possibility of thermobaric instability in regions characterized (for dynamic reasons) by shallower thermoclines, and, consequently, the occurrence of ΔT. This is exemplified by the following. In the current divergence zone (the Listvyanka-Tankhoy cross-section, Fig. 5) the temperature profiles measured at the stations 3, 5, 8, and 12 km from Tankhoy have a crossed T_{md} profile, whereas no thermobaric instability was found at the central station (18 km from Tankhoy) or the station located near Listvyanka.

Thus, TDS is essentially constant within the range of depth characterised by temperatures near T_{md}, and therefore vertical TDS gradients are too small to provide stability of stratification when the temperature profile crosses the T_{md} profile. Small TDS gradients also testify to intensive mixing within this depth range: water mixing due to thermobaric instability occurs within the layer from tens to a hundred meters in thickness (Fig. 4, 5).

Conclusion

Any study of the hydrodynamics of a deep temperate lake requires an understanding of the peculiar stability properties of waterbodies near the temperature of maximum density. In the present work we have used a simple model based on particle dynamics to illustrate the importance of nonlinear effects in determining their stability.

According to linear theory, a water column warmer than T_{md} is stable only if the temperature gradient is positive, while a water column below T_{md} requires a negative temperature gradient for stability; thus a temperature profile that crosses the T_{md} profile must change the sign of its gradient, leading to the mesothermal maximum observed during winter in Lake Baikal [*Shimaraev et al.*, 1994]. The marginal cases of homothermy and of a water column at maximum density throughout its depth are generally regarded as neutrally stable [*Eklund*, 1965]. However, the latter case is in fact unstable to the smallest disturbances, and even a water column that is stable according to linear theory may be destabilised if the temperature is close to T_{md} and sufficiently large disturbances (e.g., due to internal waves) are present. We propose this as the explanation for the convective mixing event illustrated in Fig. 3 - 5, which occurred in an area of otherwise stable stratification.

Experimental data show that thermobaric instability phenomena take place rather often during summer stratification and may effect a redistribution of nutrients in the water column. It may also be important for the maintenance of diatom algae within the photic zone.

Acknowledgements
This research was carried out within the framework of BICER activity supported by the Royal Society of the UK and INTAS grant N 96-1937. The CTD probe was purchased with the support of INTAS grant N 94-3121.

References

Abramowitz, M. and Stegun, I.A. (1965) Handbook of mathematical functions, Dover, New York.
Carmack, E.C. and Farmer, D.M. (1982) Cooling processes in deep, temperate lakes: a review with examples from two lakes in British Columbia. J. Mar. Res., 40, Suppl., 85–111.
Carmack E.C. and Weiss, R.F. (1991) Convection in Lake Baikal: An example of thermobaric instability. In: Deep convection and deep water formation in the ocean., P.C.Chu and J.C. Gascard, eds., Elsevier, 215–228.
Eklund, H. (1963) Fresh water: temperature of maximum density calculated from compressibility. Science, 142, 1457–1458.
Eklund, H. (1965) Stability of lakes near the temperature of maximum density. Science, 149, 632–633.
Farmer, D.M. (1975) Potential temperatures in deep freshwater lakes. Limnol. Oceanogr., 20, 634–635.
Farmer, D.M. and Carmack, E.C. (1981) Wind mixing and restratification in a lake near the temperature of maximum density. J. Phys. Oceanogr., 11, 1516–1533.
Foster, T.D. (1972) An analysis of the cabbeling instability in sea water. J. Phys. Oceanogr., 2, 294–301.
Gill, A.E. (1982) Atmosphere-Ocean Dynamics, Academic Press, New York.
Granin, N.G. (1995) Stability of stratification at temperature close to the temperature of maximum density. Second Vereshchagin Baikal Conference: Abstracts, Irkutsk,
p. 46.

Granin, N.G. (1996) Thermobaric instability in temperate lakes during summer temperature stratification. American Society of Limnology and Oceanography: Abstracts, University of Wisconsin-Milwaukee, p. 52.

Hohmann, R., Kipfer, R., Peeters, F., Piepke G. and Imboden, D.M. (1997) Processes of deep water renewal in Lake Baikal. Limnol. Oceanogr., 42, 841-855.

Kay, A. (1998) Particle dynamics in deep cold water. Mathematics today, 2, 11-16.

Matthews, P.C. (1998) A model for the onset of penetrative convection. J. Fluid Mech., 188, 571–583.

Millard, R.C., Owens, W.B. and Fofonoff, N.P. (1990) On the calculation of the Brunt-Vaisala frequency. Deep Sea Research, 37, 167–181.

Peeters, F., Piepke, G., Kipfer, R., Hohmann, R. and Imboden, D.M. (1996) Description of stability and neutrally buoyant transport in freshwater lakes. Limnol. Oceanogr., 41, 1711–1724.

Shimaraev, M.N. and Granin, N.G. (1991) Temperature stratification and mechanism of convection in Lake Baikal. Doklady Akademii Nauk SSSR, 321, 831–835.

Shimaraev, M.N., Verbolov, V.I., Granin, N.G. and Sherstyankin, P.P. (1994) Physical limnology of Lake Baikal: a review. BICER, Irkutsk, 81 pp.

Walker, S.J. and Watts, R.G. (1995) A three-dimensional numerical model of deep ventilation in temperate lakes. J. Geophys. Res., 100, 22711–22731.

Weiss, R.F., Carmack, E.C. and Koropalov, V.M. (1991) Deep-water renewal and biological production in Lake Baikal. Nature, 349, 665–669.

Lake Baikal
K. Minoura (editor)
2000 Elsevier Science B.V.

Study of the elemental composition of suspended particles in large continental lakes (Baikal and Khubsgul)

Potyomkina, T. G.* and Potyomkin, V. L.
Limnological Institute, Siberian Branch of the Russian Academy of Sciences
e-mail address: klimat@lin.irk.ru

* Corresponding author, Box 4199, 664033 Irkutsk, Russian Federation.

Abstract

Samples of suspended material (solid particles) from Lake Baikal (Russia) and Lake Khubsugul (Mongolia) were analyzed by using X-ray fluorescence elemental analysis with excitatiomn by monochromatized synchrotron radiation (SRXFA, Novosibirsk, Russia), and electron probe X-ray microanalysis (EPXMA, Antwerpen, Belgium). The distribution of chemi-cal composition at different depths in the lakes is demonstrated. It is shown that the concentra-tions in the central parts of the waterbody are steady, but that the concentrations are higher in the southern basin, which is liable to pollution.

Introduction

Lakes Baikal and Khubsugul are rift freshwater water bodies. In spite of their different size, they have an asymmetrical cross profile, an elongated form, rather steep underwater slopes, and rocky coasts. The maximal depth of Baikal is 1642 m, whereas that of Khubsugul is 238 m.
It is necessary to study the suspended matter in these lakes, which are in an initial sedi-mentogenesis stage, in order to determine the conditions of their formation, to identify their main sources, and to investigate the processes influencing their formation.

There have been few studies on the distribution of elements in the suspended matter of Lake Baikal (preliminary results have been published [Granina *et al.*. 1994; Potyomkina *et al.*. 1994; Shevchenko *et al.*. 1994]), and data on the elements in Lake Khubsugul have been obtained for the first time. One of the important aspects of the study of suspended matter is its geochemical composition, which depends the mineralogy and granulometry of the suspended matter, on the rocks and soils of the lake]s basins, on physical geographical conditions, and on technogenic factors. Because the concentration of suspended matter in deep lake sites is low - about 1 mg/l, it is difficult to study the geochemical composition of suspended matter from these sites by classical methods, and for this reason the SRXFA-method was used to determine the elements in the suspended matter.

Methods

In summer 1997 water sampling was conducted in the three basins of Lake Baikal and in Lake Khubsugul (Mongolia), on central, vertical cross-sections and at near-coastal sites. The water samples were collected with 10-liter bathometers. The suspensions were extracted from the water by vacuum filtration through "Nuclepore" filters 47 mm in diameter having a pore size of 0.4 mm. The mean weight of the dry suspension on the filters was about 1 mg. We performed X-ray fluorescence element analysis with excitation by monochromatized synchrotron radiation (SRXFA, Novosibirsk, Russia) to quantitatively identify the content of the elements on the filters. The station is described in Ref. (Baryshev *et al.*. 1989). To quantitatively identify the elemental content from Ca to Zr, as well as Pb on the filters, we performed analyse coupled with an external reference sample. The measurement time was 10 min per sample.

Some of the samples were analyzed in Belgium (Antwerp University, Department of Chemistry) at the EPXMA station. This method was used to determine the chemical composi-tion and granulometry (morphological characteristics) of single micrometer-sized suspended particles. This station is described in detail in Ref. (Jambers and Van Grieken 1997). The method incorporates the following advantages: high sensitivity, small sample weight, multielemental data, and rapid acquisition of results.

Results

The results of investigation of lacustrine suspended matter by the SRXFA method show-ed that the elements Fe, Ca, Mn, Ti, V, Cr, Cu, Zn, and Pb predominate in both Lake Baikal and Lake Khubsugul, and that the concentration of other elements is low. On the whole, the concen-trations of the above elements vary within the following range (μg/l):

central areas of Lake Baikal (locations of the sites are listed in Table 1)

Ca 0.00 - 0.68	Ti 0.00 - 4.42	
Fe 0.90 - 13.42	Cu 0.08 - 0.49	
Mn 0.29 - 0.66	Zn 0.05 - 0.28	Pb 0.00 - 0.34

for near-coastal areas of Lake Baikal (Selenga dem1lta river, 52°30'N-106°40' E)

Ca 1.60 - 77.61	Ti 3.65 - 42.38	
Fe 4.81 - 156.81	Cu 0.04 - 5.10	
Mn 0.25 - 21.82	Zn 0.02 - 15.62	Pb 0.62 - 1.50

for Lake Khubsugul (locations of sites are listed in Table 1)

Ca 0.00 - 8.25	Ti 0.01 - 2.26	
Fe 16.90 - 127.64	Cu 0.00 - 0.15	
Mn 0.78 - 4.91	Zn 0.21 - 1.92	Pb 0.01 - 0.32 .

According to the results of element analysis, the amounts of Ca, Fe, Ti, and Cu in the suspended matter is higher in the southern basin of Baikal than in the middle and northern Basins. The concentrations of Mn and Pb were higher in the suspended matter from northern Baikal. Rb was found only in the middle Basin. Comparison of these findings with earlier data (Potyomkina *et al.*, 1998) shows a quantitative differences between samplings in different years. This is typical of natural objects and reveals their

Table 1. The distribution of total concentrations of suspended particles (mg/l) in different lakes.

Lake	Depth, m	Concentration of suspended matter, mg/l
Baikal, southern basin 51°44' N- 105°01' E	0	1.93
	5	1.80
	15	1.18
	50	0.23
	100	0.17
	200	0.13
	500	0.22
	850	0.07
Baikal, middle basin 52°55' N- 107°47' E	0	0.28
	400	0.19
	600	0.10
	1400	0.09
	1600	0.07
Baikal, northern basin 55°10' N- 109°20' E	0	0.67
	200	0.16
	390	0.14
	500	0.11
Khubsugul 51°14' N- 100°20' E	0	0.37
	50	0.50
	84	1.22
	100	0.52

perennial dynamics. However, the ratios between the elements are maintained, suggesting a stable pattern of formation of the lake suspensions.

The vertical distribution of the total amount of suspended matter in open Baikal generally decreases from the surface to the bottom, whereas in Lake Khubsugul it increases (Table 1). The values for total suspended matter content on the vertical lines of the three basins of Lake Baikal range from 0.07 to 1.93 mg/l, with the highest concentrations being more common in the Southern Basin. The highest concentration of suspended matter in water, 3.5 mg/l, was measured at Nizhne-Angarsk City Port (northern Baikal). The situation in regard to element distribution is more complicated.

Discussion

In the coastal zone of Lake Baikal near the Selenga delta where the river waters have a great influence, the element content of the suspended material is much higher than in suspensions from open Baikal. This is because the sampling station is located in where the waters and sus-pensions of the largest tributaries of the Selenga delta. The content of V (0.2-0.5 μg/l) and Cr (0.2-0.3 μg/l) in the Selenga coastal zone is higher than that in the lake, and Br and Ni (0.015-0.04 μg/l), which have never been detected in open Baikal, have been found to be present. Thus, the river suspended matter is more elementrich and can influence the element content of the suspend-ed matter in the lake.

Granulometric measurements of particles have shown for the first time that the suspend-ed matter Lake Baikal is stable at all depths in the waterbody and is not subject to large fluctua-tions. The maximum partical diameter is no greater than 1 μm (fig.1). This distinguishes it from river suspended matter, whose particle are characteristically coarse (up to 10-100 μm). These particles are not transported for and are deposited on the bottom.

Since lacustrine suspensions are formed both by riverine suspensions and by atmospheric aerosols, and the source of both of which is the soil, it would be difficult to classify lacustrian samples on the basis of one or two elements. In this situation we should take advantage of ele-ment analysis, and just the data for some elements allow the use multi-dimensional statistical procedures (cluster and factor analysis). Cluster analysis is intended for classification of observa-tions in more or less homogeneous groups. Hierarchical clustering provides for unification of the most similar observations, and then the next closest observations, are added to them. Next, the groups are combined to those with which they are most closely connected, and this is repeated until full classification of object is achieved.

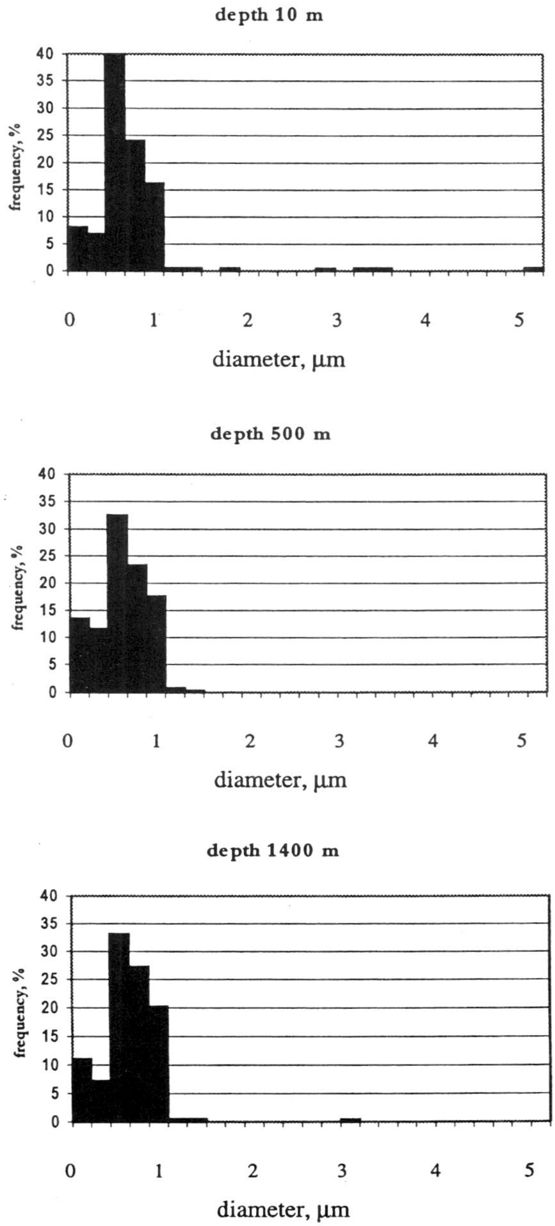

Fig.1. Probability of particle diameter (southern part of Lake Baikal).

The surface samples and the samples taken in central basin of Lake Baikal are united in separate clusters. The ports, which have additional sources of technogenic pollution, are separate.

It is possible to draw similar conclusions from the results of factor analysis. Factor analy-sis is used to reveal groups of elements that are connected with each other. When this approach is used the content of any single element in each sample is considered the result of the combined influence of several sources (factors) that are to be revealed during the analysis. The main idea is to set aside groups of interrelated elements. This means that such groups are identified from the elements having significant loads of a given factor. Factor loads are calculated from correlating matrices of elements. The matrix of factor loads is given in Table 2.

To interpret the results 4 factors have been chosen by the most widespread principle: only factors with means greater than 1 are chosen. These factors explain 89% of the total variable dispersion. The first factor is connected to elements of soil origin (Fe, Mn, Ti; Al was identified by the EPXMA technique as alumino-silicates, and Si as Si-rich particles). The second factor reflects the contribution of atmospheric aerosol (Cr enters into the composition of the aerosol). The third factor can be explained by aerosol influence, since it is almost identical to soil by its composition, but provides only 8% of the total dispersion. The fourth factor is completely defined by the contribution, therefore it describes the local anthropogenic influence (Y is a com-pound of local rocks, therefore its content is connected with the construction of the port).

Table 2. Factor matrix after rotation (4 factors).

	Factors			
Elements	I	II	III	IV
Ca	**0.951**	0.038	0.259	0.015
Ti	**0.860**	-0.063	0.484	0.015
V	0.589	0.423	0.444	-0.069
Cr	-0.136	**0.935**	-0.087	0.019
Mn	**0.923**	-0.143	0.218	-0.019
Fe	**0.797**	-0.094	0.458	0.011
Cu	0.205	-0.096	**0.956**	-0.025
Zn	0.654	0.050	**0.709**	-0.045
Br	0.598	-0.218	0.239	-0.109
Sr	**0.968**	0.024	0.196	0.024
Y	-0.014	0.012	-0.018	**0.995**
Zr	**0.757**	0.020	0.595	-0.010
Pb	0.542	-0.052	**0.810**	0.015

Conclusion

Although the pollution of Lake Baikal is still limited, further study will be necessary to monitor the impact of human activity (through atmospheric transport and river inflow) on the unique ecosystem of the lake, particularly in the southern basin.

References

Baryshev V., N. Gavrilov, A. Daryin, K. Zolotarev, G. Kulipanov, N. Mezentsev and A. Terek-hov, 1989, Nucl.Instr. and Meth. A 282, 570 pp.

Jambers W. and R. Van Grieken, 1997, Single Particle Characterization on Inorganic Suspension in Lake Baikal, Siberia. *Environ. Sci. Technol.* **31**, 1525-1533.

Granina L., V. Baryshev, A. Grachev and O. Levina, 1994, Preliminary results of the study of suspended matter of Lake Baikal and its tributaries by X-ray fluorescent analysis based on synchrotron radiation. *Baikal as a natural laboratory for global change : abstracts*. Irkutsk, **3**, 39.

Potyomkina T., V. Baryshev and A. Grachev, 1994, Chemical composition of suspension in water of Lake Baikal. *Baikal as a natural laboratory for global change : abstracts*. Irkutsk, **3**, 78.

Potyomkina T., V. Baryshev, A. Grachev and V. Potyomkin, 1998, Nucl.Instr. and Meth. A 405, 543pp.

Shevchenko V., Yu. Anokhin, T. Prokhorova, W. Jambers, R. Van Grieken, J.M. Martin and V. Makhov, 1994, Composition of suspended matter of rivers flowing into Lake Baikal according to results of chemical and microprobe analysis. *Baikal as a natural laboratory for global change : abstracts*. Irkutsk, **3**, 82.

Lake Baikal
K. Minoura (editor)
2000 Elsevier Science B.V.

Atmospheric and riverine input of nutrients and organic matter into Lake Baikal

Sorokovikova, L. M.*, Khodzher, T. V., Sinyukovich, V. N., Golobokova, L. P., Bashenkhaeva, N. D., and Netsvetaeva, O. G.
Limnological Institute of SB of RAS, Irkutsk, Russia
E-mail: root@lin.irkutsk.su
(*corresponding author)

Abstract

Long-term changes in the input of nutrients and organic matter into Lake Baikal were considered. An increase in the input of substances into the lake from both riverine waters and atmospheric depositions resulting from human activity was established. The results showed that most nutrients and organic substances have entered the lake with fluvial waters whose a maximum flux occurs during periods of extremely high riverine water discharge.

Introduction

Baikal is one of the largest lakes in the world and contains more than 80% of Russian fresh water resources. Its drainage area is 540,000 km^2, and more than 500 rivers and streams contribute to its water chemistry. The first and most complete study of the chemistry of the fluvial waters and atmospheric depositions in Lake Baikal, and the first construction of the budget of nutrients and organic matter (OM) were performed by K.K.Votintsev and co-authors in the 1950s (Votintsev *et al.*, 1965). The results obtained during this period characterized the natural composition of riverine waters and atmospheric depositions, thereby providing a base for assessing later changes. As new data on the concentrations of inorganic and organic forms of nitrogen and phosphorus in the fluvial and atmospheric waters have accumulated in subsequent years, the budgets of nutrients and OM have been revised and more precisely defined (Votintsev and Popovskaya 1974; Tarasova and Meshcheryakova 1992). This study aims to evaluate the recent input of nitrogen, phosphorus, silicon, and OM into Lake Baikal via fluvial waters and atmospheric depositions.

Material and methods

The results of long-term observations on the input of nutrients and OM into Lake Baikal conducted by the scientists of the Limnological Institute of SB of RAS during different periods, – 1950-1955 by Votintsev *et al.* (1965) and 1981-1984 by Tarasova and Meshcheryakova (1992) – and new data obtained by the authors in 1993-1997 are analyzed in this paper. The authors made seasonal observations (winter, spring, summer, autumn) on the Selenga River and six tributaries of southern Baikal, in 1993-1996, and monthly observations in 1997. Samples were collected from the Upper Angara River and Barguzin River in summer only. Surface waters were sampled on the profile through the river mouth at three sites: near the right shore, in the center of the river, and near the left shore. To analyze atmospheric depositions, the data published in Votintsev *et al.* (1965) were used for 1950-1955, and more recent data (Obolkin and Khodzher 1990; Khodzher and Obolkin 1992) were used for 1981-1984. In 1993-1997 the authors performed seasonal studies, which allowed us to evaluate the modern chemical composition of atmospheric depositions (rainfall, snow cover, dry deposition). The region of the sampling sites for atmospheric depositions was extended in recent years, while preserving the main base stations (Fig.1). The total atmospheric flux (*P*) of nutrients and OM was evaluated as the sum of the fluxes calculated for the southern, central, and northern basins:

$$P = \sum_{i=1}^{3} a_i h_i S_i ,$$

where a_i – the concentration of elements in atmospheric depositions in each lake basin, S_i – the area of each lake basin; h_i – the average amount of atmospheric depositions in each lake basin.

Samples of fluvial waters and atmospheric depositions were run through nuclepore filters having a pore size of 0.4 mm. The concentration of nutrients was analyzed by spectrophotometry: nitrate nitrogen, with disulphophenol acid; ammonium nitrogen, with the Nessler reagent; nitrite nitrogen, - with the Griss reagent; and phosphate and silicon, - with ammonium molybdate, while total organic

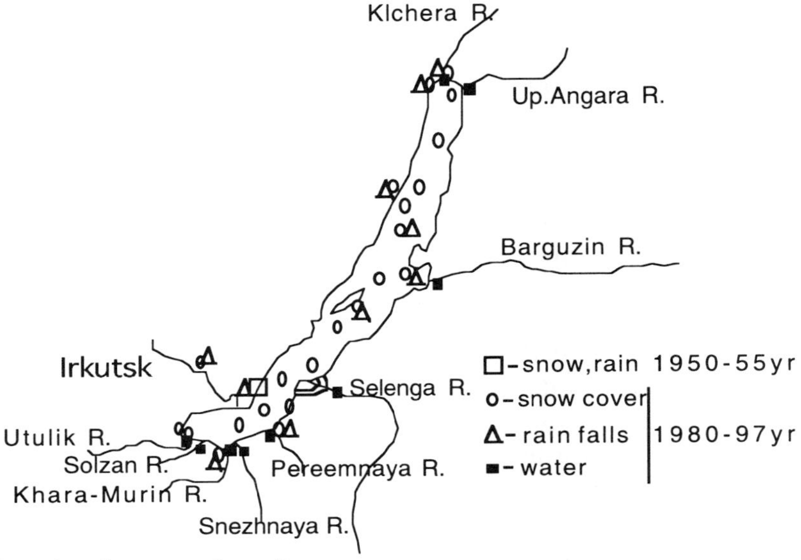

Fig.1. Map showing sampling sites.

carbon was calculated from COD (chemical oxygen demand) (Glazunov, 1963; Mannual/1977; Stroganov and Buzinova, 1980). The detection limits and standard deviations of the methods used were as follows: N-NO_3, 0,005 mgN/l and 7 %, respectively, N-NH_4, 0,02 mgN/l and 4 %; phosphorus, 3 μgP/l and 5 %; silicon, 0.1 mgSi/l and 5 %; and COD, 0.5 mgO/l and 8%.

In 1950-1955 silicon was measured by the Denizhe method (Votintsev *et al.*, 1965), however, the color developed was visually characterized by using K_2CrO_4 solution as the standard (Glazunov, 1963). The relationship between Si concentrations [mg/l] determined this way (Y) and colorimetrically (X) is as follows (Domysheva *et al.*, 1998):

$$Y = 1.389 X + 0.071 \quad (1)$$

Equation (1) was used when the data obtained in the 1950s were recalculated for comparison with those obtained in 1993-1997.

Results and Discussion

Up to 60-90% of the total input of nutrients and organic matter into the lake is provided by riverine fluxes. Baikal tributaries are characterized by quite different fluxes of components entering the lake. Most dissolved solids enter the lake with the waters of the Selenga River, which supplies about 50% of the total riverine water inflow (Votintsev *et al.*, 1965). When the water discharge was at an average level (1996-1997), the input with Selenga waters was: inorganic nitrogen, 8.7 Ktons/yr; inorganic phosphorus, 608 tons/yr; dissolved silicon, 114 Ktons/yr; and OM, 339 Ktons/yr. The input of these constituents into the lake during the year coincided well with the seasonal variability of riverine water discharge (Fig.2). About 80% of annual fluvial water discharge and riverine chemical input took place during the warm season (May-September). Minimal chemical fluxes occured in late winter (March).

Our results show that the inter-year differences in riverine fluxes of substances is caused mainly by corresponding changes in fluvial water discharge. Deviations in nitrate nitrogen, phosphorus, and silicon riverine fluxes from the averaged long-term annual fluxes to the lake were 20-30%. For OM and ammonium nitrogen, whose concentrations increase as long as water discharge increases, the fluctuations were up to 40-50%. For instance, the ammonium nitrogen flux with Selenga waters averaged 2,6 Ktons N/yr, however, it can reach 4 Ktons N/yr during a flood, as recorded in August 1993 (Sorokovikova *et al.*, 1995).

Intensive opening up of natural resources in the Lake Baikal basin in the 1960s to 1980s caused increased input of chemicals into the riverine system both as waste and through increased concentrations of substances in atmospheric depositions. In 1996-97 the weighted mean (by water discharge) concentration of inorganic phosphorus in Selenga water was 21 mgP/l (range: 5 to 38 µgP/l), whereas in the 1950s the weighted mean concentration did not exceed 13 µg/l, in accordance with Votintsev *et al.*, 1965. The phosphorus riverine flux increased during this period, from 0.6 to 1.1 Ktons P/yr (Table 1).

It should be noted, that there is a misprint in Votintsev *et al.*, 1965, on page 480: the PO_4 input with "other tributaries" was published as 1.86 Ktons/yr whereas it should be 0.186 Ktons/yr. This misprint resulted in a higher calculated input of total phosphorus into the lake. Votintsev *et al.* (1965) published it as 3.94 Ktons

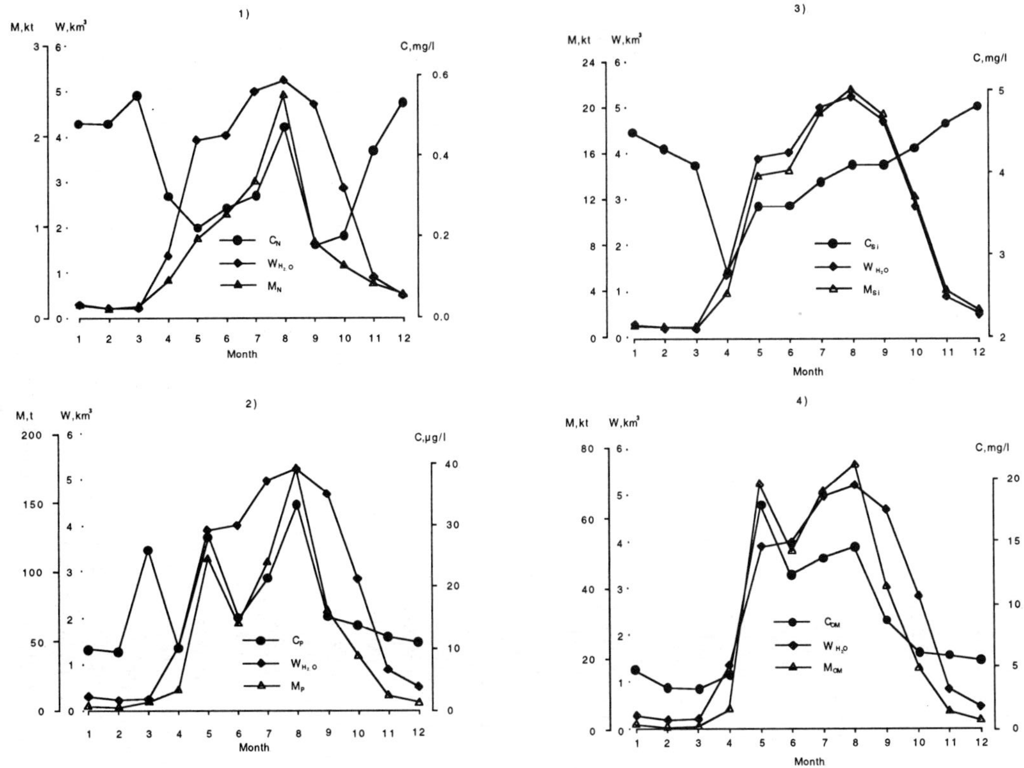

Fig 2. The Selenga River: intra-year changes in water discharge (W) and concentrations (C) and fluxes (M) of nutrients and OM into Lake Baikal.
1 - silicon, 2 - OM , 3 - phosphorus, 4 - nitrogen

Table 1. Input of nutrients and organic matter into Lake Baikal, Ktons/yr

Income source	N	P	Si	OM
From tributaries				
1950-1955[1]	5.2	0.6	292.8	584
1981-1984[2]	3.8	0.5	281.1	734
1993-1997	14.1	1.1	240	595
From atmosphere				
1950-19551	2.1	0.1	1.9	24
1981-19842	3.1	0.3	3.4	56
1993-1997	4.5	0.4	4.0	67

1-Votintsev et al. 1965

2-Tarasova and Mescheryakova 1992

PO_4/yr, whereas the correct value is 2.26 Ktons PO_4/yr. This inaccuracy was repeated in subsequent publications concerning the chemical mass budget of the lake (Tarasova and Meshcheryakova, 1992; Granina 1997). The data for phoshorus (P) are shown in Table 2.

It is clear from Table 2 that the fluvial input of nitrogen and phosphorus in 1981-84 was lower than in the 1950s, and the weighted mean concentration of nutrients in fluvial waters decreased too. For example, the concentration of nitrate nitrogen in Selenga water was 0.10 mgN/l, and the corresponding Selenga riverine flux of nitrate nitrogen was 2.6 Ktons/yr in the 1950s (Votintsev *et al.* 1965), whereas in the 1980s the concentration decreased to 0.06 mgN/l, and the input to 1.7 Ktons/yr (Tarasova and Meshcheryakova 1992). We believe that the 1981-1984 input of nitrate nitrogen and phosphorus published in Tarasova and Meshcheryakova, 1992, is underestimated, because there were no reasons for water quality improvement. Due to the absence of any sewage purification plants until 1986, the industrial waste of most of the factories and plants in the city of Ulan-Ude and other settlements located on the banks of the Selenga entered the river untreated. Moreover, the authors (Tarasova and Meshcheryakova, 1992) claim an increase in OM input in 1981-1984, when there was a decrease in the input of nutrients. This is paradoxical, because nutrients and OM are both indicators of water quality and normally change in parallel.

The 1990s were characterized by a high input of nutrients and OM into the lake, and this was to a considerable degree attributable

Table 2. Recalculated inorganic phosphorus input into Lake Baikal in 1950-1955, Ktons/yr

Input	in Votintsev et al. 1965	Recalculated from Votintsev et al., 1965
with 18 main tributaries	0.55	0.55
with other tributaries	0.61	0.06
with atmospheric depositions	0.01	0.01
Total	1.17	0.62

to increased fluvial water discharge. In 1993-1995 the latter was 40% higher than the long-term mean values. In general, the concentration of nutrients in fluvial waters in this period was 10-50% higher than in previous years of observation. The concentration of nitrate nitrogen ranged from 0.22 to 0.49 mgN/l; ammonium nitrogen, from 0.01 to 0.18 mgN/l; inorganic phosphorus, from 1.0 to 34 μgP/l; and silicon, from 1.0 to 4.8 mg/l. Nitrite nitrogen was episodically found in Selenga and Barguzin waters in summer. The inorganic nitrogen riverine flux was calculated as the sum of ammonium, nitrate, and nitrite fluxes, whereas it was previously based on nitrate nitrogen only (Votintsev *et al.* 1965).

The percentage of ammonium nitrogen flux ranged from 25% to 50% of the total nitrogen input, depending on water discharge conditions. The highest percentage (>30%) was recorded during flood events. When floods occur, total riverine nitrogen flux can reach extremely high values: for example, it was 12.5 Ktons/yr in 1973 (V.Bogdanov, personal communication).

The modern input of dissolved silicon is lower than estimated for previous periods, and this is caused by improvement in the method of silicon assay. A coefficient of 1.386 was used to compare the data obtained in the 1950s and the 1990s (Domysheva *et al.* 1998, see "Materials and Methods"). The mean silicon concentration in Selenga water in 1951-1955 was 5.4 mg Si/l (Votintsev *et al.* 1965), and recalculating it by using the proposed coefficient, yielded 3.9 mg Si/l, which perfectly matched the Si concentration measured in 1993-1997 (Sinyukovich *et al.*, 1998).

Disposal of domestic, industrial, and agricultural wastes into the Selenga and Barguzin riverbeds resulted in increased riverine fluxes of OM and nutrients into the lake. The anthropogenic contribution can be calculated by using the equation

$$\Delta R = 100(R_2 - R_1 K_w)/R_1,$$

where ΔR is the anthropogenic component in %; R_2, the input of the constituent during the accounting period; R_1, the same as in the 1950s, taken as the background value; and K_w, the correction for the difference in water discharge during the background and accounting periods.

Selenga River water flow in 1995-1997 was the same as the mean fluvial water flow used by Votintsev *et al.* (1965), and the coefficient K_w is 1 when calculating ΔR. Thus, the increase in riverine flux of nutrients and OM into the lake was caused by an increase in anthropogenic load in the Selenga River drainage basin. The increase in Selenga riverine flux compared to the 1950s is as follows: nitrate nitrogen, 57%; inorganic phosphorus, 42%; and dissolved OM, 15%. The smaller increase in OM input compared to that of the nutrients is probably attributable to the intensive process of OM destruction that occurrs within the Selenga waters (Sorokovikova and Avdeev, 1992).

Long-term observations performed in the Baikal region showed that the highest concentrations of inorganic nitrogen, phosphorus, and OM in atmospheric depositions were recorded in summer, and the lowest in the winter. Depending on the site, the concentration of ammonium nitrogen in the atmospheric depositions ranged from 0.06 to 2.0 mg/l; nitrate nitrogen, from 0.01 to 4.0 mg/l; inorganic phosphorus, from 1.0 to 20.0 μg/l; and dissolved OM, from 1.0 to 4.5 mgO/l.

Long-term observations indicated that the mean annual atmospheric flux of inorganic nitrogen in most Lake Baikal regions was from 0.05 to 0.2 tons/km^2/yr, thereby corresponding to the background values for Eastern Siberia and the Arctic regions of Russia. Higher atmospheric fluxes of nitrogen (0.5-1.0 tons/km^2/yr) are typical for the northern slopes of the Khamar-Daban Ridge, which are directed towards the lake and where the rivers of southern Baikal have their sources, as well as for the valleys of Selenga and Barguzin Rivers.

From 1971 to 1981 the atmospheric flux of inorganic nitrogen in Southern Baikal doubled (Valikova *et al.* 1985) because of industrial development in the region and increased burning of fuel. According to Tarasova and Meshcheryakova 1992, from the 1950s to the 1980s the atmospheric flux of nitrogen ranged from 39% to 81% of the riverine input into the lake, and the corresponding values for phosphorus were 11% to 60%. However, percentages of 60% - 80% seem to be too high. Our recent data show that both the atmospheric and the riverine fluxes of nutrients have increased, and that the ratio of the former to the latter is no more than 40%, the same as recorded by Votintsev *et al.* (1965) in the 1950s. Similar ratios (30-50%) are typical of the Northern Atlantic, Baltic Sea (Nering *et al.* 1981), Onega Lake (Hydrochemistry...1973), Ladoga Lake (Anthropogenic...1982), and other water bodies.

Conclusion

An analysis of data obtained from long-term observations shows that during the last decades the riverine and atmospheric fluxes of OM, nitrogen, and phosphorus compounds into Lake Baikal have increased substantially because of intensive human activity. The input of the substances studied was determined by the fluvial water discharge, and it sharply increases during floods.

Improvements in methodology, an increased number of sampling sites for atmospheric depositions within the lake basin, and implementation of a system of observations allowed us to more precisely define the input of nutrients and OM into Lake Baikal. Monitoring studies need to be continued in order to control the amount of chemicals entering the lake.

References

Anthropogenic eutrophication of the Ladoga Lake., 1982,. Nauka, Leningrad, 173 pp, (in Russian).

Domysheva V.M., M.N. Shimaraev, L.A. Gorbunova, L.P. Golobokova, I.V. Korovyakova, A.A. Zhdanov and V.V. Tsechanovsky, 1998, Silicon in Lake Baikal. Geogr. i Prirodn. Resursy, 4, 73-81, (in Russian).

Glazunov I.V., 1963, Hydrochemical Regime and Chemical Outflow of the Angara River at its Source, In: Issue of LIN SB AN SSSR,

3(23), 57-94, (in Russian).
Granina L.Z., 1997, The chemical budget of Lake Baikal: A review. Limnol. Oceanogr. 42(2), 373-378.
Hydrochemistry of Onega Lake and its Tributaries, 1973, Nauka, Leningrad, 124 pp. (in Russian).
Khodzher T.V. and V.A. Obolkin, 1992, Monitoring of Precipitation and Aerosol near Lake Baikal. Thirtheenth international conference on: "Nucleation and Atmospheric aerosols", Salt Lake City, USA, 13, 256-257.
Mannual on Chemical Analyses of Surface Waters., 1977, Leningrad: Hydrometeoizdat, 534 pp., (in Russian).
Nering D., A. Vilde and K. Rode, 1981, Studies of atmospheric input of nutrients into Baltic Sea, In: Change by chemical elements on the interfaces of marine environment, Moscow, 367pp., (in Russian).
Obolkin V.A. and T.V. Khodzher, 1990, Annual input of sulphates and mineral nitrogen from the atmosphere in the region of Lake Baikal. Meteorology & Hydrology, 7, 71-76, (in Russian).
Sinyukovich V.N., L.M. Sorokovikova, L.P. Golobokova and M.P. Chubarov, 1998, Peculiarities of dissolved silicon input into Lake Baikal. Geogr. i Prirodn. Resursy, 2, 66-70, (in Russian).
Sorokovikova L.M., V.N. Sinyukovich, V.V. Drjukker, T.G. Potyomkina, O.G. Netsvetaeva and V.A. Afanasiev, 1995, Ecological peculiarities of the Selenga River in conditions of flood. Geogr. i Prirodn. Resursy, 4, 64-71, (in Russian).
Sorokovikova L.M. and V.V. Avdeev, 1992, Primary production and distribution of organic matter of the Selenga River. Vodnye Resursy, 19, 163-165, (in Russian).
Stroganov N.S. and N.S. Buzinova, 1980, Practical mannual on water chemistry, Publ. house of Moscow University: Moscow, 193 pp., (in Russian).
Tarasova E.N. and A.I. Meshcheryakova, 1992, Modern State of Hydrochemical Regime of Lake Baikal, Novosibirsk: Nauka. 141 pp., (in Russian).
Valikova V.I., A.A. Matveev and B.B. Chebanenko, 1985, Input of some substances with atmospheric deposition in the region of Lake Baikal, In: Improvement of regional monitoring of the state of Lake Baikal, pp.58-66, (in Russian).
Votintsev K.K., I.V. Glazunov and A.P. Tolmacheva, 1965,

Hydrochemistry of Rivers in the Basin of Baikal, Moscow: Nauka, 495 pp., (in Russian).
Votintsev K.K. and G.I. Popovskaya, 1974, The role of allochtonous organic substance in Lake Baikal, In: Nature of Baikal. Izd. Geogr. Obsch, SSSR: Leningrad, pp.169-178, (in Russian).

Lake Baikal
K. Minoura (editor)
2000 Elsevier Science B.V.

Comparison of persistent organochlorine pollutant behavior in the food webs of Lakes Baikal and Superior

Kucklick, J. R.[1]* and Baker, J. E.[2]
[1] National Institute of Standards and Technology, Analytical Chemistry Division
Charleston Laboratory
219 Fort Johnson Road, Charleston, SC 29412
Phone: (843)-762-8572, Fax: (843)-762-8724
E-mail: john.kucklick@nist.gov
[2] University of Maryland, Chesapeake Biological Laboratory
PO Box 38 Solomons, MD 20688
Phone: (410)- 326-7205, Fax: (410)-326-7341
E-mail: baker@cbl.umces.edu
(*corresponding author)

Abstract

A comparative study was undertaken to examine the behavior of persistent organochlorine pollutants (POPs) in the food webs of Lake Baikal and Lake Superior. Samples of zooplankton, macrozooplankton, amphipods, as well as size or age classes of sculpins and salmonids were taken from Lake Baikal during August - September 1993 and from Lake Superior during May - July, 1994. Samples were analyzed by gas chromatography with electron-capture and mass-spectrometry detection for 74 PCB congeners, chlordanes, HCHs, HCB, DDT compounds, and toxaphene. The lipid content and stable nitrogen isotopes of the samples were also measured. Stable nitrogen isotopes were measured to provide information on an organism's trophic level. The HCB, HCHs, Σchlordanes, ΣPCBs, and toxaphene concentrations were generally lower in the samples from Lake Baikal than from Lake Superior. DDT compounds were generally higher in concentration in the samples from Lake Baikal than from Lake Superior. In both lakes, wet weight POP concentrations in the food web were mainly affected by an increase in lipid with trophic position-more so than trophic position alone. The ratios of log predator/prey for individual PCB congeners were not related to log K_{ow} in the Lake Superior food web, which conflicted with the results of laboratory feeding studies, suggesting that uptake rates of high log K_{ow} POPs from food are offset by re-equilibration with water. The reverse, however, was true of Lake Baikal, where there was a good relationship between log predator/prey POPs concentration and log K_{ow}. This is likely due to temporal variability in the concentrations of POPs in the water column arising from local and/or regional POP sources.

Introduction

Persistent organic pollutants (POPs) are a class of anthropogenic contaminants that generally have a recognized degree of toxicity and a long half-life in the environment (months to years). Certain POPs, such as DDTs, polychlorinated biphenyls (PCBs), chlordanes, hexachlorocyclohexanes (HCHs), hexachlorobenzene (HCB) and toxaphene are extremely recalcitrant, bioaccumulative, and dispersed throughout the globe. The mobility of these compounds in the environment has led to varying degrees of contamination of water bodies nearly everywhere and their subsequent incorporation into the food web. Once in the food web, POPs are passed to higher trophic levels in a process referred to as bioaccumulation (Kidd *et al.*, 1995), where they may affect wildlife and human health.

Surprisingly, there have been few investigations of POP distribution in a complete lake food web (e.g., Oliver and Niimi, 1988, Kidd *et al.*, 1995, Kiriluk *et al.*, 1995). Since there have been so few complete field studies on POP behavior in food webs, drawing up "rules" based on a just a few investigations is risky. For instance, Kidd *et al.*. (1995) and Kiriluk *et al.* (1995) demonstrated that trophic position describes more variance in POP concentrations than lipid in either the Lake Laberge or Lake Ontario food webs. The reverse, however, is true of Lake Superior (Kucklick and Baker, 1998). There have been even fewer comparative studies of POPs between two lakes that incorporate lower trophic levels and a wide array of POPs, such as large numbers of PCB congeners.

In addition to field studies, knowledge regarding the behavior of POPs in food webs, especially regarding fractionation from trophic transfer, has been gained from laboratory studies (e.g., Gobas *et al.*, 1993) and simulation models (e.g., Thomann and Connolly, 1984). These studies of trophic transfer have their limitations, such as non-steady state conditions in the laboratory and lack of calibration data sets in modeling studies. In all modeling studies, laboratory data have been used as the basis for the predictions of POP behavior in the field. For instance, laboratory investigations have demonstrated that the patterns of POPs fractionate between predator and prey, leading to the retention of more hydrophobic POPs in the predator (Gobas *et al.*, 1993 and LeBlanc, 1995). However, field studies suggest that there may be competing processes, such as re-equilibration of the predator with the surrounding water (e.g., Kucklick and Baker, 1998). Therefore, a complete predictive understanding of POP behavior in aquatic food webs requires a balance between laboratory and field studies.

The objective of this paper is to provide a detailed comparison of Lake Baikal and Lake Superior with respect to POP concentrations, fractionation, and the influence of lipid and trophic position on bioaccumulation in their food webs. The data for this comparison came from Kucklick *et al.* (1996) for Lake Baikal and from Kucklick and Baker (1998) for Lake Superior. These two studies are unique in that the data are internally consistent (having been collected and analyzed by the same investigators), include information on lower trophic levels, have a large number of POPs (85) spanning a wide hydrophobicity range, and include data on trophic position ($\delta^{15}N$).

The studies focussed on POPs in the food webs of Lake Baikal and Lake Superior. Lake Baikal (*ca.* 52°N to 56°N to 104°E to 110°E) is an extremely large lake in terms of volume (2.3 x 10^4 km^3) and has a long water residence time (*ca.* 300 years). The bulk of Lake Baikal's watershed is undeveloped, with the largest runoff sources of contaminants being the Selenga River (Iwata *et al..*, 1995) and possibly a large pulp mill on the southeast shore of the lake. Atmospheric deposition of POPs may also be an important source of contamination of the lake (Iwata *et al.*, 1995 and McConnell *et al.*, 1996). Lake Superior is also a very large lake and has a residence time of roughly 170 years. Most of Lake Superior's watershed is undeveloped, and its largest source of POPs is atmospheric deposition (Eisenreich, 1987). Lake Baikal is the deepest purely freshwater lake in the world (-1,637 m), and Lake Superior has the largest surface area (83,300 km^2).

Materials and Methods

Sample Collection

The details of sample collection, processing, and analysis are given in Kucklick *et al.* (1996) and Kucklick and Baker (1998), and they are only summarized here. The Lake Baikal and Lake Superior food webs were sampled from August 25 through September 8, 1993 and from June 25 to July 1, 1994, respectively. Samples of fish (*Coregonus autumnalis migratorious* (the omul), *Comephorus dybowskii, and Comephorus baikalensis*), benthic amphipods (*Acanthogammarus* sp.), and zooplankton (*Macrohectopus*) were collected from the central basin of Lake Baikal by trawling and deployment of a 63μm plankton net while aboard the *R/V Vereshchagin* (Tab. 1). Deepwater sculpins (*Myoxocephalus thompsonii*) and amphipods (*Dioporeia hoyi*) were collected from Lake Superior from the *R/V Edwin Link* and the *Clelia* research submersible (Tab. 2). *Mysis relicta* were collected by plankton tows from the *R/V Edwin Link.* Fish from Lake Superior, including smelt (*Osmerus mordax*), herring

Table 1. POPs concentrations (ng/g wet mass), $\delta^{15}N$, and percnt lliqid in samples from the Lake Baikal food web.

Sample	lipid (%)	$\delta^{15}N$	HCB	ΣPCBs	dieldrin	ΣChlordanes	4,4'-DDE	4,4'-DDT	toxaphene
zooplankton tows									
station 1: 25 August, 1993	1.4	6.3	<0.3 ng	4.1	0.4	1.7	0.27	<2.7 ng	<92 ng
station 1: 26 August, 1993	1.3	6.8	<0.3 ng	9.8	0.5	2.0	<3.1 ng	3.14	<92 ng
station 1: 27 August, 1993	1.3	7.2	<0.3 ng	3.7	0.4	0.5	<3.1 ng	<2.7 ng	<92 ng
Macrohectopus	3.5	9.0	<0.3 ng	17	0.7	3.9	<3.1 ng	<2.7 ng	<92 ng
benthic amphipods (Acanthogammarus)	1.6	12.0	<0.3 ng	37	1.5	11	20	2.04	31
C. dybowskii									
year 4	2.1	13.1	3.0	59	0.3	4.1	20	12	55
year 5	1.2	12.8	1.0	80	0.6	5.7	23	15	51
year 6	1.7	13.4	1.1	66	0.5	4.6	24	15	48
year 7-8	1.4	13.1	0.9	73	0.5	4.8	21	13	42
C. baikalensis									
year 3	3.1	13.1	4.1	72	0.6	9.6	25	15	32
year 4	2.4	13.7	3.0	66	0.5	9.5	18	11	37
year 5	26	13.3	15	422	6.5	44	138	107	330
year 6	41	13.4	26	712	10	69	327	791	712
year 7	40	13.2	23	633	10	66	298	610	607
year 8-10	42	9.3	25	659	11	67	313	658	643
Coregonus									
3 years	0.6	9.3	3.2	18	0.3	6.7	21	10	<92 ng
6 years	1.3	11.3	1.2	42	0.5	5.1	15	6	22
7 years	1.2	11.4	2.2	37	0.8	4.4	39	21	<92 ng
8 years	1.0	13.0	4.2	40	0.5	6.1	58	18	24
9 years	1.5	11.2	4.5	34	1.2	6.4	35	13	24
10 years	1.3	11.9	2.5	75	0.5	7.6	25	11	41
13 years	2.3	12.2	2.9	86	1.0	9.6	37	15	52

Table 2. POPs concentrations (ng/g wet mass), $\delta^{15}N$, and percnt lliqid in samples from the Lake Superior food web.

Organism	% lipid	$\delta^{15}N$	HCB	SPCBs	dieldrin	Schlordanes	4,4'-DDE	Toxaph
[1]**Smelt**								
1-49 mm	2.9	--	0.8	21	4.6	6.0	9.0	99
50-99 mm	2.6	--	1.3	28	7.5	7.0	12	140
100-149 mm	4.0	6.27	1.3	41	22	16	14	210
150-199 mm	4.6	6.98	1.8	53	16	22	33	200
[1]**Herring**								
200-249 mm	5.7	7.61	3.0	120	19	43	34	560
250-300 mm	9.6	7.81	4.3	150	35	67	63	720
[1]**Bloater**								
150-199 mm	11	6.99	8.4	130	54	70	69	1360
200-249 mm	11	7.20	7.1	140	52	81	102	1170
250-300 mm	11	8.34	4.7	180	42	87	71	840
Sculpins								
deepwater sculp1383	2.6	10.17	1.1	52	12	30	30	260
deepwater sculp NOAA 3	4.0	--	2.2	--	31	46	48	460
1deepwater sculp 50-99 mm	3.2	--	1.4	42	32	37	35	360
1slimy sculp 1-49 mm	6.1	--	4.1	46	42	35	20	450
1slimy sculp 100-149 mm	3.0	7.22	2.8	42	18	31	16	390
1spoonhead sculp	6.9	--	2.6	47	54	33	9.2	360
Mysis								
large site1383	1.2	6.92	<0.3 ng	9.2	3.5	5.9	6.0	30
large site NEO 80	2.1	7.18	0.1	11	7.1	8.4	8.4	40
large site NOAA3	1.4	7.10	<0.3 ng	9.2	3.3	3.9	5.2	21
small site 1383	1.2	4.93	<0.3 ng	10	2.9	2.6	4.8	<92 ng
small site NEO 80	2.1	--	--	7.5	--	--	--	--
small site NOAA 3	1.1	--	0.2	5.6	2.4	2.0	1.6	37
Limnocalanus								
site 1383	2.5	4.93	--	11	--	--	--	--
site NEO 80	4.4	4.99	1.4	14	5.6	17	1.1	180
site NOAA 3 (tow 1)	2.2	4.97	0.8	9.0	5.6	5.8	3.1	110
site NOAA 3 (tow 2)	2.4	--	0.9	7.7	5.9	5.7	3.7	120
Amphipod								
site NEO 80	1.3	4.58	0.8	11	18	10	3.7	100
[1]**Lake Trout**								
1	7.2	8.81	1.7	82	12	29	35	250
2	10	8.91	2.6	120	21	37	46	440
3	7.0	9.35	2.8	160	19	58	87	540
4	9.4	8.74	2.7	110	16	33	47	370
5	11	8.83	3.1	113	21	35	44	360

[1]Samples collected by U.S. Fish and Wildlife Service

(*Coregonus artedii*), bloater (*Coregonus hoyi*), slimy sculpin (*Cottus congatus*) and spoonhead sculpin (*Cottus ricei*) were obtained from trawls near the vicinity of the Kenweenaw Peninsula. Lake trout (*Salvelinus namaycush*) were obtained gutted from local fishermen. All samples were composited based on age or size class, then homogenized (Tekmar). Filets of lake trout and omul were homogenized, then analyzed individually.

Extraction and analysis

All samples were analyzed for organochlorine compounds, and most for stable isotopes (Tabs. 1 and 2). Between 2 to 20 g (wet weight) of the homogenized composite sample were Soxhlet extracted with dichloromethane, and the lipids were first quantified gravimetrically, then removed by gel permeation chromatography (Kucklick *et al.* 1996). The sample extracts were purified by a two-fraction Florisil technique (Kucklick *et al.* 1996). PCBs were quantified (74 congeners) in the first fraction by gas chromatography with electron-capture detection (GC-ECD) using a Hewlett Packard 5890 equipped with a 60 m x 0.25 µm x 0.25 mm DB-5 capillary column (J&W Scientific). The remaining organochlorines were quantified by negative-chemical ionization mass spectrometry (GC-NCI/MS) after the two fractions were recombined. The instrument was a Hewlett Packard 5890 coupled to a 5989 mass spectrometer. Details of the OC quantification (internal standards, GC and MS conditions) and quality control (spike recoveries and standard reference material analyses) are given in Kucklick *et al.* (1996). Total toxaphene is reported as the sum of the hepta-, octa-, and nona-chlorobornanes in the sample that was quantified relative to technical toxaphene using 2,2',3,4,4',5,6,6'-octachlorobiphenyl, congener 204, as the internal standard. Stable nitrogen isotopes were analyzed as described in Kucklick *et al.* (1996). Quality control information for organochlorine and stable isotope analyses is given in Kucklick *et al.* (1996).

Results and Discussion

The concentrations of organochlorines measured in the Lake Baikal and Lake Superior food webs are given in Tables 1 and 2, with the data here presented on a wet mass basis. HCB concentrations in Lake Superior were generally lower than in Lake Baikal. Concentrations ranged from below the limit of detection (3 ng/extract) in Lake Baikal zooplankton tows to 16.0 ng/g ± 10.4 ng/g in *C. baikalensis*. Levels of HCB in Lake Superior ranged from below the limit of detection in some *Mysis relicta* samples to only 6.7 ng/g ± 1.9 ng/g in bloaters. ΣPCBs were generally lower in Lake Baikal than Lake Superior, except for the 5-10 year classes of *C. baikalen-*

sis, where ΣPCBs ranged from 422 ng/g to 712 ng/g (Tabs. 1 and 2). ΣPCB concentrations in omul from Lake Baikal relative to a trophically comparable fish, the lake trout from Lake Superior, were 47 ± 24 ng/g versus 117 ± 28 ng/g, respectively. Dieldrin concentrations ranged from 0.48 ng/g ± 0.13 ng/g to 6.4 ng/g ± 4.8 ng/g in *C. baikalensis* from Lake Baikal and 3.8 ng/g ± 1.9 ng/g in *Mysis relicta* to 49 ng/g ± 6.4 ng/g in bloater from Lake Superior (Tabs. 1 and 2). Overall Σchlordane (*cis*- and *trans*-chlordane, *cis*- and *trans*-nonachlor, heptachlor epoxide and oxychlordane) concentrations were higher in the food web of Lake Superior than of Lake Baikal. Concentrations of Σchlordanes in Lake Superior ranged from 4.6 ng/g ± 2.6 ng/g in *Mysis relicta* to 79.3 ng/g ± 8.5 ng/g in bloaters, as opposed to only 1.41 ng/g ± 0.79 ng/g in zooplankton tows to 44.1 ng/g ± 28.3 ng/g in *C. baikalensis* from Lake Baikal. 4,4'-DDE concentrations in Lake Superior ranged from 2.6 ± 1.1 ng/g in Limnocalanus to 81 ng/g ± 19 ng/g in bloaters versus below the limit of detection (3.1 ng/extract) in some zooplankton tows to 187 ± 145 ng/g in *C. baikalensis* from the Lake Baikal food web. 4,4'-DDT was not detected in Lake Superior biota, but concentrations in the Lake Baikal food web ranged from below the limit of detection (2.7 ng/extract) in two zooplankton tows to 365 ng/g ± 358 ng/g in *C. baikalensis* (Tab. 2). Toxaphene concentrations were much higher in the Lake Superior food web than in the Lake Baikal food web (Tabs. 1 and 2). Toxaphene levels in Lake Superior ranged from just below the detection limit in Mysis relicta to 1120 ng/g ± 263 ng/g in bloaters, and below the detection limit in zooplankton tows and some omul, to only 394 ng/g ± 307 ng/g in *C. baikalensis* from Lake Baikal.

The concentrations of organochlorines found in the Lake Baikal and Lake Superior fishes were in reasonable agreement with those found by others (comparisons made on a wet mass basis). For instance, levels of Σchlordanes, 4,4'-DDE, 4,4'-DDT and ΣPCBs in *C. baikalensis* were 43 ng/g, 125 ng/g, 429 ng/g and 660 ng/g, respectively (Nakata *et al*.., 1995). In the present study the concentrations of Σchlordanes, 4,4'-DDE, 4,4'-DDT and ΣPCB in *C. baikalensis* were 41.4 ng/g ± 28.2 ng/g, 186 ngg/ ± 145 ng/g, 365 ng/g ± 358 ng/g and 427 ng/g ± 294 ng/g, respectively. The levels of HCB, dieldrin, Σchlordanes, 4,4'-DDE, and ΣPCB found by Gerstenberger *et al*. (1997) in Lake Superior Siscowet lake trout were 3.2 ng/g, 19 ng/g, 65.6 ng/g, 66.5 ng/g, and 370 ng/g, respectively. In the current study, the values in lake trout were 2.6 ng/g ± 0.51 ng/g, 17.8 ng/g ± 3.7 ng/g, 38.2 ng/g ± 11.4 ng/g, 51.9 ng/g ± 20.4 ng/g, 117 ng/g ± 28 ng/g, respectively. Toxaphene levels measured in Lake Superior lake trout in this study were lower than those reported by Glassmeyer *et al*. (1997), 390 ± 110 ng/g versus 4900 ± 1400 ng/g, and in better agreement with those of Newsome *et al*. (1993; 936 ng/g).

Evident from Tables 1 and 2 is the variability of POP concentrations within an individual species. This is most striking in *C. baikalensis* where, for instance, ΣPCBs on a wet mass basis range from 66 ng/g in the 4-year class to 712 ng/g in 6-year class (Fig. 1). Lipid content in this species is also highly variable, ranging from 2.4% in the 4-year class to 42% in the 8-10 year class. Higher lipid appears to relate to higher PCB concentrations when expressed on a wet mass basis. This is also evident to a lesser extent in the Lake Superior food web, where bloaters had the highest lipid content and the highest ΣPCB levels (Tab. 2 and Fig. 2). Obviously, lipid plays a role in organochlorine variance, as demonstrated by Kiriluk *et al.* (1995) in Lake Ontario and Kucklick and Baker (1998) in Lake Superior.

The variability of POP concentrations on a wet weight basis is shown graphically in Figures 1 and 2, using ΣPCBs as an example. ΣPCBs are highly correlated with other POPs in the Lake Superior data set (Kucklick and Baker, 1998) and with other POPs in the Baikal data set ($r^2 > 0.90$), and are therefore used to simplify the following discussion. The bars in Figures 1 and 2 are arranged in order of $\delta^{15}N$, which has been shown to be a proxy for trophic position (Peterson and Fry, 1987). Figures 1 and 2 indicate that there is a general increase in ΣPCB concentrations on a wet mass basis with trophic position. In Lake Baikal and Lake Superior the r^2s for $\delta^{15}N$ versus log lipid on a wet mass basis were 0.70 and 0.53, respectively, indicating that trophic position has a clear influence on POP concentrations.

Figures 3 and 4 show the results of normalizing the ΣPCBs in the two food webs to lipid. Much of the increase in the ΣPCB concentrations with trophic level is masked when the data are normalized to lipid, indicating that trophic position and lipid content strongly covary. The relationships (r^2) between $\delta^{15}N$ and log ΣPCB concentration on a lipid mass basis for Lake Baikal and Lake Superior, 0.54 and 0.53, respectively, were also significant ($p<0.05$), showing a persistent effect of trophic position after accounting for lipid.

There has been general agreement in the literature that POP concentrations on a wet mass basis increase with food web length (Rasmussen *et al.*., 1990; Broman *et al.*, 1992; Kidd *et al.*, 1995). However, whether the observed POPs increases are due solely to trophic position or increases in lipid content in the food web is less understood. We used two approaches to help elucidate the relative importance of these two factors (Kucklick and Baker, 1998). The first approach uses the results of the simple regression of ΣPCBs (wet mass) versus lipid and ΣPCBs (wet mass) versus $\delta^{15}N$. This defines the maximum variance explained by the two independent variables ($\delta^{15}N$ and %lipid). The minimum variance is given by the ratio of the partial sum of squares to the total sum of squares in the multiple regression of

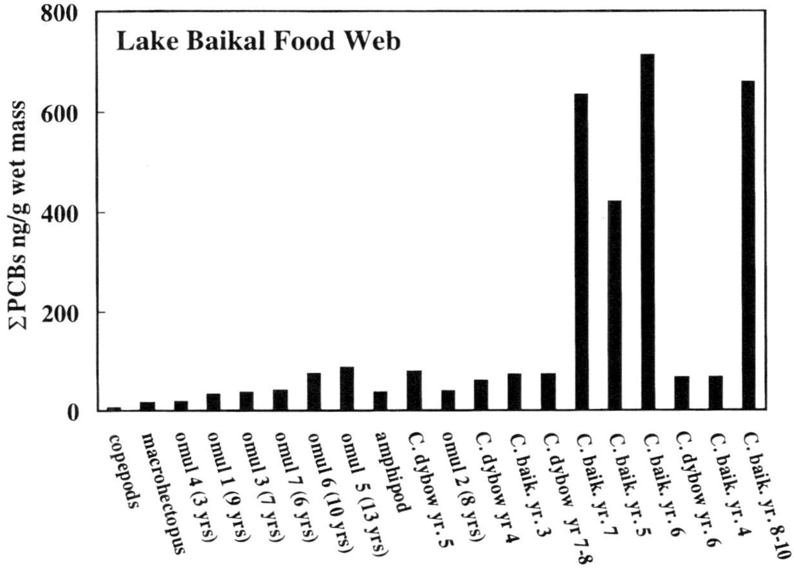

Figure 1. Concentrations of ΣPCBs in Lake Baikal's food web in ng/g wet mass. The bars are arranged from low to high trophic position based on their $\delta^{15}N$ values (parts per thousand).

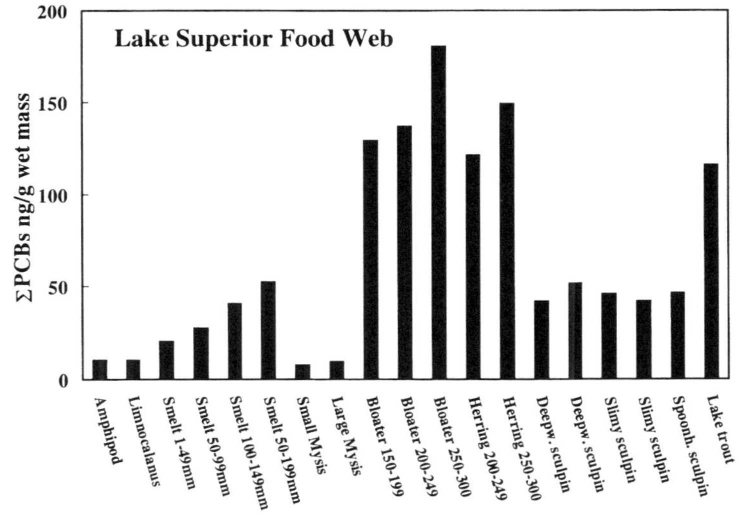

Figure 2. Concentrations of ΣPCBs in Lake Superior's food web in ng/g wet mass. The bars are arranged from low to high trophic position based on their $\delta^{15}N$ values (parts per thousand).

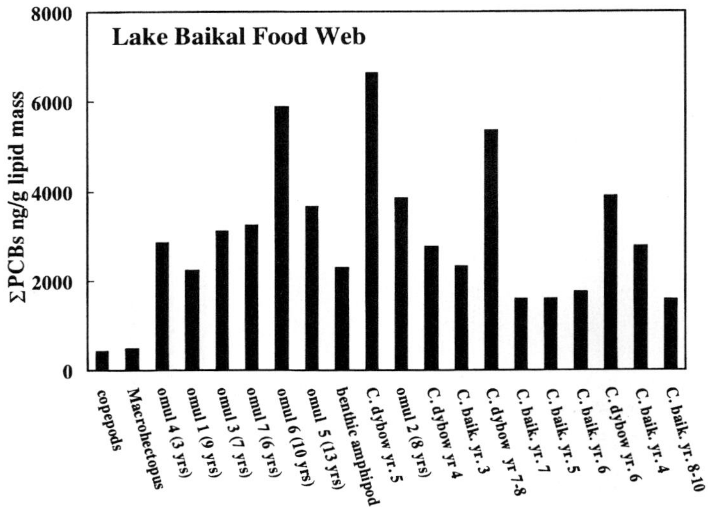

Figure 3. Concentrations of ΣPCBs in Lake Baikal's food web in ng/g lipid mass. The bars are ordered as in Figures 1 and 2.

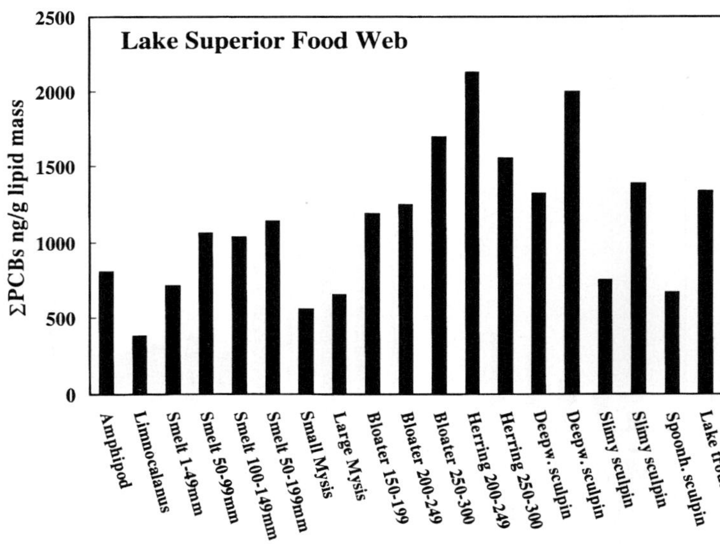

Figure 4. Concentrations of ΣPCBs in Lake Superior's food web in ng/g lipid mass. The bars are ordered as in Figures 1 and 2.

ΣPCBs, %lipid and $\delta^{15}N$. For Lake Superior, trophic position contributed between 3.7% and 42% to the variance in ΣPCBs, whereas lipid contributed between 42% and 81% (Kucklick and Baker, 1998). In the Lake Baikal food web, $\delta^{15}N$ explained between 0.7% and 22% of the variance in ΣPCB, while lipid alone explained between 78% and 99% of the variance in wet mass ΣPCB.

Path analysis was also used to examine the influence of trophic position and lipid on ΣPCB (ng/g wet mass). With path analysis, the variables are placed in a logical cause-effect relationship (Pedhazur, 1982). In this case, trophic position exerts a direct effect on ΣPCB concentrations (i.e., trophic transfer) and an indirect effect through increasing %lipid with trophic position (Kucklick and Baker, 1998). Path analysis uses the correlation matrix of the variables to quantify the relative importance of the two paths. There is a direct path in which wet mass ΣPCBs are only influenced by trophic position and an indirect path in which trophic position influences %lipid that in turn influences wet mass ΣPCBs (see Kucklick and Baker, 1998). In the Lake Superior food web, 65% of the control of trophic position on ΣPCBs was by the indirect effect, and 35% was due to the direct effect of trophic position on ΣPCBs. In Lake Baikal, the indirect path had a greater influence, 83%, versus 17% for the direct path, 17%. In summary, the main effect of trophic position on ΣPCB in both lakes is through its influence in increasing lipid content.

The large number of compounds measured spanning a relatively large hydrophobicity range (log K_{ow} ca. 5.5 to 8.0) in these two investigations also allowed examination of partitioning behavior in the food web. In this analysis, values of individual PCB congeners are represented as the log of the ratio of the wet mass concentrations in predator and prey (biomagnification factor or BMF) versus log K_{ow} (Figures 5 and 6). Laboratory investigations have suggested that, for non-metabolized compounds, this relationship should be linear or curvilinear, since the partitioning from food (*i.e*, fugacity) increases with hydrophobicity (Gobas *et al.*, 1993; Leblac 1995; Fisk *et al.*, 1998). Plots of log predator/prey versus log K_{ow} of PCB congeners show conflicting results between the two lakes (Figures 5 and 6). In Lake Baikal, the plot is linear ($r^2 = 0.55$; Figure 5), but in Lake Superior, there is no relationship ($r^2 = 0.041$; Figure 6). In both systems, we would have expected to see the curvilinear relationship with log K_{ow} reported in laboratory studies of Fisk *et al.* (1998) or the linear relationship suggested by Leblanc (1995). We interpret the absence of a relationship between log BMF and log K_{ow} in Lake Superior as evidence that diffusive exchange with the water occurs at a faster overall rate than incorporation from food (Kucklick and Baker, 1998). In other words, if the predator and

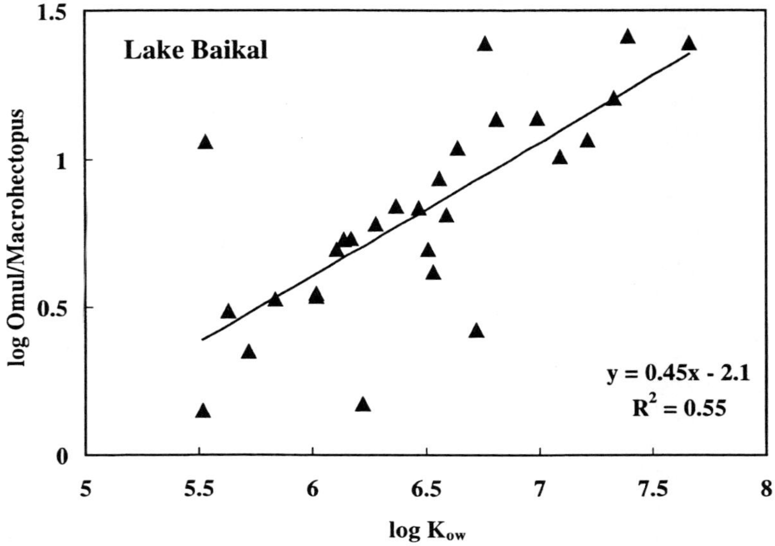

Figure 5. Plot of PCB congener concentrations (ng/g wet mass) expressed as log Omul/Macrohectopus (BMF) versus log K_{ow} for Lake Baikal. Values for log K_{ow} are from Franz (1990).

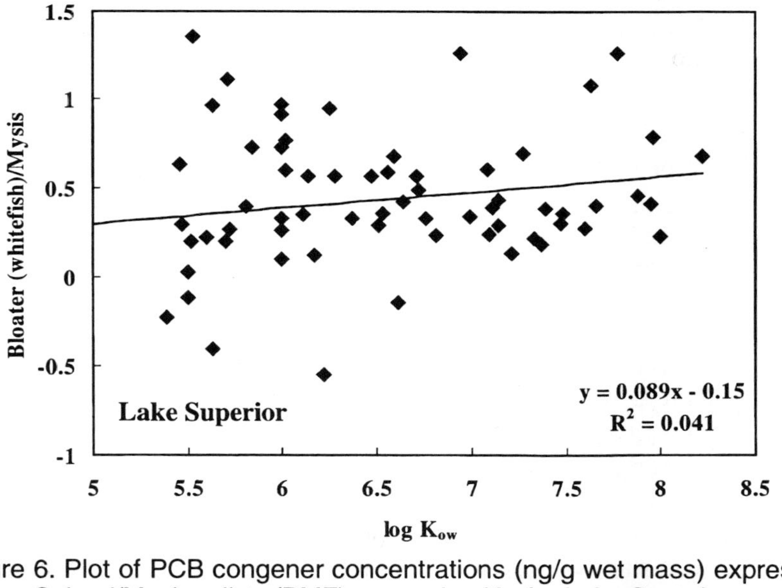

Figure 6. Plot of PCB congener concentrations (ng/g wet mass) expressed as log *C. hoyi/Mysis relicta* (BMF) versus log K_{ow} for Lake Superior.

the prey are both in equilibrium with the water, then there will be no relationship between the ratio of the two and log K_{ow}. Biotic equilibrium with water in Lake Superior is supported by a significant relationship between the log bioconcentration factor (BCF; ng/kg lipid organism / ng/L water) and the log K_{ow} for both lake trout and *Mysis relicta*; $r^2 = 0.81$ and 0.72, respectively (water data for Lake Superior from Jeremiason *et al.* (1994)). If this holds true, then the relationship between log BMF and log K_{ow} in Lake Baikal results from the predator or prey not being in equilibrium with the water, which was the case during this sampling, as shown by the poor relationship between the log BCF and log K_{ow} (see Kucklick *et al.* (1996) Figure 9). Kucklick *et al.* (1996) also showed that water samples collected from the southern and central basins had patterns enriched in highly chlorinated PCB congeners, suggesting a local or regional source of PCBs that may be causing non-steady state conditions. Dissolved ΣPCBs were also nearly twice as high in the southern basin in 1993 than in 1991 (up to 1,800 pg/L versus 1100 pg/L) suggesting temporal variability in surface water PCBs (Kucklick *et al.*, 1996; Kucklick *et al.*, 1994).

In conclusion, several similarities and differences were found in this comparative study between POPs in Lake Baikal and Lake Superior. The concentrations of most organochlorine compounds were generally lower in Lake Baikal than in Lake Superior, with ΣDDTs being an exception. ΣDDTs concentrations were higher in Lake Baikal, and much was in the form of 4,4'-DDT, suggesting regional usage of this compound. The two lakes also showed different POP patterns when expressed as log BMF versus log K_{ow}. The lack of a relationship between these two parameters in Lake Superior may be a result of a diffuse and constant regional source of POPs, which has allowed the food web to reach steady state with other compartments in the lake, such as the water. The good relationship between log BMF and log K_{ow} in Lake Baikal is likely due to temporal variability in concentrations of POPs in the surface water generated by a non-steady POPs source. This may arise from stored POPs delivered to the lake during the spring thaw, or atmospheric or riverine sources. The two lakes appeared similar in regard to the relation between trophic position, lipid content, and wet mass POP concentrations in the food web. Higher trophic level organisms have higher wet-mass POP concentrations, mainly because they are fatter. The direct effect of trophic position on wet mass POP concentrations in both food webs was minimal and may have been masked by other processes, such as re-equilibration of the organism with the surrounding water. This does not mean that trophic transfer is not important, surely it is. Assessing the relative importance of POP trophic transfer versus water re-equilibration is a rate question and would require the development and calibration of a bioenergetic model for POPs in these two lakes.

References

Broman D., C. Näf, C. Rolff, Y. Zebuhr, B. Fry and J. Hobbie, 1992, Using ratios of stable nitrogen isotopes to estimate bioaccumulation and flux of polychlorinated dibenzo-p-dioxins (PCDDs) and dibenzofurans (PCDFs) in two food chains from the northern Baltic. Environmental Toxicology and Chemistry, 11, 331- 345.

Eisenreich S.J., 1987, The chemical limnology of nonpolar organic contaminants: polychlorinated biphenyls in Lake Superior. Sources and Fates of Aquatic Pollutants, American Chemical Society. Washington, D.C. pp. 393-467.

Fisk A.T., R.J. Norstrom, C.D. Cymbalisty and D.C.G. Muir, 1998, Dietary accumulation and depuration of hydrophobic organochlorines: Bioaccumulation parameters and their relationship to the octanol/water partition coefficient. Environmental Toxicology and Chemistry, 17, 951-961.

Franz T.P., 1990, PCBs in rural Minnesota precipitation. Masters Thesis. University of Minnesota.

Gerstenberger S.L., M.P. Gallinat and J.A. Dellinger, 1997, Polychlorinated biphenyl congeners and selected organochlorines in Lake Superior fish, USA. Environmental Toxicology and Chemistry, 16, 2222-2228

Glassmeyer S.T., D.S. DeVault, T.R. Myers and R.A. Hites, 1997, Toxaphene in Great Lakes Fish: A temporal, spatial, and trophic study. Environmental Science and Technology, 31, 84-88.

Gobas F.A.P.C., J.R. McCorquodale and G.D. Haffner, 1993, Intestinal absorption and biomagnification of organochlorines. Environmental Toxicology and Chemistry, 12, 567- 576.

Iwata H., S. Tanabe, K. Ueda and R. Tatsukawa, 1995, Persistent organochlorine residues in air, water, sediments, and soils from the Lake Baikal region, Russia. Environmental Science and Technology, 29, 792-801.

Jeremiason J.D., K.C. Hornbuckle and S.J. Eisenreich, 1994, PCBs in Lake Superior, 1978-1992: Decreases in water concentrations reflect loss by volatilization. Environmental Science and Technology, 28, 903-914.

Kidd K.A., D.W. Schindler, D.C.G. Muir, W.L. Lockhart and R.H. Hesslein, 1995, High concentrations of toxaphene in fishes from a subarctic lake. Science, 269, 240-242.

Kiriluk R.M., M.R. Servos, D.M. Whittle, G. Cabana and J.B. Rasmussen,

1995, Using ratios of stable nitrogen and carbon isotopes to characterize the biomagnification of DDE, mirex, and PCB in a Lake Ontario pelagic food web. Canadian Journal of Fisheries and Aquatic Sciences, 52, 2660-2674.

Kucklick J.R., H.R. Harvey, P.H. Ostrom, N.E. Ostrom and J.E. Baker, 1996, Organochlorine dynamics in the pelagic food web of Lake Baikal. Environmental Toxicology and Chemistry, 15, 1388-1400.

Kucklick J.R. and J.E. Baker, 1998, Organochlorines in Lake Superior's Food Web Environmental Science and Technology, 32, 1192-1198.

Kucklick J.R., T.F. Bidleman, L.L. McConnell, M.D. Walla and G.P. Ivanov, 1994, Organochlorines in the water and biota of Lake Baikal, Siberia. Environmental Science and Technology, 28, 31-37.

Leblanc G.A., 1995, Trophic-level differences in the bioconcentration of chemicals: Implications in assessing environmental biomagnification. Environmental Science and Technology, 29, 154-160.

McConnell L.L., J.R. Kucklick, T.F. Bidleman, G.P. Ivanov and S.M. Chernyak, 1996, Air-water gas exchange of organochlorine compounds in Lake Baikal, Russia. Environmental Science and Technology, 30, 2975-2983.

Nakata H., S. Tanabe, R. Tatsukawa, M. Amano, N. Miyazake and E.A. Petrov, 1995, Persistent organochlorine residues and their accumulation kinetics in Baikal Seal (*Phoca sibirica*) from Lake Baikal, Russia. Environmental Science and Technology, 29, 2877-2885.

Newsome W.H. and P. Andrews, 1993, Organochlorine pesticides and polychlorinated biphenyl congeners in commercial fish from the Great Lakes. Journal of AOAC International, 76, 707-710.

Oliver B.G. and A.J. Niimi, 1988, Trophodynamic analysis of polychlorinated biphenyl congeners and other chlorinated hydrocarbons in the Lake Ontario ecosystem. Environmental Science and Technology, 22, 388-397.

Pedhazur E.J., 1982, Multiple Linear Regression in Behavioral Research: Explanation and Prediction. Holt, Rinehart and Winston, New York, pp. 577-635.

Peterson B.J. and B. Fry, 1987, Stable isotopes in ecosystem studies. Annual Review of Ecological Systems, 18, 293-320.

Rasmussen J.B., D.J. Rowan, D.R.S. Lean and J.H. Carey, 1990, Food chain structure in Ontario lakes determines PCB levels in lake trout (*Salvelinus namaycush*) and other pelagic fish. Canadian Journal of Fisheries and Aquatic Sciences, 47, 2030-2038.

Thomann R.V. and J.P. Connolly, 1984, Model of PCB in the Lake Michigan lake trout food chain. Environmental Science and Technology, 18, 65-71.

Lake Baikal
K. Minoura (editor)
2000 Elsevier Science B.V.

Carbon and nitrogen isotope studies of the pelagic ecosystem and environmental fluctuations of Lake Baikal

Ogawa, N. O.[1,4]*, Yoshii, K.[1], Melnik, N. G.[2], Bondarenko, N. A.[2], Timoshkin, O. A.[2], Smirnova-Zalumi, N. S.[2], Smirnov, V. V.[3], and Wada, E.[1]

[1] Center for Ecological Research, Kyoto University, Otsu 520-0105, Japan
[2] Russian Academy of Science, Limnological Institute, Irkutsk 664033, Russia
[3] Russian Academy of Science, Baikal Museum, Listvyanka 666016, Russia
[4] Present address:
Graduate School of Environmental Earth Science, Hokkaido University, Sapporo, 060-0810, Japan
Department of Marine Chemistry and Geochemistry, Woods Hole Oceanographic Institution, Woods Hole, MA 02543 USA

Correspondence:
Nanako Ohkouchi Ogawa
Department of Marine Chemistry and Geochemistry, Woods Hole Oceanographic Institution, Woods Hole, MA 02543 USA
FAX: +1-508-289-2164
E-mail: nanako@ees.hokudai.ac.jp

Abstract

Carbon and nitrogen isotope ratios were measured in pelagic organisms collected in Lake Baikal, in 1992-1994 and in scale specimens of omul fish (*Coregonus autumnalis migratorius*) collected from 1947 to 1995. The nitrogen isotope ratio ($\delta^{15}N$) of the pelagic organisms showed a clear trend toward step-wise enrichment with trophic level, which can be described by the following equation:

$$\delta^{15}N = 3.3\ (\text{trophic level} - 1) + 3.8\ (‰).$$

The carbon isotope data ($\delta^{13}C$) suggested that pelagic phytoplankton are the primary carbon source of the pelagic food web, because the $\delta^{13}C$ values of the consumers were close to those of pelagic phytoplankton. The $\delta^{13}C$ values recorded in omul scales showed a trend to gradually decrease from 1947 to 1995. The trend reflected the decreasing $\delta^{13}C$ in atmospheric CO_2 which was caused by the release of fossil-fuel carbon into the atmosphere. Periodic variations were observed in the $\delta^{13}C$ and $\delta^{15}N$ records of omul scales. They may be related to variations in ecological or physiological factors.

Introduction

An anthropogenic impact has been evident in the 20th century both on a global and a regional scale. Unfortunately, however, reconstruction of these phenomena involved has often been difficult because of lack of direct observational records. Under these circumstances, other approaches are required to more precisely reconstruct the perturbations and to elucidate the mechanisms of the processes in question.

Lake Baikal is a valuable site for monitoring the effect of global change on aquatic ecosystems involving isotopic food web structures because of following conditions:
(1) Anthropogenic perturbation is currently minimal.
(2) The turnover rate of the lake water column is rapid (6.2-11.2 yrs; Peeters *et al.*, 1997).
(3) ΣCO_2 is abundant in the lake water (up to 10 mgC/L).
(4) A thermal-bar system prevents transport of organic matter from coastal to pelagic regions (Shimaraev *et al.*, 1993).
(5) The pelagic food chain is simple (Mazepova, 1998).
In this study we determined the carbon and nitrogen isotope ratios of various pelagic organisms and fish scale specimens to elucidate the food web structure in Lake Baikal and to reconstruct fluctuations in global and lacustrine environments in the recent past. We wish to emphasize that our fish scale specimens can serve as a valuable medium for recording changes in paleo-lacustrine environments.

Materials and Methods

Present Baikal baiota

Phytoplankton, zooplankton, and pelagic fish were collected at several sites in Lake Baikal during the cruises of R/V Obruchev from June to August during 1992 to 1996 (Yoshii *et al.*, 1999, Ohkouchi, 1999). Seal samples collected in the central and south basins in May 1992 were supplied by Dr. N. Miyazaki (The University of Tokyo). All samples were desiccated in a dry oven at 60°C for a week and were ground into powder. The powdered samples were extracted with a mixture of and methanol (2:1; v:v) to remove lipids, and they were stored in glass bottles until the isotope analyses.

Omul scales from 1947-1996

Omul (*Coregonus autumnalis migratorius*), Baikalian pelagic whitefish, are classified as pelagic, benthic-deepwater, and coastal-epipelagic groups based on differences in feeding habits and ecological-morphological characteristics (Smirnov, 1992). In this study only benthic-deepwater omul that belong to the Posolskaya subpopulation were used. Omul live in the subsurface to deep layer of water (water depth of 50-350 m, Fig. 1) and mainly feed on the macrozooplankton Macrohectopus branickii (approximately 50%, Table 1). The life span of omul is around 16 years. Omul fish of various ages had been caught in the mouth of tributaries of south basin almost every year from 1947 to 1995, and scale specimens had been collected from them. The age of the specimens was determined by counting the scale rings (Smirnov and Smirnova-Zalumi, 1993). Scales collected from omul of 12 different ages were supplied for isotope analysis in this study. The fork length (length from fish mouth tip to the caudal depression) ranged from 304 to 450 mm. To eliminate organic substances on the surface of the scales, before analysis the samples were treated with 5% NaOH and 5% H_2O_2 (1:1, v/v) solution, washed with distilled water, and then dried at 60°C.

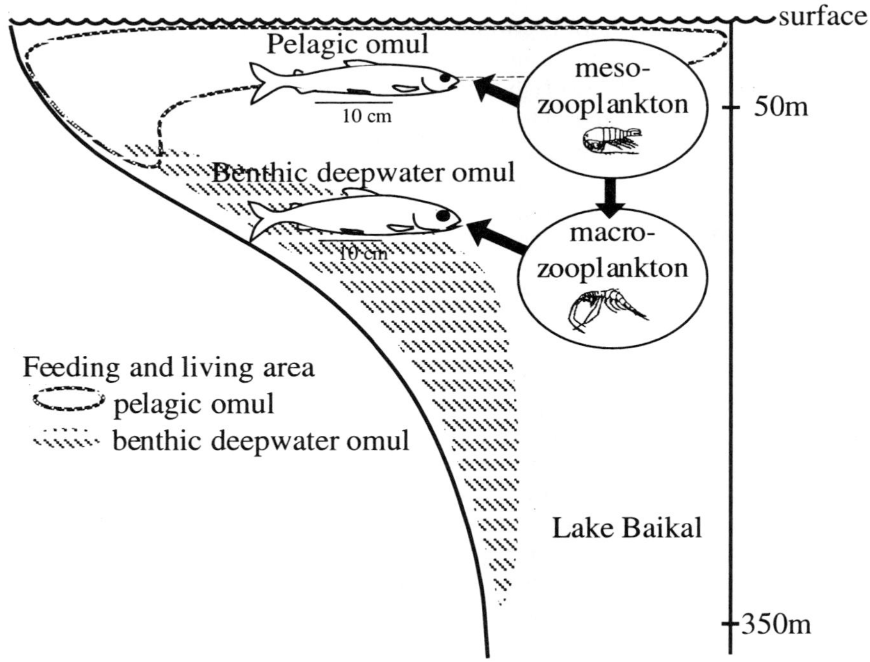

Figure 1. Schematic diagram of the ecological characteristics of the Baikalian omul, Coregonus autumnalis migratorius, Georgi (After Smirnov, 1992). The arrows indicate the predation relationships between omul, mesozooplankton, and macrozooplankton (zooplanktonic gammarids).

Table 1. Characteristics of three morhpo-ecological groups of Omul fish in Lake Baikal*

	Coregonus autumnalis miratorius (Georgi)		
	Pelagic	Coastal epipelagic	Benthic deepwater
Living zones	Epipelagic zone	Coastal pelagic zone	Benthic layer of slope area (50-350m depth)
Spawning place in rivers (distance from the lake)	Upper reach (>100km)	Middle reach (150-700km)	Lower reach (3-30km)
Morphologicy			
Head size	Short	Longer than pelagic	Longest
Eye size	Small	-	Large
Caudal peduncle height	Narrow	-	Wide
Body height	Small	Greater than pelagic	High
Body height/ Caudal-peduncle height	Small	-	Large
Number of gill rayker	Many (44-55)	Fewer than palagic	Smallest (36-44)
Shape of gill rayker	Long, thin, and densely-placed	Spacer	Coase
Diet **			
Mesozooplankton	41	23	10
Macrozooplankton	23	34	52
Pelagic cottidei (fry)	27	26	-
(medium-old aged)	-	-	25
Benthic gammmarids	-	-	12
Others	9	17	-

* Modified from Smirnov (1992).

** Relative amounts (%) obtained from gut content analyses (Smirnov and Ustyuzhanina-Gurova, 1969)

Isotope analyses

Carbon and nitrogen isotope ratios were determined by either standard the combustion method as described by Minagawa *et al.* (1984) or with an elemental analyzer-stable isotope ratio monitoring mass spectrometer (EA-IRMS). In the combustion method, each of the dry samples was sealed into a pre-combusted quartz tube with copper oxide, copper, and a piece of silver foil and evacuated down to 10^{-3} mmHg. The sample tube was combusted at 850°C for 2 hours, and allowed to cool overnight. The N_2 and CO_2 gas produced was cryogenically purified and collected into pyrex tubes with liquid nitrogen and dry-ice ethanol traps. A molecular sieve (Wako Chemicals, 5A 1/16) was used to collect the N_2 gas. In the EA-IRMS method, each of the dry samples was wrapped in a tin capsule and introduced to the elemental analyzer. Fisons EA1108 and NA1500 elemental analyzers and Finnigan MAT Delta-S and 252 mass spectrometers were used. An L-alanine laboratory standard ($\delta^{13}C$ = -21.4‰, $\delta^{15}N$ = -5.3‰) was used as a running standard for the isotope measurements. The analytical errors for both the $\delta^{13}C$ and $\delta^{15}N$ analyses are better than ±0.2‰.

Results and Discussion

Isotope food web structure in pelagic Baikal

Figure 2 shows the results of analysis of $\delta^{15}N$ and $\delta^{13}C$ in various kinds of organisms collected from Lake Baikal. The $\delta^{15}N$ value showed a trend toward step-wise enrichment of $\delta^{15}N$ along the predation process. The foodweb structure and trophic levels of pelagic organisms have been studied by ecological methods (e.g, Kozhov, 1963). Pelagic organisms can be divided into four groups, according to the estimated TL: phytoplankton (TL=1), zooplankton (meso- and macro-, TL=2)), omul (TL=3), and seal (TL=4). Yoshii *et al.* (1999) estimated the $\delta^{15}N$ enrichment factor during a single feeding process from a simple linear function between the estimated TL and mean $\delta^{15}N$ values of these 4 groups and suggested that step-wise enrichment of $\delta^{15}N$ in Lake Baikal can be expressed by the following equation;

$$\delta^{15}N\ (‰) = 3.3\ (TL - 1) + 3.8.$$

This enrichment pattern conforms well with many data obtained for trophic structures in lakes and oceans (e.g., Minagawa and Wada, 1984; Yoshioka *et al.*, 1994). The $\delta^{15}N$ variation that exists within a group is caused by variations in diet. For example, large Cyclops and Macrohectopus individual are sometimes carnivorous and they have relatively high $\delta^{15}N$ values. Several factors may contribute to the simple and clear increase in $\delta^{15}N$ throughout the pelagic food web in Lake Baikal: 1) pelagic phytoplankton being the major food base, with little contribution by other primary producers, such as benthic plants, 2) the small seasonal variation in the $\delta^{15}N$ of phytoplankton during our sampling periods (Yoshii *et al.*, 1999), 3) the simple species composition (Bondarenko *et al.*, 1996), and 4) the small geographic variation in phytoplankton $\delta^{15}N$ (Ogawa *et al.*, unpublished data).

The ^{13}C enrichment along the food web was not as clear as it was for of

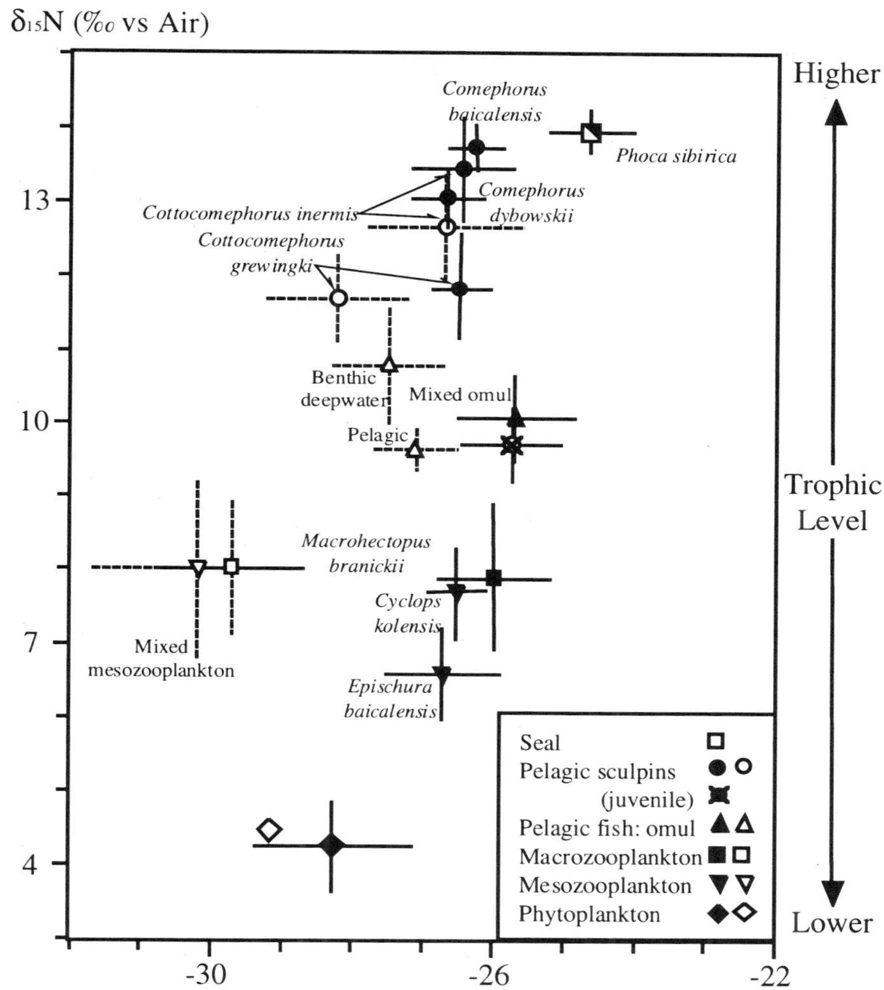

Figure 2. Carbon vs. nitrogen isotope diagram ($\delta^{13}C$-$\delta^{15}N$ diagram) of pelagic organisms in Lake Baikal (Closed symbols: Yoshii *et al.*, 1999; Open symbols (include lipid fraction): Ohkouchi, 1999). Each symbol and bar denotes a mean value and standard deviation, respectively.

^{15}N (Fig. 2), for instance, consistent ^{13}C enrichment was not observed for the four pelagic sculpin species. However, the results suggested that the carbon isotope ratio of pelagic consumers reflects and is affected by that of the phytoplankton in the lake.

Carbon isotope record in omul fish scales over the past 50 years

Figure 3 shows the $\delta^{13}C$ record of the scales of benthic deepwater omul fish ($\delta^{13}C_{scale}$) collected in the south basin of Lake Baikal from 1947 to 1995. Although some variability exists even within a single year (up to 1.4 ‰), the $\delta^{13}C_{scale}$ tended to gradual decrease from 1947 to 1995. In 1947 the $\delta^{13}C_{scale}$ value was -21.7±0.4 ‰ (n=3), whereas in 1995 it was -23.3‰ (n=1). Thus, the overall amplitude of this trend reached approximately 1.6‰. If a simple linear function is applied to all $\delta^{13}C_{scale}$ results, calculation of the rate of decrease of $\delta^{13}C_{scale}$ yields 0.021‰/yr. Interestingly, the tendency for $\delta^{13}C_{scale}$ to decrease closely conforms to the trend for the $\delta^{13}C$ of atmospheric CO_2 ($\delta^{13}C_{CO2}$), which is the ultimate source of the carbon in

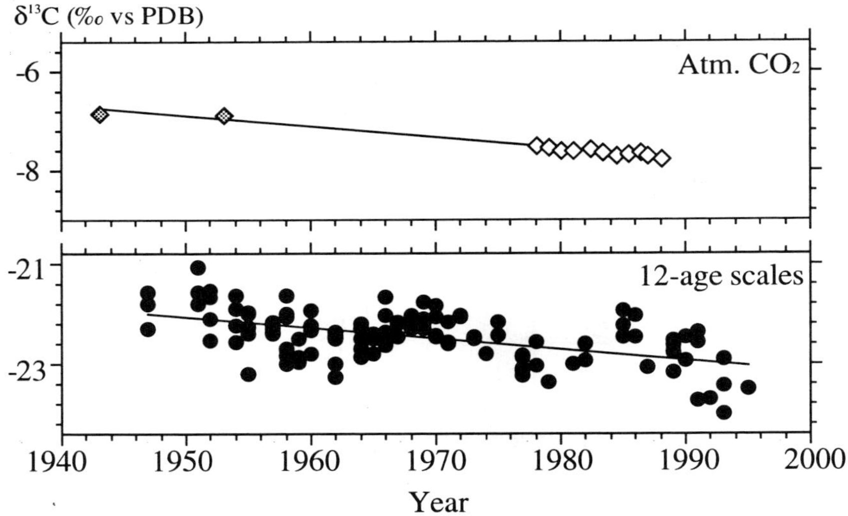

Figure 3. Carbon isotope record of atmospheric CO_2 (top) and scales of omul of 12 ages (bottom; Ogawa *et al.*, in prep.). The $\delta^{13}C$ values for atmospheric CO_2 are seasonally adjusted mean values of directly measured $\delta^{13}C$ at Mauna Loa and the South Pole (open squares: Keeling *et al.*, 1989) and in an Antarctic ice core (shaded squares: Friedli *et al.*, 1987). Omul fish belonging to the benthic-deepwater group collected from the south basin of Lake Baikal were used.

fish scales via phytoplankton. Specifically, the $\delta^{13}C_{CO2}$ values measured at Mauna Loa and in an ice core from Antarctica decreased approximately 1 ‰ during the past 40 years (approximately 0.022 ‰/yr), a change caused by the additional entry of fossil fuel carbon into the atmosphere (Friedli *et al.*, 1987; Keeling *et al.*, 1989; Fig. 3). Our findings suggest that the recent shift in atmospheric $\delta^{13}C_{CO2}$ not only affects autotrophic organisms, but deeply penetrates into the entire ecosystem of the pristine Lake Baikal.

In addition to the trendency to decrease, the $\delta^{13}C_{scale}$ values showed cyclic variation, with relatively high values around 1950, 1970, and 1985, and low values around 1960 and 1980, and thus, the period of the cycle was estimated to be slightly less than 20 years. Similar cyclic variation was also observed in the $\delta^{15}N$ record ($\delta^{15}N_{scale}$), with variations from 10.5% to 12.5‰ (Fig. 4). This cyclicity may be explained by ecological effects, such as changes in the phytoplankton biomass or the foodweb structure in the lake. One of the factors that may be related to the change in $\delta^{13}C_{scale}$ values is isotopic fractionation by phytoplankton. An increase in phytoplankton biomass has been observed every 3-4 years, with periodic (11-22 years) cyclicity (Rychkov *et al.*, 1989, Bondarenko, 1999). Theoretically, the $\delta^{13}C$ of phytoplankton is controlled by both the $\delta^{13}C$ value of ΣCO_2 (related to atmospheric CO_2) and kinetic fractionation during photosynthetic carbon assimilation (e.g., Farquhar *et al.*, 1982). Factors with the potential to change this fractionation are temperature, *p*H, growth rate, and cell size (e.g., Takahashi *et al.*, 1991; Hinga *et al.*, 1994). Thus, oscillations in phytoplankton biomass could affect changes in their $\delta^{13}C$ values through changes in their growth rate and cell size. On the other hand, a shift in the omul diet is another factor that may have contributed to the variation in

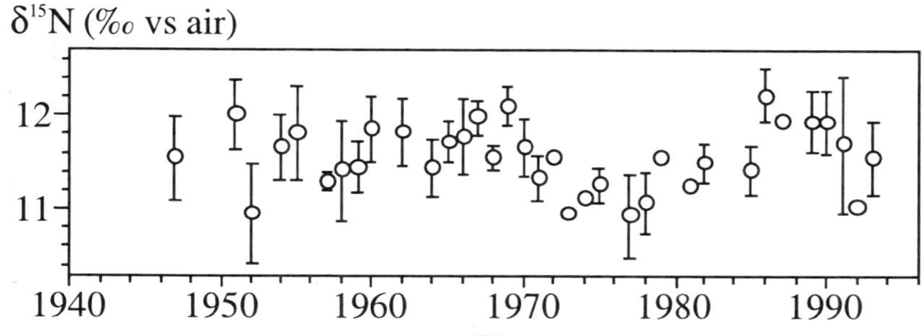

Figure 4. Nitrogen isotopic record of scales of omul fish of 12 ages belonging to the benthic-deepwater group and collected from the south basin of Lake Baikal. Symbols and bars indicate mean values and standard deviations, respectively.

$\delta^{15}N_{scale}$. As described above, the $\delta^{15}N$ value of omul fish is governed by the $\delta^{15}N$ value of its diet. The $\delta^{15}N_{scale}$ values are higher during the period when benthic deepwater omul more freqently depend for food on fish larvae, which have highest $\delta^{15}N$ value in the omul diet (Table 1, Fig. 2). Further observations are required to fully understand the relationship between the oscillations of $\delta^{13}C_{scale}$ and $\delta^{15}N_{scale}$ and ecological factors, especially in regard to isotopic distributions related to changes in the plankton biomass in the lake.

Conclusion

Based on the above arguments, we conclude that the whole pelagic foodweb structure is clearly described by the simple $\delta^{15}N$-TL equation and $\delta^{13}C$ values of the phytoplankton. This suggests that the Baikal ecosystem is isotopically ordered, and thus stable isotope monitoring is a potential method to access foodweb fluctuations in the lake. We also conclude that the decrease in the carbon isotope ratio of omul scales principally reflects changes in the $\delta^{13}C$ of atmospheric CO_2, the ultimate source of the carbon in the omul body. We do not have a convincing explanation for the periodic variations in $\delta^{13}C_{scale}$ and $\delta^{15}N_{scale}$, but they may be linked to some ecological effect, such as changes in the phytoplankton biomass related to isotope fractionation of phytoplankton and variations in the omul diet. Our omul scale $\delta^{13}C$ data also suggest that the recent shift in atmospheric $\delta^{13}C$ not only affects autotrophic organisms, but deeply penetrates throughout the entire ecosystem of Lake Baikal.

Acknowledgements

The authors very much thank Dr. Michael A. Grachev, Director of the Limnological Institute of the Russian Academy of Science for his invitation to Lake Baikal. The authors also thank Dr. N. Miyazaki for all his equipment for seal sampling and Dr. N. Ohkouchi for helpful discussion. This study was partly supported by Research Fellowships of the Japan Society for the Promotion of Science for Young Scientists to N.O. Ogawa.

References

Bondarenko N.A., N.E. Guselnikova, N.F. Logacheva and G.V. Pomazkina, 1996, Spatial distribution of phytoplankton in Lake Baikal, Spring 1991. Freshwater Biology, 35, 517-523.

Bondarenko N.A., 1999, Floral shift in the phytoplankton of Lake Baikal, Siberia: Recent dominance of Nitzschia acicularis. Plankton Biology and Ecology, 46, 18-23.

Farquher G.D., M.H. O□Leary and J.A. Berry, 1982, On the relationship between carbon isotope discrimination and the intercellular carbon dioxide concentration in leaves. Aust. J.

Friedli H., H. L LUötscher, H. Oeschger, U. Siegenthaler and B. Stauffer, 1987, Ice core record of the $^{13}C/^{12}C$ ratio of atmospheric CO_2 in the past two centuries. Nature, 324, 237-238.

Hinga K.R., M.A. Arthur, M.E.Q. and D. Whitaker Pilson, 1994. Carbon isotope fractionation by marine planton in culture: The effect of CO_2 concentration, pH, Temperature and scecies. Global Biogeochem. Cycles 8, 91-102.

Keeling C.D., R.B. Bacastow, A.F. Carter, S.C. Piper and T.P. Whorf, 1989, A three-dimensional model of atmospheric CO_2 transport based on observed winds: 1. Analysis of observational data, In: Aspects of climate variability in the Pacific and the Western Americas, eds. D.H. Peterson, American Geophysical Union, Washington, DC.

Kozhov M., 1963, Lake Baikal and its life. Dr. W. Junk. Publishers, Netherlands.

Mazepova G.F., 1998, The role of copepods in the Baikal ecosystems. Journal of Marine Systems, 15, 113-120.

Minagawa M. and E. Wada, 1984, Stepwise enrichment of ^{15}N along food chains: Further evidence and the relation between $\delta^{15}N$ and animal age. Geochimica et Cosmochimica Acta, 48, 1135-1140.

Minagawa M., D.A. Winter and I.R. Kaplan, 1984, Comparison of Kjeldahl and combustion methods for measurement of nitrogen isotope ratios in organic matter. Analytical Chemistry, 56, 1859-1861.

Ogawa N.O., N.S. Smirnova-Zalumi, V.V. Smirnov, N.G. Melnik, M.N. Shimaraev, T. Yoshioka and E. Wada Carbon isotopic composition of atmospheric CO_2 recorded in scale specimens of Omul fish in Lake Baikal. (in preparation).

Ohkouchi N., 1999, Fluctuations of lacustrine environments during the last several decades revealed from stable isotope ratios of fish specimens, Doctoral Thesis. Kyoto University, Kyoto.

Peeters F., R. Kipfer, R. Hohmann, M. Hofer, D.M. Imboden, G.G. Kodenev and T. Khozder, 1997, Modeling transport rates in Lake Baikal: gas exchange and deep water renewal. Environmental Science and Technology, 31, 2973-2982.

Rychkov I.L., E.N. Kuzevanova and G.I. Pomazkoya, 1989, Long-term natural changes in plankton of the southern Baikal. Sym. Biol. Hung, 38, 361-366.

Shimaraev M.N., N.G. Granin and A.A. Zhdanov, 1993, Deep ventilation of Lake Baikal waters due to spring thermal bars. Limnology and

Oceanography, 38, 325-333.
Smirnov V.V., 1992, Intraspecific structure of Baikal Omul, *Coregonus autumnalis migratorius* (Georgi). Pol. Arch. Hydrobiol., 39, 325-333.
Smirnov V.V. and N.S. Smirnova-Zalumi, 1993, Divelopment of annual growth zone on the scale of Baikal Omul *Coregonus autumnalis migratorius*. Voprosy Ikhtiologii, 33, 121-129.
Takahashi K., E. Wada and M. Sakamoto, 1991, Relationship between carbon isotope discrimination and the specific growth rate of green alga *Chlamydomonas reinhardtii*. Jpn. J. Limnol. 52, 105-112.
Yoshii K., N.G. Melnik, O.A. Timoshkin, N.A. Bondarenko, P.N., Yoshioka T. Anoshko and E. Wada, 1999, Stable isotope analyses of the pelagic food web in Lake Baikal. Limnology and Oceanography, (in press).
Wada T., E. Yoshioka and H. Hayashi, 1994, A stable isotope study on seasonal food web dynamics in a eutrophic lake. Ecology, 75, 835-846.

Lake Baikal
K. Minoura (editor)
2000 Elsevier Science B.V.

Some speculations on the possibility of changes in deep-water renewal in Lake Baikal and their consequences

Kipfer, R.* and Peeters, F.
Environmental Physics, Swiss Federal Institute of Technology (ETH) and Swiss Federal Institute of Environmental Science and Technology (EAWAG), 8600 Dübendorf, Switzerland
Tel: (41) 1 823 55 30/31, fax: (41) 1 823 52 10, e-mail: kipfer@eawag.ch, peeters@eawag.ch.
*Correspondence to Rolf Kipfer

Abstract

Most of the processes that have been identified as responsible for deep-water renewal in Lake Baikal require that the surface water of the lake have higher salinity than the water at a depth of 250 to 400 m. Moderate changes in the temperature of the surface water do not seem to significantly affect the occurrence of vertical advective transport into the deep water. Thus, deep-water renewal in Lake Baikal is probably more strongly affected by changes in salt levels, particle load, and discharge of inflows than by moderate temperature variations due to climate change. Increasing salt levels in the deep water might lead to an increase in the stability of the water column and thus to lower rates of vertical turbulent diffusive transport of, for example, oxygen.

Changes in the rate of deep-water formation would not only have an impact on the ecological state of Lake Baikal but would also affect the biota and its evolution. Deep-water formation may be the 'missing link' connecting internal biological processes in Lake Baikal, especially speciation, with external forcing, linked, for instance, to enhanced salt input due to tectonic activity and climatic change.

Despite the great depth of Lake Baikal (maximum depth. 1632 m), oxygen is present throughout the entire water column, indicating rapid deep-water exchange. Tracer measurements demonstrate that up to 10% of the deep water is renewed annually, and several processes contribute to the deep-water exchange (Fig.1, Weiss *et al.*, 1991; Shimaraev *et al.*, 1993; Imboden, 1994; Kipfer *et al.*, 1996; Hohmann *et al.*, 1997).

There are only two brief periods, in the course of the year early spring and late autumn, in which significant deep-water exchange can occur. Only

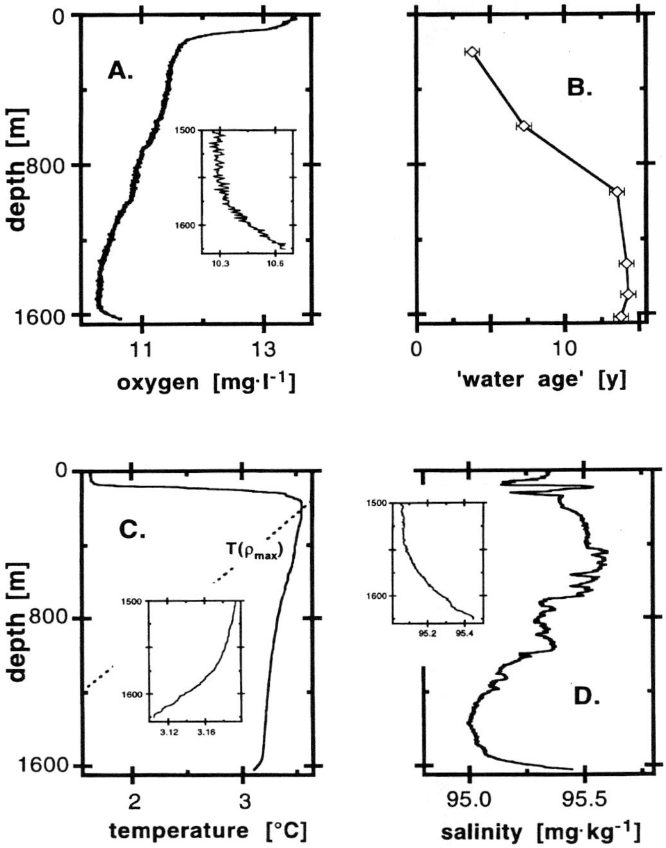

Figure 1. Tracer profiles indicating deep-water exchange in the central basin of Lake Baikal.

A: Lake Baikal exhibits high oxygen levels (above 80% surface saturation) throughout its entire water column. Small inset: close to the lake bottom the concentration of dissolved oxygen ($[O_2]$) increases with increasing depth. B: 'Water age' (WA) - a measure of the time elapsed since the water was last in contact with the atmosphere - increases with increasing depth. Close to the lake bottom, however, WA decreases with increasing depth. High $[O_2]$ and low WA in the bottom water indicate that the lower, deepest water is being rapidly replaced by surface water (which is saturated with O_2, WA ≈ 0) via advective transport along the lake boundaries. C/D: Each of the processes that renews deep water leaves characteristic traces in the distribution of temperature and ions in the water column. The young and oxygen-rich bottom water of the central basin of Lake Baikal is colder (C) and contains more particles and ions (D) than the water above it. Water with similar properties can be found in an underwater canyon where the Selenga leaves its delta. In spring, the water in the canyon is colder and richer in particles and ions than the open lake water at equivalent depth. The water in the canyon, therefore, has a greater density than the open lake water, and it sinks along the canyon's bottom boundary to greater depths, where it spreads out to the deepest regions of the central basin.

during these seasons are the physical properties of the surface water that control water density (temperature and salinity) similar to the conditions commonly found in the deep water. Moreover, it is only during these periods that the density of surface water is sufficiently large to generate deep-water convection. During the course of five joint Russian-Swiss expeditions several processes were identified as being responsible for deep-water renewal (Tab.1), and these mixing processes lead to significant vertical transport, which may be sufficient to explain the fast deep-water exchange indicated by the tracer mea-surements (Weiss *et al.*, 1991; Peeters *et al.*, 1997; Hohmann *et al.*, 1998).

The temperature of the deep water in Lake Baikal is always close to the temperature at which the surface water reaches its maxi-mum density (Fig.

Table 1. Prominent deep-water formation processes in Lake Baikal and their impact on the physical state of the deep water below 500m.

Process leading to Deep-water formation	Site	Basin affected	Impact on temperature	Impact on salinity
River inlets	Selenga Delta	CB,(SB)	–	+
Thermal bar near Boldakova (Shimaraev et al., 1993)	Entire shore region	CB	+, –	+
Flow over sills, exchange between basins(Peeters et al., 1996; Hohmann et al., 1997)	Academician Ridge	NB	–	0
Hydrothermal activity (Kipfer et al., 1996)	Frolikha Bay	NB	+	+
Vertical turbulent diffusion	everywhere	all basins	+	?
Total			'zero'	+

+ / – : process tends to increase or to decrease the parameter
Total : expected integrated effect on Lake Baikal
NB, CB, SB : northern, central and southern basin

1 (C)), and thus salinity should significantly contribute to the density stratification. The density structure in the region close to the so-called mesothermal maximum (Fig. 1 (D), MTM, ~ 200 m deep), which separates the surface water from the deep-water, in particular, is mainly controlled by salinity. Even small salinity differences in this zone regulate the transport of surface water through the MTM towards greater depths during the inverse stratification of the water column that occurs in early spring (Peeters *et al.*, 1996; Hohmann *et al.*, 1997). During this period spatial salinity gradients are the driving force of the verti-cal exchange in the two most active regions of largescale deep-water formation; the Selenga Delta and the Academician Ridge (Fig. 2, Tab. 1).

In the central basin the deepest water has a temperature of about 3.1°C. Buoyancy-driven transport of surface water at 3.1°C down to the greatest depth requires that the salinity of the surface water be about 2 mg kg^{-1} greater than the salinity of the ambient water at the MTM (200 m depth). Surface water with sufficiently increased salinity and low temperature is provided by the Selenga River discharging into the southern and central basin. In the northern basin deep-water temperatures are above 3.3°C. To generate buoyancy-driven transport of 3.3°C water from the surface down to the greatest depth, the salinity of the surface water need only be 0.8 mg kg^{-1} greater than at the MTM. Surface water in the central basin has slightly higher salinity than in the northern basin (Fig. 2). At Academician Ridge horizontal transport of central basin water to the north leads to vertical salinity gradients in the northern basin sufficient to drive deep-water convection. Because the difference in salinity between central and northern basin water in the top 300 m is at most 0.8 mg kg^{-1}, buoyancy driven convection in the northern basin occurs only for water temperatures of 3.3°C or above, which explains why the deep water temperature in the northern basin is 3.3°C or higher.

The expected overall effects of the different deep-water formation processes on the physical state of the deep water are shown in the last two columns of Table 1. Deep-water formation associated with mixing processes due to hydrothermal activity and thermal bar fronts increases the temperature of the deep-water. However, deep-water renewal related to river inlets and intrabasin mixing causes cold surface water to sink to greater depths. Russian measurements dating back to the beginning of this century suggest that the temperature in the deep water is in a steady state (Shimaraev *et al.*, 1994). Thus, the processes responsible for deep-water renewal must compensate for each other with respect to heat transport.

The situation in regard to salinity is different. All of the pro-cesses that lead to deep-water renewal by large-scale convection seem to increase the salinity of the deep water in Lake Baikal, meaning that salinity should

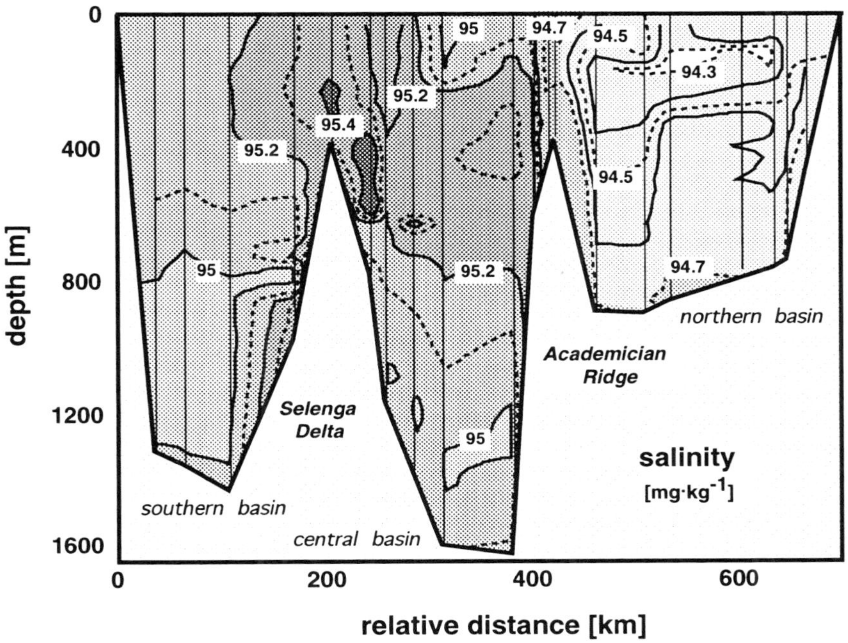

Figure 2. Spatial salinity gradients in Lake Baikal

The density structure (ρ) of Lake Baikal is not determined by temperature (T) alone but by the ion distribution as well (salinity S). Both vertical and horizontal ρ-, T-, and S-gradients develop, and these gradients drive the formation of deep water and lead to the development of a thermo-saline circulation.

Selenga, the major tributary feeding Lake Baikal, drains into the central basin and exhibits an approximately 50% higher degree of mineralisation than the lake. In contrast, Upper Angara, the second largest tributary, adds freshwater having a low mineralisation level into the northern basin. As a result, the water of the central basin is the most strongly mineralised in Lake Baikal.

The changes in density (Δρ) resulting from the slight variations in temperature (ΔT <0.3°C) in the deep water are extremely small because the gradient $\partial\rho/\partial T$ disappears for temperature close to 4°C. If ions are added to such a water body, however, the density gradients induced by gradients in the ion concentrations might be significant. The calculation of water density in Lake Baikal, is also complicated by the fact that T(ρmax) is a function of hydrostatic pressure and decreases rapidly with increasing depth (approx. 2°C per 1000 m, Fig. 1).

Dissolved ions play a central role in renewing the deep water of Lake Baikal. While the variation in ion concentrations within each basin is extraordinarily low (e.g. 0.5 mg.kg^{-1} in the northern basin, Fig. 1) the salinity gradients between basins and between the lake and river water play a key role in transporting cold surface water down to a depth of about 500 m. At this depth the low temperature of the convecting water drives it down to its greatest depth (Weiss et al., 1991; Kipfer et al., 1996; Hohmann et al., 1997; Peeters et al., 1997).

steadily increase with time. Because the salt levels in Baikalian waters are extremely low, the imbalance in the recent salt budget may be only a transient phenomenon.

Any persistent change in the salt budget of Lake Baikal will directly influence its mixing processes and its deep-water formation, and therefore ultimately affect the water quality and the ecosystem of the lake (Fig. 2, Hohmann *et al.*, 1997).

Despite the generally pristine state of Lake Baikal, signs of subtle changes have recently been detected. For several years river-borne algae adapted to nutrient-rich environments have been migrating from delta regions to the open water. This invasion seems to be indicative of the slowly increasing availability of nutrients. The enhanced nutrient level and the suspected salt non-equilibrium of the lake water could be interpreted as an expression of the rising pres-sure of civilisation around Lake Baikal. Intensification of urbanisation and agriculture may increase soil erosion in the catchment and lead to an increase in nutrient and salt input into the lake through the river discharge. Since the water residence time of Lake Baikal, i.e., the time required for all inflows to refill the lake, is more than 350 years, slight disturbances in water quality are difficult to detect, but remain in the system for centuries. Therefore, the anticipated recent changes in the salt budget of Lake Baikal must be regarded as irrever-sible, and they will affect the ecology of Lake Baikal over several hundred years.

In contrast to salt balance, which is sensitive to small changes in the salt concentration in the inflows, the thermal conditions of the deep water of Lake Baikal can be expected to be fairly insensitive to changes in external heat fluxes, e.g., to variations in climatic conditions. As long as the surface water freezes and thaws regularly, deep-water convection is likely to occur just as it does now. The density anomaly of freshwater, and especially the pressure dependence of the temperature of maximum density ($T[\rho_{max}]$, Fig. 1. (C)), in combination with low water temperatures in the deep-water region lead to a barrier of potential energy at the depth of the MTM, between 200 m and 450 m, which prohibits convection of cold surface water to the deep-water region of Lake Baikal (Peeters *et al.*, 1996). It should be noted that this barrier to convection decreases if the temperature in the deep-water region rises. The potential energy barrier at intermediate depth can only be overcome by water that has greater density, due to increased salinity or an increased load of suspended particles, than the ambient water at the depth of the barrier, or if strong winds cause thermobaric instabilities. The density excess required to allow buoyancy-driven deep-water convection of water at 3.1°C and 3.3°C is about $1.8 \cdot 10^{-3}$ kg m^{-3} and $5.8 \cdot 10^{-4}$ kg m^{-3}, and corresponds, for instance, to an increased salinity of 2 mg kg^{-1} and 0.8 mg kg^{-1}, respectively. Salinity, suspended particle load, and the occurrence of strong winds are the most important parameters responsible for convection, as long as the surface water gets colder than the lowest temperature in the deep water (about 3.1°C), and low enough surface water temperatures are guaranteed if freezing and thawing occurs.

The high oxygen levels in the deep water of Lake Baikal result from the interplay between the fast vertical transport of oxygen linked to the rapid deep-water renewal (about 10% per year) and very low oxygen consumption, on the order 0.10 - 0.15 $mgO_2\ l^{-1}\ yr^{-1}$, in the deep-water (e.g. Weiss *et al.*, 1991; Peeters *et al.*, 1997). Because deep-water renewal is not sensitive to moderate changes in surface water temperature, we speculate that the vertical transport of oxygen in Lake Baikal is not severely affected by moderate climatic changes in temperature. However, changes in the hydrological regime, and especially changes in the salinity and the load of suspended particles in the inflows, might first lead to increased vertical convective transport, and then to a reduction in vertical turbulent diffusion due to increased stability in the deep water, similar to what occurs in Lake Lugano (Wüest *et al.*, 1992). Under such circumstances, one would expect transport of oxygen by turbulent diffusion to be reduced, which would lead to lower dissolved oxygen concentrations in the deep-water regions of Lake Baikal as a result of oxygen consumption by biodegradation.

If deep-water renewal were to stop completely, it would require at least 100 years to reach anoxic conditions in the deep water of Lake Baikal, assuming that oxygen depletion continues to occur at the current rate, which is on the order 0.10 - 0.15 mg $l^{-1}\ yr^{-1}$. Such a situation is conceivable only under severe climatic conditions leading, for example to an ice shield covering the entire lake, or to a drastic change in water discharge or salt and particle concentrations in the river inflows.

Data on the abundance of species in Lake Baikal and on their genetic evolution imply that a sudden and distinct mass extinction occurred about 3 million years ago and triggered rapid species radiation (Timosh-kin, personal communication).

Reduced deep-water exchange causing a decline in oxygen concentrations may have a severe impact on the biota of Lake Baikal. If deep-water exchange is reduced drastically during severe climatic conditions, anoxic conditions may develop and possibly lead to mass extinction on a time scale on the order of 100 years or more.

References

Hohmann R., R. Kipfer, F. Peeters, G. Piepke, D. M. Imboden and M. N. Shimaraev, 1997, Processes of deep water renewal in Lake Baikal. Limnol.

Oceanogr., 42, 841-855.
Hohmann R., M. Hofer, R. Kipfer, F. Peeters and D. M. Imboden, 1998, Distribution of helium and tritium in Lake Baikal. J. Geophys. Res. 103, 12823-12838.
Imboden D. M., 1994, Deep water formation: The physical mystery of Lake Baikal. Baikal as a natural laboratory for global change, 1, 21-22 (abstr.).
Kipfer R., W. Aeschbach-Hertig, M. Hofer, R. Hohmann, D. M. Imboden, H. Baur, V. Golubev and J. Klerkx, 1996, Bottomwater formation due to hydrothermal activity in Frolikha Bay, Lake Baikal, eastern Siberia. Geochim. Cosmochemi. Acta 60, 961-971.
Peeters F., G. Piepke, R. Kipfer, R. Hohmann and D. M. Imboden, 1996, Description of stability and neutrally buoyant transport in freshwater lakes. Limnol. Oceanogr., 41, 1711-1724.
Peeters F., G. Piepke, R. Kipfer, R. Hohmann and D. M. Imboden, 1997, Modelling transport rates in Lake Baikal: gas exchange and deep water renewal. Environ. Sci. Technol., 31, 2973-2982.
Shimaraev M. N., N. G. Granin and A. A. Zhadanov, 1993, Deep ventilation of Lake Baikal due to spring thermal bars. Limnol. Oceanogr. 38, 1068-1072.
Shimaraev M. N., V. I. Verbolov, N. Granin and P. P. Sherstayankin, 1994, Physical limnology of Lake Baikal: a review. Irkutsk, Okayama, 80 pp.
Weiss R. F., E. C. Carmack and V. M. Koropalov, 1991, Deep-water renewal and biological production in Lake Baikal. Nature 349, 665-669.
Wüest A., A. Aeschbach-Hertig, H. Baur, M. Hofer, R. Kipfer and M. Schurter, 1992, Density structure and tritium-helium age of deep hypolimnetic water in the northern basin of Lake Lugano. Aquat. Sci. 54, 205-218.

Lake Baikal
K. Minoura (editor)
2000 Elsevier Science B.V.

Contamination of the ecosystems of Lake Baikal by persistent organochlorines

Nakata, H.[1,#], Tanabe, S.[1]*, Iwata, H.[2], Amano, M.[3], Miyazaki, N.[3], Petrov, E.A.[4], and Tatsukawa, R.[5]
[1] Center for Marine Environmental Studies, Ehime University, Tarumi, 3-5-7, Matsuyama 790-8566, JAPAN (Fax: +81-89-946-9904, E-mail: shinsuke@agr.ehime-u.ac.jp)
[2] Department of Environmental Veterinary Sciences, Graduate School of Veterinary Medicine, Hokkaido University, Kita-Ku Kita 18 Nishi 9, Sapporo 060-0818, JAPAN (Fax: +81-11-717-7569, E-mail: iwatah@vetmed.hokudai.ac.jp)
[3] Otsuchi Marine Research Center, Ocean Research Institute, The University of Tokyo, Akahama, Otsuchi-cho, Iwate 028-1102, JAPAN (Fax: +81-193-42-3715, E-mail: miyazaki@wakame.ori.u-tokyo.ac.jp)
[4] Limnological Institute of the Siberian Branch of the Academy of Science of Russia, 664033 Irkutsk, Uran-Batorskaya 3, RUSSIA (Fax: +7-3952-290-551)
[5] Kochi University, Akebono-cho 2-5-1, Kochi 780-8520, JAPAN (Tel: +81-888-44-0111)

[#] Present address: Department of Environmental Sciences, Faculty of Science, Kumamoto University, Kurokami 2-39-1, Kumamoto 860-8555, JAPAN (Fax: +81-96-342-3380, E-mail: nakata@aster.sci.kumamoto-u.ac.jp)

* To whom correspondence should be addressed.

Abstract

Contamination by persistent organochlorines (OCs), including PCBs, DDTs, HCHs, and chlordanes, was assessed in air, water, sediment, soil, fish, and seal samples collected from Lake Baikal. DDT concentrations in air and water were one order of magnitude higher than in the Arctic. HCH and PCB concentrations in water samples were found to be higher in the lower reaches of the Selenga River and the southern basin of the lake, suggesting the presence of local sources of these contaminants. Higher concentrations of PCBs and DDTs were found in the blubber of Baikal seals. A positive age-accumulation of PCBs, DDTs, and CHLs was found in males, whereas a steady state was observed in females, implying the transfer of these chemicals from mothers to their pups. Isomer-specific analysis of PCBs suggested that Baikal seals have higher or comparable capacity to metabolize toxic contaminants than marine mammals, but a clearly lower capacity than terrestrial mammals, which seems to be a causative factor in the higher accumulation of OCs in this species.

Introduction

During the last decade, many investigations have been conducted to elucidate the global distribution of persistent organochlorines, such as polychlorinated biphenyls (PCBs) and DDTs, and the reports have documented that these contaminants are widely transportable through atmosphere and have eventually contaminated the entire world, including the polar regions (Iwata *et al.*, 1993; Simonich and Hites, 1995, AMAP, 1998). In contrast to several items of evidence indicating increased OC contamination in the Arctic, the sources and pathways of contaimination remain obscure. This is partly due to the existence of unsurveyed areas of OC pollution. While reports on the recent status of OC contamination have been increasing in the area of tropical Asia, Oceania (Iwata *et al.*, 1994a), and Europe (Jones *et al.*, 1995; Lead *et al.*, 1996), only a few reports are available in northeastern areas, especially Siberia.

In 1987 and 1988, an acute disease struck the Baikal seal, and 8-10 thousands of animals died (Grachev *et al.*, 1989). Although the immediate cause of the outbreaks was attributed to morbillivirus infection, it is suspected that some stressor, such as chronic exposure to persistent organochlorine contaminants, may have played an important role in triggering severe manifestations in epizootics by causing immunosuppression in mammals (Kannan *et al.*, 1993; Aguilar and Borrell, 1994).

Against this background, air, water, sediment, soil, fish, and seal samples were collected from Lake Baikal in 1992, and their levels of contamination by DDTs, PCBs, chlordane compounds (CHLs), and HCHs were determined. The data were then compared with data for various other regions previously reported in order to better understand the status of OC contamination in the Lake Baikal ecosystems. In addition, differences in levels in relation to sex and age, and the specific features of OC metabolism by the Baikal seal were assessed in comparison with aquatic and terrestrial mammals.

OC contamination in air, water, sediment, and soil samples

The OC concentrations in air, water, sediment, and soil samples collected from the Lake Baikal region are summarized in Table 1. The Aerial concentrations of CHLs and PCBs were comparable to those in the Arctic (CHLs: a few pg/m^3, PCBs: < 100 pg/m^3) as a 'remote site' from the source, but the DDT concentrations in air appeared to be 1 order of magnitude higher than the Arctic concentrations, which are in the several pg/m^3 range (AMAP, 1998).

Table 1. Range of OC concentrations in air, water, sediment and soil samples collected from the Lake Baikal region [a)]

OCs	air (pg/m3)	water (pg/L)	sediment (ng/g dry wt.)	soil (ng/g dry wt.)
DDTs	10 — 29	< 2.0 — 15	0.014 — 2.7	0.34 — 28
PCBs	8.7 — 23	18 — 590	0.08 — 6.1	1.4 — 92
CHLs	1.3 — 18	< 2.0 — 41	< 0.001 — 0.003	< 0.001 — 0.009
HCHs	230 — 960	56 — 960	0.019 — 0.12	0.043 — 16

a) Cited from Iwata *et al.* (1995)
DDTs: *p,p'*-DDE + *p,p'*-DDD + *p,p'*-DDT
CHLs: trans-chlordane + cis-chlordane + trans-nonachlor
HCHs: $\alpha + \beta + \gamma$ isomers

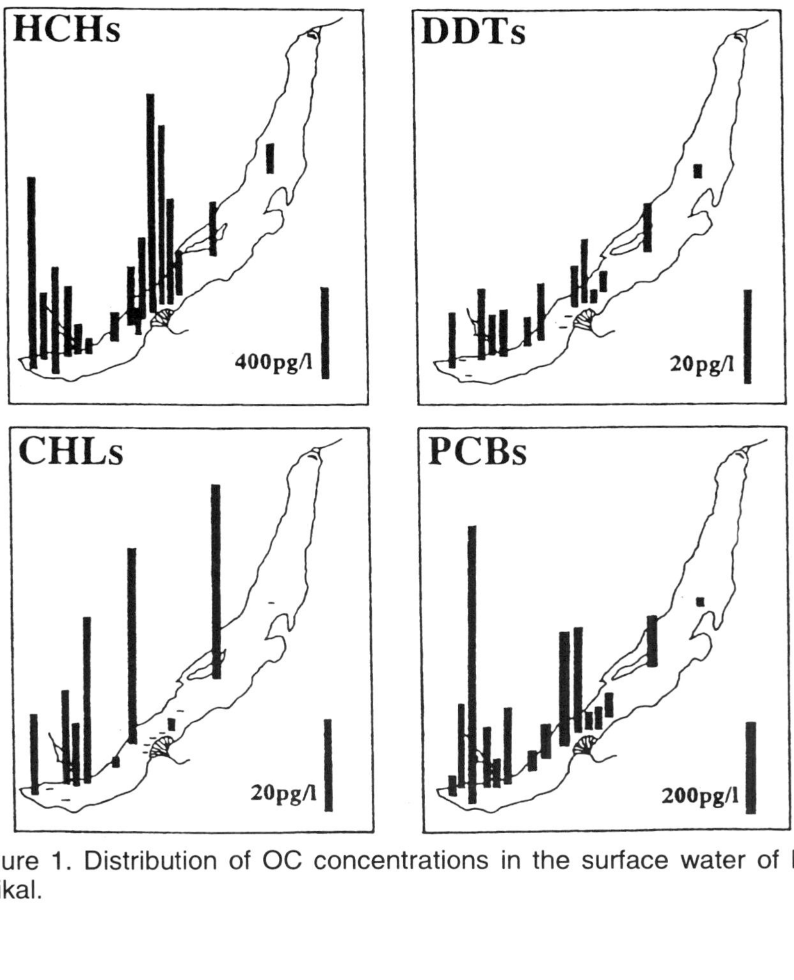

Figure 1. Distribution of OC concentrations in the surface water of Lake Baikal.

The data for OCs in surface water showed higher HCH and PCB concentrations in the lower reaches of the Selenga River and the southern basin of the lake (Fig. 1). PCBs levels in the Selenga River estuary of Lake Baikal (36-240 pg/L) were 1-2 orders of magnitude higher than observed in the Arctic Ocean (Hargrave *et al.*, 1988, Iwata *et al.*, 1993). These findings are in agreement with those obtained in a previous study (Kucklick *et al.*, 1996), suggesting an inflow of PCBs through the river and tributaries of Lake Baikal.

The data for OC levels in sediment samples showed prominent contamination by PCBs (6.1 ng/g dry) and DDTs (2.7 ng/g dry wt.) in a sample collected from the southern basin. The lake sediments are more contaminated by PCBs than Chukchi Sea and Bering Sea sediments (Iwata *et al.*, 1994b), but were clearly less contaminated than heavily polluted sites such as Baltimore harbour (Ashley and Baker, 1999). The OC levels in soil varied with the sampling locations (Table 1). A soil sample collected from a potato field was highly contaminated by HCHs (16 ng/g dry wt.) and DDTs (28 ng/g dry wt.), suggesting possible use of these compounds around Lake Baikal at present or in the past.

OC concentrations and the kinetics of their accumulation in the Baikal seal

Status of contamination

OCs were detected in all seal and fish samples from Lake Baikal (Table 2). DDT compounds were the OCs detected in the highest concentration in Baikal seals, ranging from 4.9 to 160 μg/g on a lipid weight basis, and were followed by PCBs (3.5 to 64 μg/g), CHLs (0.22 to 1.9 μg/g), and HCHs (0.028 to 0.14 μg/g). OC residues in mature males were significantly higher than in mature females ($p < 0.05$), and this may have been attributable to transfer of OCs from females to pups through gestation and lactation, as documented in other pinnipeds (Addison & Smith, 1974). PCBs and DDTs were the dominant components in Baikal fishes, ranging in concentration from 0.82 to 3.2 μg/g and from 0.57 to 2.0 μg/g, respectively (Table 2).

The mean concentration of DDTs in adult Baikal seals was 64 μg/g in males and 22 μg/g in females, that is, about one order of magnitude higher than in North Sea seals (Hall *et al.*, 1992, Law *et al.*, 1989) and comparable the concentrations in harbour seals in the Wadden Sea (Reijnders, 1980) and grey seals in the Baltic Sea (Blomkvist *et al.*, 1992). The Baltic Sea and Wadden Sea are known to be highly contaminated areas, and previous studies have shown an association between the presence of high levels of OCs in seals and disease, such as uterine occlusion and consequent

Table 2. Organochlorine concentrations (mean±SD, μg/g lipid wt.) in the blubber of Baikal seals and their fish diets (whole body) from Lake Baikal [a)]

	N	Body length (cm)	Body weight (kg)	Age (yr)	fat (%)	DDTs*	PCBs*	CHLs*	HCHs*	p,p'-DDE/DDTs*
Baikal seal										
Male	27	116±20 (79.5-145.2)	51±19 (18.9-90.8)	12±9.2 (0.5-35.5)	87±2.8 (81-91)	45±36 (5.3-160)	22±15 (4.3-64)	0.82±0.43 (0.26-1.9)	0.077±0.027 (0.038-0.14)	0.61±0.14 (0.26-0.86)
Female	31	111±15 (79.7-133.2)	52±17 (22.1-90.0)	9.9±6.2 (0.3-24.5)	87±2.3 (80-91)	20±9.9 (4.9-46)	11±4.4 (3.5-19)	0.46±0.15 (0.22-0.83)	0.054±0.020 (0.028-0.11)	0.52±0.12 (0.29-0.72)
Fish										
Comephorus baikalensis	8	15±1.5	0.020±0.008	ND	33	2.0	2.0	0.13	0.018	0.19
Comephorus dybowskii	10	12±0.67	0.007±0.002	ND	3.8	2.0	3.2	0.25	0.019	0.38
Cottocomephorus inermis	5	16±0.86	0.038±0.007	ND	8.2	0.57	0.82	0.10	0.018	0.33

a) Cited from Nakata *et al.* (1995)
* Concentrations and the ratio were significantly different between male and female. ($p<0.05$: Mann Whitney U-test)
DDTs: *p,p'*-DDE + *p,p'*-DDD + *p,p'*-DDT + *o,p'*-DDD + *o,p'*-DDT
CHLs: oxychlordane + trans-chlordane + cis-chlordane + trans-nonachlor + cis-nonachlor
HCHs: $\alpha + \beta + \gamma$ isomers
ND: Not determined

reproductive failure (Bergman & Olsson, 1985).

p,p'-DDE is a stable compound that accounts for a major proportion of the DDT metabolites, in the Baikal seal, whereas *p,p'*-DDT predominated in fishes. The ratio of the concentration of *p,p'*-DDE to total DDTs (*p,p'*-DDE/DDTs) can be used to differentiate between past and present input of technical DDT into the ecosystem. The *p,p'*-DDE/DDTs ratio in Baikal seals was 61±14% in males and 52±12% in females, (the average for males and females was 56±14%; Table 1). These ratios are lower than those observed in the harbour seals affected by a phocine distemper epizootic that started in 1988 in the North Sea around the UK coast, where ratios of 66±20% (Hall *et al.*, 1992), 66±5.7% (Law *et al.*, 1989), and 64±19% (Mitchell & Kennedy, 1992) were recorded. Agricultural use of technical DDT and its production were banned in the former USSR in the 1970s and 1980s, respectively (Barrie *et al.*, 1992), however, higher concentrations of DDTs and lower *p,p'*-DDE/DDTs ratios in Baikal seals and their fish diet imply recent input of technical DDT into the watershed of Lake Baikal. Fedorov (1999) reported finding higher DDT concentrations and persistent frequent usage in various food materials in Russia when measurements were made in 1990. The larger variations in the composition of DDT compounds in air, water, sediment, and soil samples collected around Lake Baikal also indicated current use of technical DDTs near the lake (Iwata *et al.*, 1995).

The PCB residue level in adult Baikal seals was 31 μg/g in males and 13 μg/g in females, about half the concentration of DDTs. The observed levels were comparable to those reported for harbour seals in the North Sea (Hall *et al.*, 1992, Law *et al.*, 1989) and grey seals on the east coast of Canada (Addison & Brodie, 1987), but several times lower than in ringed seals and grey seals in the Baltic sea (Blomkvist *et al.*, 1992). It was estimated that approximately 100 tons of technical PCBs and 25 thousand tons of Sovol and trichlorodiphenyls (TCD) were produced in Russia (Ivanov & Sandell, 1992). The higher levels of PCB residues in Baikal seals seem to suggest use or leakage comparable to the levels in the developed nations in the region.

Age- and sex-dependent accumulation

The OC residue levels varied with the age and sex of the Baikal seals. In both sexes, the concentrations of PCBs, DDTs, and CHLs increased until 7-8 years of age, which is the age at which Baikal seals mature (King, 1983). Even after maturity, male animals showed a positive correlation between OC levels and age, whereas the levels in females remained constant (Fig. 2). This pattern is common in cetaceans and pinnipeds, indicating that OC uptake exceeds excretion in these animals and that consider-

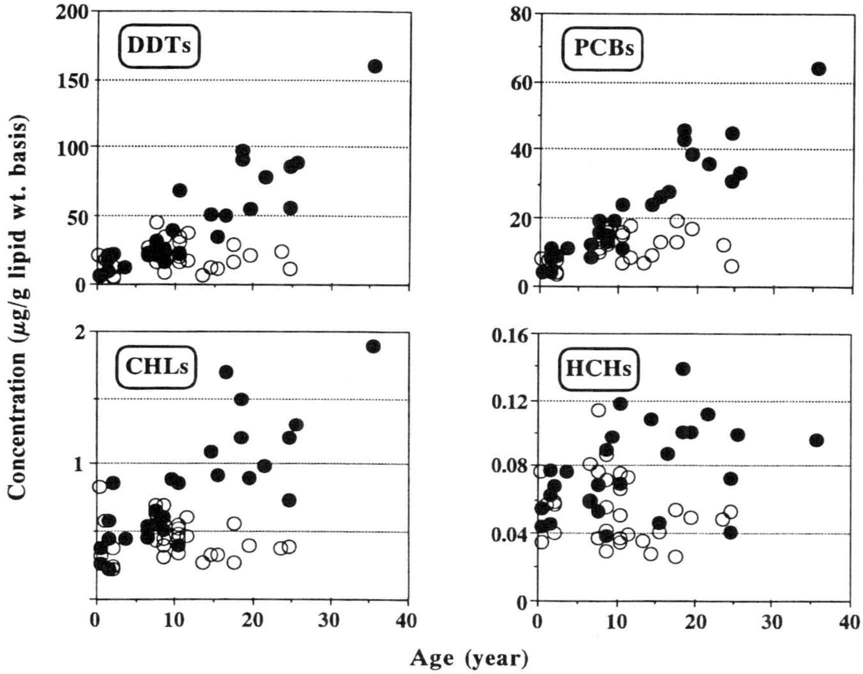

Figure 2. Age trends of organochlorine residue levels in male (●) and female (○) Baikal seals.

able quantities of OCs are transferred from mothers to their pups during the gestation and lactation period (Subramanian *et al.*, 1987, Aguilar *et al.*, 1988, Tanabe *et al.*, 1994). Age-dependent accumulation of residue levels of HCHs was less pronounced, and similar observations were made in harbour seals, grey seals (Ofstad & Martinsen, 1983), and northern fur seals (Tanabe *et al.*, 1994), presumably due to the biodegradability and less persistent nature of HCHs.

In grey seals, about 15% of the PCB burden and 30% of the DDT burden in mothers were transferred to their pups during the reproduction process (Addison & Brodie, 1977). To estimate the transfer rates in Baikal seals, we tried calculating the PCB and DDT body burdens of mature male and female animals (males: ≥8.0 yr; females: ≥6.0 yr [King *et al.*, 1983]). More than 90% of the PCB and DDT body burden of adult marine mammals is present in the blubber (Hidaka *et al.*, 1983), and thus in the present study the body burden estimate were based on the concentrations in blubber and total blubber weight measured at the time each animal was dissected on board. The relationship between age and PCB burden in the two sexes is shown in Fig. 3. The point of intersection obtained on the male and female regression lines is in the 1st reproductive year, and at 8.4 years of age the seals contained 298 mg of PCBs. It has been reported that 88% of adult female Baikal seals breed every spring (Thomas *et al.*, 1982), and thus the PCB body burden would be 349 mg in males and 301 mg in females the next reproductive year. This suggests that 48 mg of PCBs are transferred from mothers to pups every reproductive season. The rate of elimination of PCBs was then calculated by using the following formula:

$$\text{Female burden} = \text{Male burden } (1-0.01\ P)$$

Where P is the transfer rate (%). Consequently, 14% of the total PCBs in mother seals is estimated to be transferred to their pups. When the same approach was used for DDTs, a 20% transfer rate (135 mg burden) was obtained. The PCB and DDT body burden in four Baikal seals (≤0.5 years old) was 62±22 mg and 127±83 mg, respectively, and these values were rather close to the quantities transferred from mother to pup (48 mg of

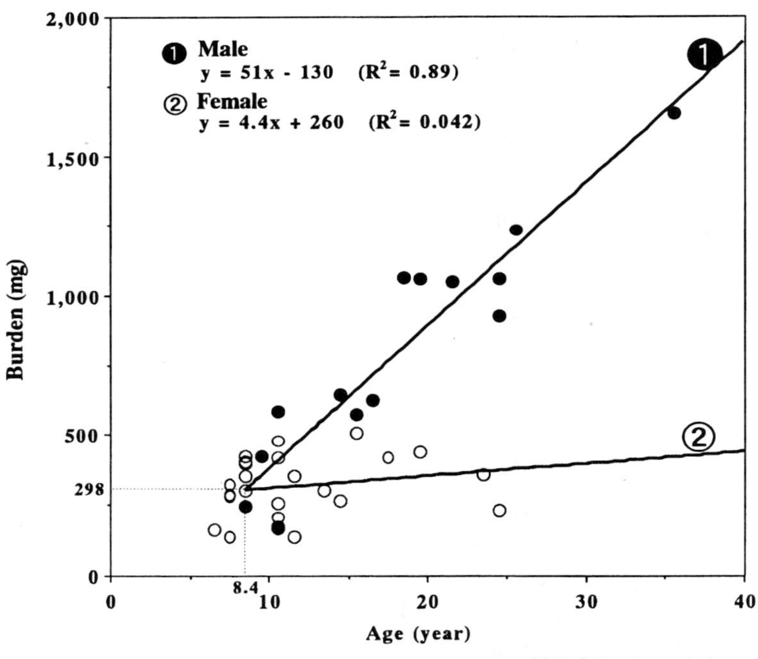

Figure 3. Estimated transferred body burdens of PCBs in adult males (●) and females (○) from mothers to pups in Baikal seals.

PCBs and 135 mg of DDTs). The estimated transfer rates in Baikal seals were comparable to those reported for grey seals (Addison & Brodie, 1977), but clearly lower than those of striped dolphins (Tanabe *et al.*, 1981). The lactation period of Baikal seals and grey seals is known to be 2.5 and 0.5-0.7 months respectively, (King, 1983), whereas that of the striped dolphin is 18 months (Miyazaki, 1981). The smaller transfer rates of OCs in seals than in dolphins may reflect the shorter lactation period.

PCBs congeners and their metabolism

The mean profile of PCB isomers and congeners in Baikal seals and their fish diets is shown in Fig. 4. The levels of tri-, tetra-, and penta-chlorinated congeners in seals were lower than in fishes, indicating that Baikal seals are capable of metabolizing some of the lower chlorinated congeners present in their diet. Hexachlorinated congeners predominated in Baikal seals, followed by penta-, hepta-, octa- and tetrachlorinated homologues. The proportion of total PCBs accounted for by tetrachlorinated congeners in Baikal seals was 1.5%, considerably lower than observed in ringed seals from the Canadian Arctic (Muir *et al.*, 1988) and Ganges river dolphins from India (Kannan *et al.*, 1994). This suggests that Baikal seals may have a greater metabolic capacity to degrade lower chlorinated biphenyls than other aquatic mammals.

The metabolism of PCB isomers and congeners has been found to be associated with hepatic mixed function oxides, such as cytochrome P-450 (Safe *et al.*, 1980; Shimada & Sato, 1980), while those having the vicinal non-chlorinated meta-para carbons and ortho-meta ones are metabolized by phenobarbital- (PB-type) and 3-methylcholanthrene- (MC-type) induced microsomal enzymes, respectively (Kato *et al.*, 1980, Shimada & Sawabe, 1983, Mills *et al.*, 1985). To investigate the metabolic capacity of the Baikal seal in greater detail, the metabolic index (MI) values of PCB isomers and congeners were calculated according to the model proposed by Tanabe *et al.* (1988):

$$\text{metabolic index } (MI_i) = \log CR_{180}/CR_i$$

Where CR180 is the ratio of the concentration of PCB isomer #180 (IUPAC No.) in fish diets to its concentration in the diet of mammals, and CRi is the ratio of the concentration of other isomers. PCB isomers with higher MI values are more biodegradable than those with lower MI values close to zero. The MI values of specific PCB isomers in different species of higher animals (Tanabe *et al.*, 1988) were then compared (Fig. 5). Interestingly, the MI values for PCB isomers metabolized by a PB-type

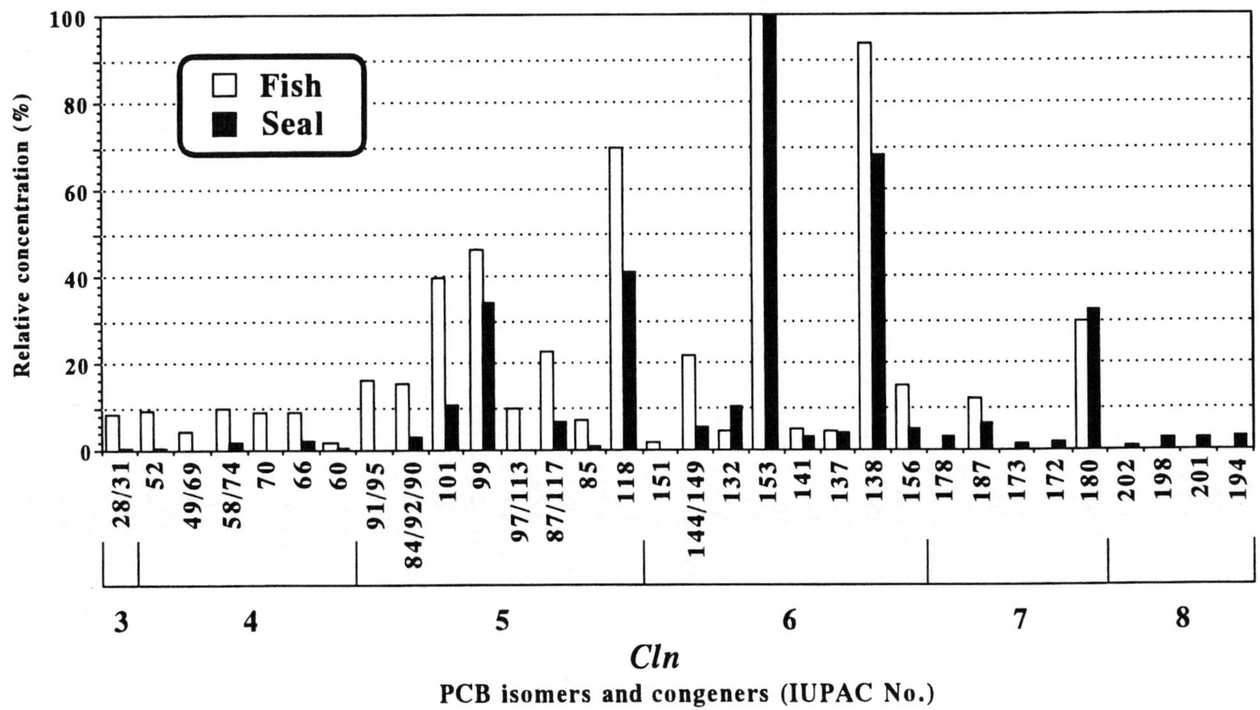

Figure 4. PCB isomer and congener composition in Baikal seal and their fish diet. “Relative concentration” means the ratio of the concentration of an individual PCB to that of its peak (IUPAC No. 153), which was assigned a value of 1.

enzyme (PCB #52) were higher in Baikal seals than in marine mammals, but clearly lower than in terrestrial mammals. The MI values for PCB isomers metabolized by a MC-type enzyme (PCB #66) on the other hand, was lower in Baikal seals, and comparable to the value in cetaceans. Earlier studies demonstrated that cetaceans have less capacity to degrade toxic contaminants because of poor function of both their PB- and MC-type enzymes, and cetaceans therefore are recognized as one of the animal groups having high concentrations of persistent OCs (Tanabe *et al.*, 1988, Watanabe *et al.*, 1989). Baikal seals have a higher capacity to degrade OCs than cetaceans, but their capacity is generally lower than that of terrestrial mammals (Fig. 5). This may be linked to high levels of accumulation of OC residues in Baikal seals. A comparison of MI values in pinnipeds revealed that Baikal seals have higher PB-type enzyme activity than marine species, but clearly lower MC-type enzyme activity (Fig. 5). These differences in MI values imply the functioning of different types of drug-metabolizing enzyme systems in the degradation of xenobiotics and also suggest the occurrence of different types of toxic effects in seals inhabiting freshwater and marine water.

Toxicological Implications

Table 3 shows the ranges and mean concentrations of non- and mono-*ortho* coplanar congeners and total PCBs in Baikal seal and their fish diets. The residue levels of non-*ortho* coplanar congeners were 3-4 orders of magnitude lower than those of total PCBs, and in adult males ranged from 12-41 ng/g wet wt. for CB 77, 2.3-5.9 ng/g for CB 126, and 0.20-0.70 ng/g for CB 169. Among the mono-*ortho* congeners, CB 118 predominated (1,100 to 5,800 ng/g wet wt. in adult males), and was followed by CB 105 (420 to 2,700 ng/g), and CB 156 (130 to 620 ng/g).

The estimated 2,3,7,8-TCDD toxic equivalents of non-, and mono-ortho coplanar congeners in Baikal seal are shown in Table 3 and compared with those of marine mammals reported previously. The mean TCDD equivalent of coplanar congeners in Baikal seal was 550 pg/g on a wet weight basis (Table 3). This level was clearly lower than in diseased/stranded striped dolphin (3,300 pg/g wet, Kannan *et al.*, 1993) and in Risso's dolphin and the bottlenose dolphin in the Mediterranean Sea (2,500-3,100 pg/g wet wt., Corsolini *et al.*, 1995), but higher than in the larga seal in the Okhotsk Sea (40 pg/g, Nakata *et al.*, 1998) and the harbour seal in the North Sea (59 pg/g, Storr-Hansen and Spliid, 1993). By contrast the ratio of concentrations TEQs to PCBs in Baikal seal was 47 x 10^{-6}, and approximately 5 - 10 times higher than in cetaceans (4.2-9.7 x 10^{-6}) and 6 times higher than in harbour seals. These findings indicate that the contribution of toxic PCB congeners to total TEQs is more prominent in Baikal seal than in marine

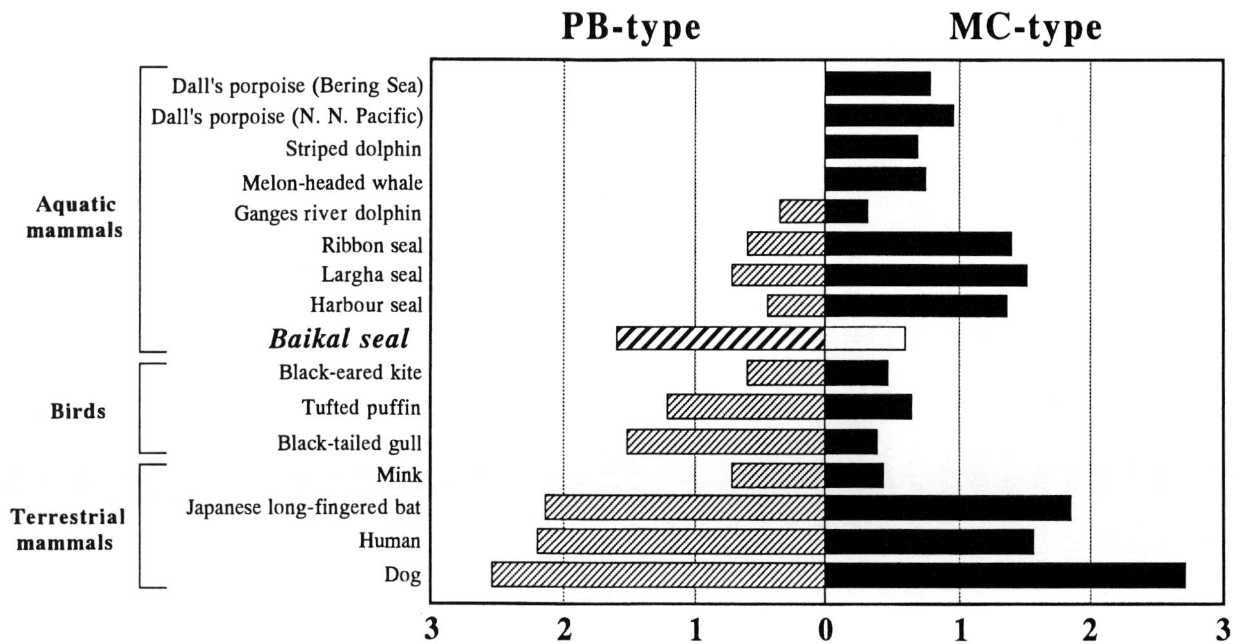

Figure 5. Estimated PB-type and MC-type enzyme activities in higher trophic animals based on the metabolic index of 2,2',5,5'- (PB-type) and 2,3',4,4'- (MC-type) tetrachlorobiphenyls.

Table 3. Concentrations of non- and mono-ortho coplanar congeners and total PCBs (ng/g wet wt.), and their 2,3,7,8-TCDD toxic equivalents (pg/g) in the blubber of Baikal seals and their fish diets

		non-*ortho*			mono-*ortho*			Total
	n	77	126	169	105	118	156	PCBs
Male								
Immature	7	6.3±4.0	2.4±1.3	0.19±0.10	250±130	650±370	87±57	5,000±2,700
		(2.5-14)	(0.6-4.7)	(0.02-0.3)	(65-410)	(180-1,200)	(26-120)	(1,400-9,500)
Mature	8	23±9.8	3.5±1.2	0.37±0.19	1,200±830	2,800±1,800	260±170	26,000±17,000
		(12-41)	(2.3-5.9)	(0.20-0.70)	(420-2,700)	(1,100-5,800)	(130-620)	(11,000-59,000)
Female								
Immature	6	8.8±7.6	2.5±2.6	0.30±0.10	260±160	740±470	74±41	5,000±3,200
		(3.1-23)	(1.0-7.7)	(0.20-0.40)	(140-560)	(420-1,700)	(41-140)	(3,000-11,000)
Mature	19	14±8.1	3.4±2.0	0.30±0.20	360±160	960±500	150±67	10,000±3,800
		(5.6-33)	(1.0-9.0)	(0.06-0.8)	(110-650)	(61-1,700)	(28-300)	(4,100-17,000)
Fish	23	1.0±1.2	0.30±0.30	0.07±0.08	12±13	30±30	5.1±4.7	350±350
		(0.2-2.4)	(0.09-0.6)	(0.01-0.20)	(8.3-65)	(3.4-27)	(1.6-11)	(98-750)
Average TEQ (seal)		1.3	300	2.9	52	130	71	550

TEQs were estimated using TEF values cited from Van den Berg et al. (1998)

mammals, and the reason for this may lie in species-specific capacity for the biotransformation of these congeners. In view of the above observations, it seems plausible to assume that even if the total PCB concentration in Baikal seals is comparable to that in other mammals, the risk of TCDD-like toxicity tends to be more serious because of the high accumulation of toxic PCB congeners. Among the wide range of toxicities of TCDD-related compounds, the adverse effects caused by enzymatic induction and immunological dysfunction are of great concern in this species.

Conclusions

This investigation yielded evidence of OC pollution in the lake Baikal region. The DDT concentrations in air and water were one order of magnitude higher than in the Arctic and the HCH and PCB concentrations in water samples were found to be higher in the lower reaches of the Selenga River and the southern basin of the lake, suggesting the presence of local sources of these contaminants. Reflecting the contamination of air and water, by DDTs and PCBs, higher residues of these contaminants were found in Baikal fish and seals. The PCB concentrations in Baikal seals were comparable to those in certain species of pinnipeds in polluted waters such as the North Sea. This may be attributable to lower drug-metabolizing capacity of Baikal seals and partly to their reproductive transfer of the contaminants. Evidence has been found recently that chronic exposure to organochlorine contaminants, particularly toxic PCB congeners, is responsible for immunosuppression in harbour seals, and this may have triggered the virus-induced mass mortality in the North Sea in 1988 (Swart *et al.*, 1994; Ross *et al.*, 1996). Further studies are needed in regard to the relationship between the enrichment of toxic PCB congeners and possible toxicities, in particular immunosuppression, since mass seals deaths due morbillivirus infection occurred in Lake Baikal in 1987-88.

References

Addison, R. F. and Smith, T. G. (1974) Organochlorine residue levels in Arctic ringed seals: variation with age and sex, Oikos, 25, 335-337.
Addison, R. F. and Brodie, P. F. (1977) Organochlorine residues in material blubber, milk, and pup blubber from grey seals (Halichoerus grypus) from Sable Island, Nova Scotia, J. Fish. Res. Board Can., 34, 937-939.
Addison, R. F. and Brodie, P. F. (1987) Transfer of organochlorine residues from blubber through the circulatory system to milk in the lactating grey seal Halichoerus grypus, Can. J. Fish. Aquat. Sci., 44, 782-786.
Aguilar, A. and Borrell, A. (1988) Age- and sex-related changes in organochlorine compound levels in Fin whales (Balaenoptera physalus)

from the Eastern North Atrantic, Mar. Environ. Res., 25, 195-211.
Aguilar, A. and Borrell, A. (1994) Abnormal high polychlorinated biphenyl levels in striped dolphins (Stenella coeruleoalba) affected by the 1990-1992 Meditterranean epizootic, Sci. Total Environ., 154, 237-247.
AMAP (1998) AMAP Assessment report: Arctic pollution issues. Arctic monitoring and assessment Programme (AMAP) Oslo, Norway, xii+859 pp.
Ashley, J. T. F. and Baker, J. E. (1999) Hydrophobic organic contaminants in surficial sediments of Baltimore harbour: Inventories and sources, Environ. Toxicol. Chem., 18, 838-849.
Barrie, L. A., Gregor, D., Hargrave, H., Lake, R., Muir, D., Shearer, R., Tracey, B. and Biblemam, T. (1992) Arctic contaminants: sources, occurrence and pathways, Sci. Tot. Environ., 122, 1-74.
Bergman, A. and Olsson, M. (1985) Pathology of Baltic grey seal and ringed seal females with special reference to adrenocortical hyperplasia: Is environmental pollution the cause of a widely disturbed disease syndrome?, Finnish Game Res., 44, 47-62.
Blomkvist, G., Roos, A., Jensen, S., Bignert, A. and Olsson, M. (1992) Concentration of sDDT and PCB in seals from Swedish and Scottish waters, Ambio, 8, 539-545.
Corsolini, S., Focardi, S., Kannan, K., Tanabe, S., Borrell, A. and Tatsukawa, R. (1995) Congener profile and toxicity assessment of polychlorinated biphenyls in dolphins, sharks and tuna collected from Italian coastal waters, Mar. Environ. Res., 40, 33-53.
Fedorov, L. A. (1999) Persistent organic chemicals in the former Soviet Union, Environ. Pollut., 105, 283-287.
Grachev, M. A., Kumarev, V. P., Mamaev, L. V., Zorin, V. L., Baranova, L. V., Denikina, N. N., Belikov, S. I., Petrov, E. A., Kolesnik, V. S., Kolesnik, R. S., Dorofeev, V. M., Beim, A. M., Kudelin, V. N., Nagieva, F. G. and Sidorov, V. N. (1989) Distemper virus in Baikal seals, Nature, 38, 209.
Hall, A. J., Law, R. J., Wells, D. E., Harwood, J., Ross, H. M., Kennedy, S., Allchin, C. R., Campbell, L. A. and Pomeroy, P. P. (1992) Organochlorine levels in common seals (Phoca vitulina) which were victims and survivors of the 1988 phocine distemper epizootic, Sci, Tot. Environ., 115, 145-162.
Hargrave, B. T., Vass, W. P., Erickson, P. E. and Fowler, B. R. (1988) Atmospheric transport of organochlorine to the Arctic ocean, Tellus, 40B, 480-493.
Hidaka, H., Tanabe, S. and Tatsukawa, R. (1983) DDT compounds and PCB Isomers and congeners in weddell seals and their fate in The Antarctic marine ecosystem, Agr. Biol. Chem., 47, 2009-2017.
Ivanov, V. and Sandell, E. (1992) Characterization of polychlorinated biphenyl isomers in Sovol and trichlorodiphenyl formulations by high-res-

olution gas chromatography with electron capture detection and high-resolution gas chromatography-mass spectrometry techniques, Environ. Sci. Technol., 26, 2012-2017.
Iwata, H., Tanabe, S., Sakai, N. and Tatsukawa, R. (1993) Distribution of persistent organochlorines in the Oceanic air and surface seawater and the role of ocean on their global transport and fate, Environ. Sci. Technol., 27, 1080-1098.
Iwata, H., Tanabe, S., Sakai, N., Nishimura, A. and Tatsukawa, R. (1994a) Geographical distribution of persistent organochlorines in air, water and sediments from Asia and Oceania, and their implications for global redistribution from lower latitudes, Environ. Pollut., 15-33.
Iwata, H., Tanabe, S., Aramoto, M., Sakai, N. and Tatsukawa, R. (1994b) Persistent organochlorine residues in sediments from the Chukchi Sea, Bering Sea and Gulf of Alaska, Mar. Pollut. Bull., 28, 746-753.
Iwata, H., Tanabe, S., Ueda, K. and Tatsukawa, R. (1995) Persistent organochlorine residues in air, water, sediments and soils from Lake Baikal regions, Russia, Environ. Sci. Technol., 29, 792-801.
Jones, K. C., Duarte-Davidson, R. and Cawse, P. A. (1995) Changes in the PCB concentration of United Kingdom air between 1972 and 1992, Environ. Sci. Technol., 29, 272-275.
Kannan, K., Tanabe, S., Borrell, A., Aguilar, A., Focardi, A. and Tatsukawa, R. (1993) Isomer-specific analysis and toxic evaluation of polychlorinated biphenyls in striped dolphins affected by an epizootic in the Western Mediterranean Sea, Arch. Environ. Contam. Toxicol., 25, 227-233.
Kannan, K., Tanabe, S., R, S. and Tatsukawa, R. (1994) Biodegradation capacity and residue pattern of organochlorines in Gange river dolphin, Toxicol. Environ. Chem., 42, 249-261.
Kato, S., McKinney, J. D. and Matthews, H. B. (1980) Metabolism of symmetrical hexachlorobiphenyl isomers in the rat, Toxicol Appl. Pharmacol., 53, 389-398.
King, J., E. (1983) Seal of the World, British Museum, London, 240.
Kucklick, J. R., Harvey, H. R., Ostrom, P. H., Ostrom, N. E. and Baker, J. E. (1996) Organochlorine dynamics in the pelagic food web of Lake Baikal, Environ. Toxicol. Chem., 15, 1388-1400.
Law, R. J., Allchin, C. R. and Harwood, J. (1989) Concentrations of organochlorine compounds in the blubber of seals from Eastern and North-eastern England, 1988, Mar. Pollut. Bull., 20, 110-115.
Lead, W. A., Steinnes, E. and Jones, K. C. (1996) Atmospheric deposition of PCBs to moss (Hylocomium splendes) in Norway between 1977 and 1990, Environ. Sci. Technol., 30, 524-530.
Mills, R. A., Millis, C. D., Dannan, G. A., Guengerich, F. P. and Aust, S.

D. (1985) Studies on the structure-activities relationships for the metabolism of polychlorinated biphenyls by rat liver microsomes, Toxicol. Appl. Pharmacol., 78, 96-104.
Mitchell, S. H. and Kennedy, S. (1992) Tissue concentrations of organochlorine compounds in common seals from the coast of Northern Ireland, Sci. Tot. Environ., 115, 163-177.
Miyazaki, N. (1981) An outline of the biological studies on Sterelia coeruleoalba, In: Studies on the levels of organochlorine compounds and heavy metals in the marine environment, Fujiyama, T. (eds) University of Ryukyus. Ryukyus.1-5.
Muir, D. C. G., Norstrom, R. J. and Simon, M. (1988) Organochlorine contaminants in arctic marine food chain: Accumulation of specific polychlorinated biphenyls and chlordane-related compounds, Environ. Sci. Technol., 22, 1071-1079.
Muir, D. C. G., Wagemann, R., Hargrave, B. T., Thomas, D. J., Peakall, D. B. and Norstrom, R. J. (1992) Arctic marine ecosystem contamination, Sci. Total Environ., 122, 75-134.
Nakata, H., Tanabe, S., Tatsukawa, R., Amano, M., Miyazaki, N. and Petrov, E. A. (1995) Persistent organochlorine residues and their accumulation kinetics in Baikal seal from Lake Baikal, Russia, Environ. Sci. Technol., 29, 2877-2885.
Nakata, H., Tanabe, S., Tatsukawa, R., Koyama, Y., Miyazaki, N., Belikov, S. and Boltunov, A. (1998) Persistent organochlorine contaminants in ringed seals (Phoca hispida) from the Kara Sea, Russian Arctic, Environ. Toxicol. Chem., 17, 1745-1755.
Ofstad, E. B. and Martinsen, K. (1983) Persistent organochlorine compounds in seals from Norwegian coastal waters, Ambio, 12, 262-264.
Reijnders, P. J. H. (1980) Organochlorine and heavy metal residues in harbour seals from the wadden sea and their possible effects on reproduction, Netherlands Journal of Sea Research, 14, 30-65.
Ross, P., Swart, R. D., Addison, R., Loveren, H. V., Vos, J. and Osterhaus, A. (1996) Contaminant-induced immunotoxicology in harbour seals: wildlife at risk ?, Toxicology, 112, 157-169.
Safe, S., Wyndham, C., Parkinson, A., Purdy, R. and Crawford, A. (1980) Haloginated biphenyl metabolism, Environ. Sci. Res., 16, 537-544.
Shimada, T. and Sato, R. (1980) Covalent binding of polychlorinated biphenyls to rat liver microsomes in vitro: nature of reactive metabolites and target macromolecules, Toxicol. Appl. Pharmacol., 55, 490-500.
Shimada, T. and Sawabe, Y. (1983) Activation of 3,4,3',4'-tetrachlorobiphenyl to protein-bound metabolites by rat liver microsomal cytochrome P-448-containing monooxygenase system, Toxicol. Appl. Pharmacol., 70, 486-493.

Simonich, S. L. and Hites, R. A. (1995) Global distribution of persistent organochlorine compounds, Science, 269, 1851-1854.

Storr-Hansen, E. and Spliid, H. (1993) Distribution patterns of polychlorinated biphenyls congeners in harbor seal (Phoca vitulina) tissues: statistical analysis, Arch. Environ. Contam. Toxicol., 25, 328-345.

Subramanian, A., Tanabe, S. and Tatsukawa, R. (1987) Age and size trends and male-female differences of PCBs and DDE in dall-type dall's porpoises, Phocoenoides dalli of Northwestern north pacific, Proc. NIPR Symp. Polar Biol., 1, 205-216.

Swart, R. L. d., Ross, P. S., Vesser, L. J., Timmerman, H. H., Heisterkamp, S., Loveren, H. V., Vos, J. G., Reijnders, P. J. H. and Osterhaus, A. D. M. E. (1994) Imparement of immune function in harbour seals (Phoca viturina) feeding on fish from polluted waters, Ambio, 23, 155-159.

Tanabe, S., Tanaka, H., Maruyama, K., Tatsukawa, R. (1981) Elimination and chlorinated hydrocarbons from mother striped dolphin (Stenella coeruleoalba) through parturition and lactation. In: Studies on the levels of organochlorine compounds and heavy metals in the marine environment, Fujiyama, T. (eds) University of Ryukyus. Ryukyus, 115-121.

Tanabe, S., Watanabe, S., Kan, H. and Tatsukawa, R. (1988) Capacity and mode of PCB metabolism in small cetaceans, Mar. Mam. Sci., 4, 103-124.

Tanabe, S., Sung, J.-K., Choi, D.-Y., Baba, N., Kiyota, M., Yoshida, K. and Tatsukawa, R. (1994) Persistent organochlorine residues in northern fur seal from the Pacific coast of Japan since 1971, Environ. Pollut., 85, 305-314.

Thomas, J., Pastukhov, V., Elsner, R. and Petrov, E. (1982) Phoca sibiica, Mammalian Species, 188, 1-6.

Van den Berg, M., Birnbaum, L., Bosveld, A. T. C., Brunstrom, B., Cook, P., Feeley, M., Giesy, J. P., Hanberg, A., Hasegawa, R., Kennedy, S. W., Kubiak, T., Larsen, J. C., Leeuwen, F. X. R. v., Liem, A. K. D., Nolt, C., Peterson, R. E., Poellinger, L., Safe, S., Tillitt, D., Tysklind, M., Younes, M., Warn, F. and Zacharewski, T. (1998) Toxic equivalency factor (TEFs) for PCBs, PCDDs, PCDFs for humans and wildlife, Environ. Health Perspect., 106, 775-792.

Watanabe, S., Shimada, T., Nakamura, S., Nishiyama, N., Yamashita, N., Tanabe, S. and Tatsukawa, R. (1989) Specific profile of liver microsomal cytochrome P-450 in dolphin and whale, Mar. Environ. Res., 27, 51-65.

Lake Baikal
K. Minoura (editor)
2000 Elsevier Science B.V.

Genetic differentiation of gammarid (*Eulimnogammarus cyaneus*) populations in relation to past environmental changes in Lake Baikal

Mashiko, K.[1]*, Kamaltynov, R.[2], Morino, H.[3], and Sherbakov, D. Yu.[4]
[1] Laboratory of Basic Life Science, Teikyo University, Japan.
Fax: (+81) 426-78-3430, E-mail: kmashiko@main.teikyo-u.ac.jp
[2] Limnological Institute, Russian Academy of Science, Russia.
Fax: (+95) 420-21-06, E-mail: ravil@lin.irk.ru
[3] Department of Environmental Sciences, Ibaraki University, Japan.
Fax (+81) 29-228-8404, E-mail: morino@mito.ipc.ibaraki.ac.jp
[4] Limnological Institute, Russian Academy of Science, Russia.
Fax: (+95) 420-21-06, E-mail: sherb@lin.irk.ru
(*corresponding author)

Abstract

Individuals of the gammarid Eulimnogammarus cyaneus in Lake Baikal are genetically differentiated into two major groups, largely corresponding to the bathymetric division of lake between the north and south-central basins. As there appears to be no extrinsic (physical) barrier to genetic mixing between them in the present state of the lake, the two groups are considered to have been separated into different parts of the lake, and thereafter came to contact at a narrow marginal zone of their distribution, around the Olkhon Strait. The southern group is differentiated into further two groups at the Angara River outlet, which is perhaps caused by the formation of the Angara River as a reproductive barrier during the Late Pleistocene.

Introduction

A surprisingly great number of gammarid species and subspecies, well over than 300, occur in Lake Baikal, 98% of them being indigenous (Kamaltynov, 1992; Kozhova and Izmest'eva, 1998). They are regarded as having diversified in this geologically long-persistent lake from a few ancestral species (Bazikalova, 1945; Kozhov, 1963). However, the mechanism by which so many species have diversified remains controversial (Brooks, 1950; Fryer, 1991; Martens et al., 1994), even with recent molecular phylogenetic researches (e.g. Ogarkov et al., 1997; Sherbakov, 1999). In general, the process of speciation occurs through the interruption of gene flow among conspecific individuals, which is most effectively achieved by geographic (spatial) isolation (e.g. Mayr, 1963; Futuyma and

Mayer, 1980). This principle was put forth by Brooks (1950) regarding organisms in ancient lakes such as Baikal and Tanganyika, although Kozhov (1963) stressed the possibility of sympatric speciation through adaptation to various depths in Baikalian gammarids. The central problem concerning speciation in ancient lakes is, therefore, how genetic mixing among conspecific individuals is hampered within a closed lacustrine system. Population genetic approaches to gammarids in Lake Baikal will offer an important clue to this subject.

Differentiation of gammarid populations in the lake

The gammarid *Eulimnogammarus* (= *Philolimnogammarus*) *cyaneus*, endemic to Lake Baikal, occurs exclusively in the littoral zone of the stony or rocky shore of the lake (Weinberg and Kamaltynov, 1998). Body length ranges from 5 to 10 mm. We collected samples of this species from all around the lake (Fig. 1), and performed allozymic analysis. As shown in Figure 2, individuals in the lake were genetically divided into major two groups, the southern populations (populations 1-13 plus 24 and 25 on Olkhon Island) and the northern populations (populations 14-23 and 26-29). The two groups were abutting at a narrow zone along the west coast of the lake, around the Olkhon Strait. In particular, locations 13 and 14 are less than 10 km apart. As well, populations on Olkhon Island were divided into the two groups.

No present geographic or spatial barrier to gene flow between the two groups can be recognized. However, the Akademichesky Ridge transects the lake bottom 300 m below the surface from Olkhon Island towards the Suviatoy Nos Peninsula. This ridge is regarded to have been the northern shore of Paleo-Baikal (Khlystov et al., 1998). The ridge gradually subsided, and lake water invaded from south to north, forming the present northern basin by the end of the Miocene (Zonenshain et al., 1992; Mats, 1993). In this geological change, some islands still remained on the southwestern basement of the ridge until 150,000-200,000 years ago, like the Ushkanyi Islands on the north-eastern basement (Khlystov et al., 1998). In those circumstances, the northern basin might have been easily separated from the other lake areas whenever the lake level dropped by 300 m or more.

The genetic distance of 0.035 between the two groups (Fig. 2) suggests that they began to differentiate 180,000 years ago, based on the standardized allozymic molecular clock (Nei, 1987). This corresponds to the Quaternary glacial epoch. If the gauging is adequate, what then happened in the lake during the glaciation?

The geological view of the Quaternary fluctuations in lake level has not

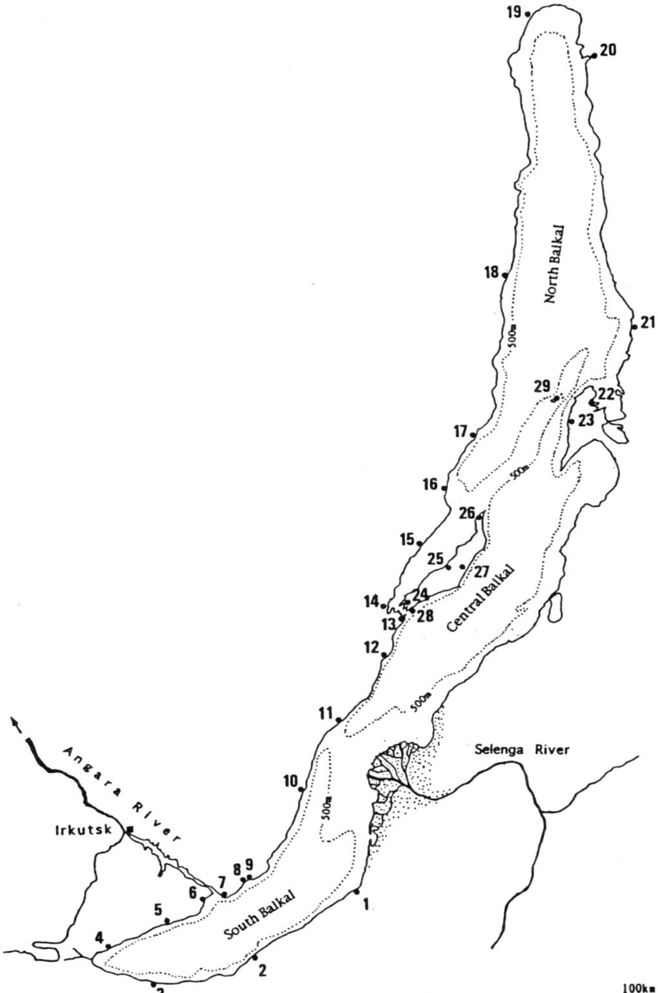

Fig. 1. The 29 locations from which specimens were collected. Locations 1-23 along the lake shore are numbered clockwise, starting from southern Baikal. Sites 24-29 are on Olkhon and Ushkanyi islands.

been completely settled, varying between the two extremes of significant drop (300-400 m) and subtle changes (Mats et al., prepared for this volume). For instance, moraines found at 300 m depth in the Frolikha Bay of northern Baikal suggest a significant drop in lake level during glaciation (Lut, 1964; cited in Mats, 1993), though alternative explanations might be

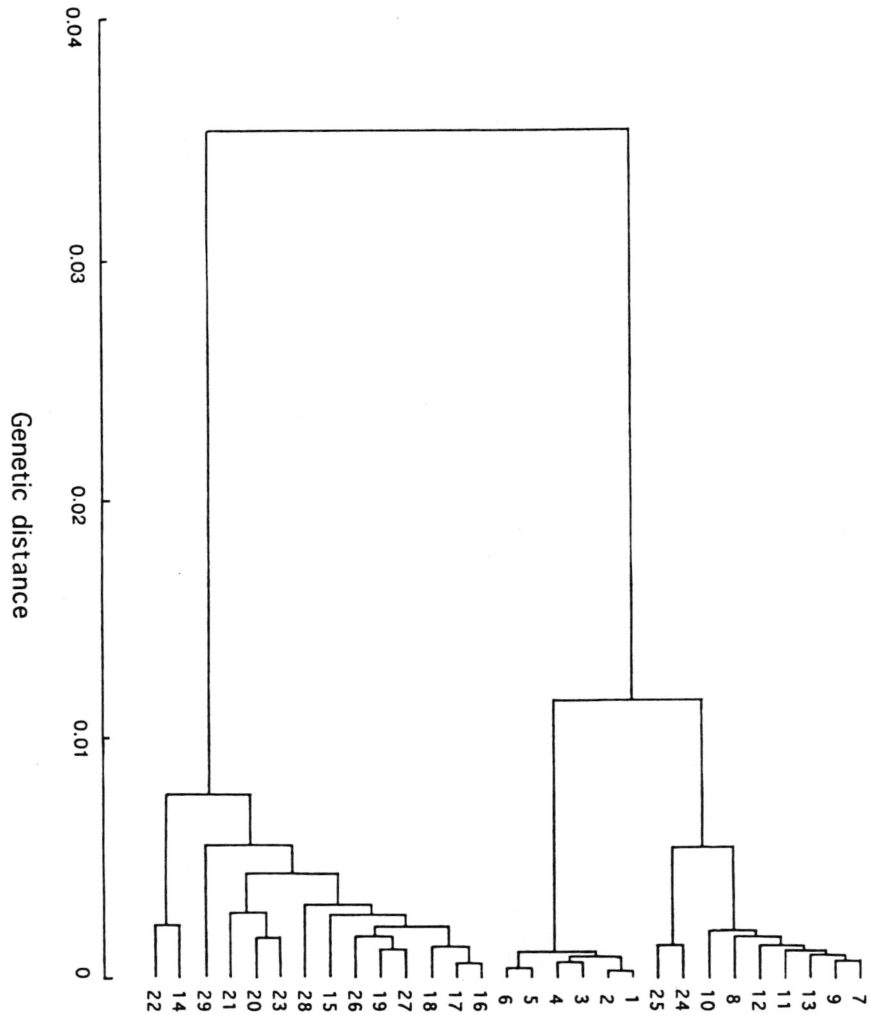

Fig. 2. UPGMA dendrogram obtained by allozyme analyses of 21 gene loci in the 29 populations (drawn from Mashiko et al., 1997, and from their unpublished data).

possible (Back et al., 1998). Our study amplifies the view that the lake level was much lower than now for a considerable period during the glacial epoch, when *E. cyaneus* populations differentiated between the northern basin and the southern and central basins. Otherwise, some other mechanisms to separate them into different parts within the lake must have functioned during that period.

The situation of populations on Olkhon Island is complicated. Two populations (24 and 25) on the southwest coast were closely related to the southern group, whereas those on other areas were akin to the northern group. It is likely that those populations came into secondary contact after the breakdown of the previously existed extrinsic barrier. In the contact zone, some special mechanism such as hybrid breakdown (e.g., Hewitt, 1988) may be operating; However, this needs to be substantiated in future studies.

Separation by the Angara River

The southern group of populations is divided into further two groups-northern (populations 7-13, 24, 25) and southern (populations 1-6)-by the Angara River outlet. Between the two groups, gene frequencies at some loci were distinctly different: For instance, the gene Ark^{82} did not appear at all in the southern populations but occurred at high frequencies in the northern populations (Mashiko et al., 1997). This means that the Angara River serves as an effective barrier to genetic mixing between the two groups of individuals. This river is the only river draining from the lake, and water flows out rapidly (ca. 1 m/s on average; estimated from Verbolov et al., 1989 and Sokolnikov, 1960) throughout the year. For this essentially lentic, shallow-water species, the rapid current must be an insurmountable obstacle to the migration of individuals across the outlet.

According to recent geological investigations (Kononov and Mats, 1986; Ryazanov, 1993), lake water formerly flowed out from the southernmost side of the lake through the ancient Kultuk River and, earlier than that, from the mid-western part (ancient Manzurka River). The present Angara River outlet is regarded to have opened during the second half of the Late Pleistocene (80,000-20,000 years ago). From the genetic distance between the two groups divided by the river, an absolute differentiation time in them is estimated to be 60,000 years. This estimate coincides well with the geological assessment of the formation of the outlet. It is therefore highly likely that the opening of the outlet in the Late Pleistocene induced the separation of those populations.

We consider that the spatial separation of populations by localized topographic changes, as indicated here for the Angara River outlet, is more important in increasing species diversity than large-scale environmental fluctuations such as the splitting of the entire lake basin. This is because regionally limited environmental changes do not throw so catastrophic impact on the whole biotic community as to cause numerous species to vanish. On the contrary, too harsh environmental changes must result in less richness of species, even if they partly contribute to speciation.

Acknowledgements
We are grateful to Drs. M. Grachev, K. Numachi, N. Miyazaki and T. Kawai, who supported our BICER research activities from various aspects. We are also thankful to the captain and crew of R/V Obrachev and to staff of the Limnological Institute, SD RAS, for their help and hospitality.

References

Back S., M. De Batist, P. Kirillov, M. R. Strecker and P. Vanhauwaert, 1998, The Florikha fan: a large Pleistocene glaciolacustrine outwash fan in Northern Lake Baikal, Siberia. J. Sedimentary Research, 68, 841-849.

Bazikalova A. Y., 1945, Amphipods of Lake Baikal. Trudy Baikal. Limnol. St. Akad. Nauka SSSR, 11, 5-440. (In Russian).

Brooks J. L., 1950, Speciation in ancient lakes. Quart. Rev. Biol., 25, 30-60.

Fryer G., 1991, Comparative aspects of adaptive radiation and speciation in Lake Baikal and the great rift lakes of Africa. Hydrobiologia, 211, 137-146.

Futuyma D. J. and G. C. Mayer, 1980, Non-allopatric speciation in animals. Syst. Zool., 29, 254-271.

Hewitt G. M., 1988, Hybrid zones-Natural laboratories for evolutionary studies. Trends Ecol. Evol. 3, 158-167.

Kamaltynov R. M., 1992, On the present state of systematics of the Lake Baikal amphipods (Crustacea, Amphipoda). Zool. Zhur., 71 (6), 24-31. (In Russian).

Khlystov O., V. D. Mats, S. Ceramicola, M. De Batist, T. K. Lomonosova and M. Grachev, 1998, Tectonic evolution of and depositional processes on Akademichesky Ridge, Lake Baikal (Siberia). Active Tectonic Continental Basins: Abstract of International Conference INTAS, Gent, Belgium, 37-38 pp.

Kononov E. E. and V. D. Mats, 1986, History of Baikal water runoff. Izv. Vyss. Ucheb. Zaved. Geol. Raved. 1986, 6, 91-98. (In Russian).

Kozhov M., 1963, Lake Baikal and its Life. Junk, The Hague, 344 pp.

Kozhova O. M. and L. R. Izmest' eva, 1998, Lake Baikal. Evolution and Biodiversity. Backhuys, Leiden, 447 pp.

Martens K., G. Coulter and B. Goddeeris, 1994, Speciation in ancient lakes —40 years after Brooks. Arch. Hydrobiol. Beih. Ergebn. Limnol., 44, 75-96.

Mashiko K., R. Kamaltynov, D. Yu. Sherbakov and H. Morino, 1997, Genetic separation of gammarid (*Eulimnogammarus cyaneus*) populations by localized topographic changes in ancient Lake Baikal. Arch.

Hydrobiol., 139, 379-387.
Mats V. D., 1993, The structure and development of the Baikal rift depression. Earth-Science Reviews, 34, 81-118.
Mayr E., 1963, Animal Species and Evolution. Harvard University Press, Cambridge, Massachusetts, 797 pp.
Nei M., 1987, Molecular Evolutionary Genetics, Columbia University Press, New York, 512 pp.
Ogarkov O. B., R. M. Kamaltynov, S. I. Belikov and D. Yu. Shcherbakov, 1997, Phylogenetic relatedness of the Baikal Lake endemial amphipodes (Crustacea, Amphipoda) deduced from partial nucleotide sequences of the cytochrome oxidase subunit III genes. Mol. Biol., 31, 24-29.
Ryazanov G. V., 1993, The unique localities around Lake Baikal. Nauka, Novosibirsk, 60 pp. (In Russian).
Sherbakov D. Yu., 1999, Molecular phylogenetic studies on the origin of biodiversity in Lake Baikal. Trends Ecol. Evol., 14, 92-95.
Sokolnikov V. M., 1960, On the water currents under ice cover in Lake Baikal and at the Angara River outlet. Trudy Baikal. Limnol. St. Akad. Nauka SSSR, 18, 264-285. (In Russian).
Verbolov V. I., V. N. Sinyukovich and N. L. Karpysheva, 1989, Water and mass exchange in the Lake Baikal and storage reservoirs of the Angara cascade. Arch. Hydrobiol. Beih. Ergebn. Limnol., 33, 35-40.
Weinberg I. W. and R. Kamaltynov, 1998, Zoobenthos communities at a stony beach of Lake Baikal. Zool. Zhur., 77, 158-165. (In Russian).
Zonenshain L. P., V. G. Kazmin and M. I. Kuzmin, 1992, New data on the Baikal Rift Geology: Results of submarine geological observations. International Project on Paleolimnology and Late Cenozoic Climate (IPPC-CE) News Letter, 6, 10-20.

Lake Baikal
K. Minoura (editor)
2000 Elsevier Science B.V.

Myological peculiarities of the comephoridae: an endemic fish taxon in Lake Baikal (Pisces: Teleostei)

Yabe, M.[1]* and Sideleva, V. G.[2]
[1] Laboratory of Marine Zoology, Faculty of Fisheries,
Hokkaido University, Hakodate, Hokkaido, 041-8611, Japan
E-mail: myabe@fish.hokudai.ac.jp
[2] Laboratory of Ichthyology, Zoological Institute,
Russian Academy of Sciences, Universitetskaya nab., 1,
St. Petersburg, 199034, Russia
(*corresponding auther)

Abstract

Examination of the myology of the Comephoridae, an endemic fish family in Lake Baikal, has revealed their morphological peculiarities. The unique muscle features of *Comephorus* found in the cheek region (the adductor mandibulae A1 and the levator arcus palatini), hyoid region (the protractor hyoidei and the hyohyoidei abductores), and branchial region (the recti ventrales and the transversi ventrales) are regarded as highly derived structures related to the pelagic feeding behavior of comephorid fishes. The muscle anatomy of the occipital region of *Comephorus* (presence of the obliquus superioris inserting on the cranium and the absence of the extrinsic muscle of the swimbladder) is distinctly different from that of the Cottoid fishes, including the Cottocomephorinae and Abyssocottidae, and is regarded as plesiomorphy in the Acanthomorpha. The myological features of *Comephorus* suggest that their origin could not be limited to the cottoids endemic to Lake Baikal.

Introduction

The thirty endemic species of sculpins inhabiting Lake Baikal are classified in three family-group taxa of the suborder Cottoidei (Sideleva, 1982, 1994, 1999; Nelson, 1994). The cottid subfamily Cottocomephorinae, comprising three genera and eight species, shows almost the same morphology as non-Baikalian cottids. The family Abyssocottidae is highly diverged only in Lake Baikal. It includes six genera and 20 species, and their morphology is similar to that of the cottids except for having unique seismosensory systems on the head and body (Sideleva, 1982). The family Comephoridae contains a single genus, *Comephorus*, and two species, *C. baicalensis* (Pallas) and *C. dybowskii* Korotneff. Comephorid fishes, known by their Russian name "golomyanka", are extremely different from other sculpins in their morphology and ecology. Both species of *Comephorus* are morphologically characterized by a translucent,fatty body, long pectoral fin and no pelvic fin, and also characterized by pelagic ecology and ovoviviparous reproduction. The phylogeny and systematics of the sculpins in Lake Baikal have been studied by Russian biologists (Berg, 1940; Taliev, 1955; Sideleva, 1982, 1994, etc.). Taliev (1995) studied the morphology and ecology of the sculpins in Lake Baikal, and suggested that the endemic sculpins, including *Comephorus*, are of monophyletic origin. Sideleva (1982), however, studied their cephalic sensory systems and concluded that *Comephorus* may have an independent origin and be distantly related to other sculpins in Lake Baikal. Nevertheless some recent molecular studies of mtDNA support the monophyly of the sculpins in Lake Baikal (Grachev *et al.*, 1992; Slobodyanyuk *et al.*, 1994, 1995), and the morphological gap between *Comephorus* and other endemic sculpins has been inexplicable. In order to clarify the phylogeny of the endemic sculpins of Lake Baikal, it is necessary to obtain more information on their biodiversity. Recently, the morphological features of somatic muscles are frequently being used to infer the phylogenetic relationships of fishes, however, the morphology of the culculature of *Comephorus* is still poorly known. The purpose of this study was to describe the myological features of *Comephorus*, and to discuss their significance in the phylogenetic study of the sculpins in Lake Baikal.

Materials and Methods

For the anatomical studies, the fish specimens were stained with alizarin red-S, dissected, and examined with a binocular microscope. The morphological terminology is chiefly according to Winterbottom (1974) and Yabe and Uyeno (1996). The specimens examined have been deposited in the Laboratory of Marine Zoology, Faculty of Fisheries, Hokkaido University (HUMZ), in Hakodate, Japan. The anatomical studies of Baikal sculpins were performed on the following specimens: Comephoridae - *Comephorus baicalensis*, HUMZ 113873 (152.5 mm SL), HUMZ 113884 (132.0 mm SL); *Comephorus dybowskii*, HUMZ 113832 (78.0 mm SL), HUMZ 113883 (101.0 mm SL). The comparison specimens were: Cottocomephorinae - *Batrachocottus multiradiatus*, HUMZ 113880 (154.4 mm SL); *Cottocomephorus inermis*, HUMZ 113864 (80.2 mm SL); *Paracottus kneri*, HUMZ 113879 (83.0 mm SL); Abyssocottidae - *Asprocottus herzensteini*, HUMZ 113874 (68.7 mm SL); and *Limnocottus megalopus*, HUMZ 113865 (153.0 mm SL). All specimens were collected during the program of the Baikal International Center for Ecological Research in 1992 and 1993.

Myological Description

Both species of the genus *Comephorus* possess almost the same myological morphology, described below.

Cheek muscles (Fig 1). - The adductor mandibulae is divided into four sections. Section A1 of the adductor mandibulae (A_1) originates from the dorsolateral margin of the preopercular. It is converted into a tapering tendon to the ventromedial head of the maxillary anteriorly, and connects the angular through a thin tendon ventrally. Section A2 of the adductor mandibulae (A_2) originates from the anterolateral margin of the preopercular and shares a myocomma with section Aw anteroventrally; some fibers of section A2 insert on the ligamentum primordium (Lp) anterolaterally. Section A3 of the adductor mandibulae (A_3) originates from the lateral surfaces of the hyomandibular and the metapterygoid, and inserts to the ligamentum primordium. Section Aw of the adductor mandibulae (Aw) occupies the meckelian fossa of the medial surface of the lower jaw. The levator arcus palatini (LAP) is clearly divided into two sections that have a common origin from the lateral surface of the hyomandibular. The anteromedial section of the levator arcus palatini has a tapering tendon that inserts on the sphenotic dorsally. The posterolateral section of the levator arcus palatini broadly inserts on the lateral process of the pterotic. The adductor arcus palatini (AAP) extends to the medial surface of the entopterygoid, metapterygoid, and hyomadibular, and lateral surface of the parasphenoid, prootic, and pterotic, and occupies the posterior half of the orbital floor. The dilatator operculi (DO) originates from the lateral margin of the pterotic and inserts on the anterodorsal tip of the opercle. The levator operculi (LO) forms a broad muscular sheet between the pterotic and the opercle. The adductor hyomandibulae (AH) originates from the pterotic and inserts on the dorsomedial surface of the hyomandibular.

Muscles of the hyoid region (Fig. 2, A). - The protractor hyoidei (PH) is a slender muscle with a myocomma. It originates from the ceratohyal immediately anterior to the base of the 3rd branchiostegal ray and extends under the intermandibularis (IM) to insert on the anteromedial surface of the dentary. There are two sections of the hyohyoidei abductores. The first section (HAB_1) originates from the distal part of the 1st branchiostegal ray and extends forward to fuse with its antimere at ventral midline and connects to anterior part of the hyoid arch through thin tendinous tissue. The second section of the hyohyoidei abductores (HAB_2) connects the medial portion of each branchiostegal ray to the ventromedial margin of the hyoid arch. The hyohyoidei adductores (HAD) interconnect the distal

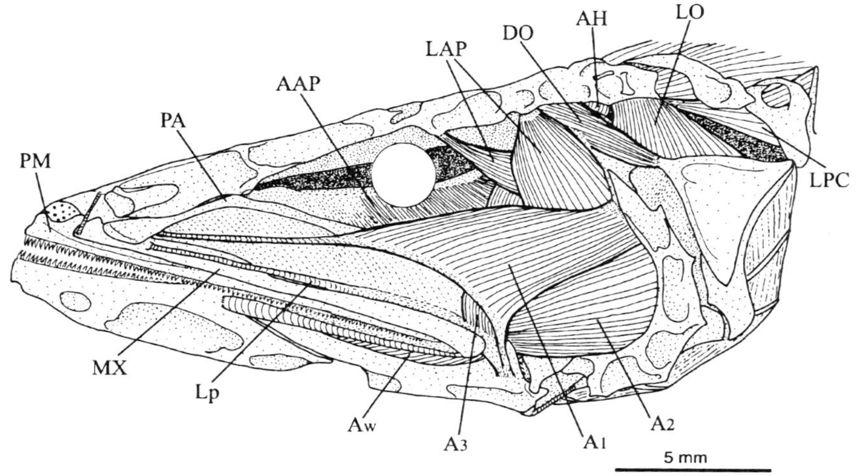

Fig.1. Left lateral view of the cheek musculature of Comephorus dybowskii(HUMZ 113882). AAP - adductor arcus palatini; AH- adductor hyomandibulate; A1, A2, A3, Aw-sections of the adductor mandibulae; DO-dilatator operculi; LAP-levator arcus palatini; LO-levator operculi; LP-ligamentum primordium; LPC-levator pectoralis; MX-maxillary; PA-palatine; PM-premaxillary.

portions of the branchiostegal rays, and fiber arising from the last branchiostegal ray insert on the medial surface of the opercle. The hyohyoides inferioris is absent. The sternohyoideus (SH) originates from the lateral surface of the urohyal, and extends to the ventral part of the cleithrum.

Muscles of the ventral branchial arch region (Fig. 2, B). - The obliqui ventrales (OV1-3) connect the hypobranchial to the ceratobranchial on the 1st to 3rd branchial arches. The recti ventrales consist of two sections. The first(RV_1) connects the 3rd hypobranchial to the 4th ceratobranchial, and the second (RV_2) connects the urohyal to the 3rd hypobranchial. The transversi ventrales consist of three elements. The anterior element (TVa) obliquely interconnects the 4th ceratobranchials of both sides; the middle element (TVm) connects the medial margin of the 4th ceratobranchial to the anterior tip of the 5th ceratobranchial; and the posterior element (TVp) interconnects the posteroventral surfaces of the 5th ceratobranchials of both sides. The rectus communis (RC) originates from the urohyal and tendinously inserts on the lateral margin of the 5th ceratobranchial. The pharyngoclavicularis externus originates from the ventrolateral margin of the 5th ceratobranchial and inserts on the anteroventral surface of the clei-

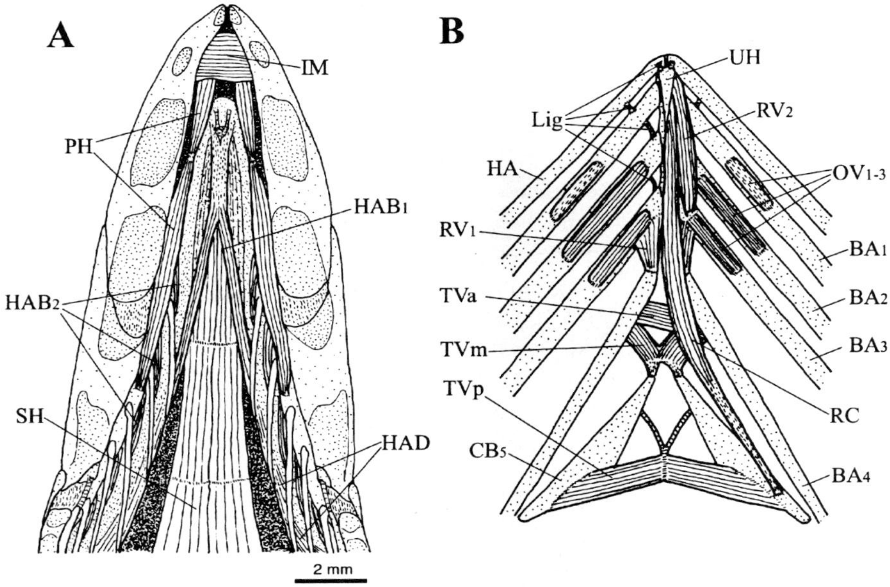

Fig. 2. Ventral view of the hyoid region (A) and schematic illustration of the ventral branchial arch muscles (B) of Comephorus dybowskii (HUMZ 113882). Elements of the pharyngoclavicularis on the fifth ceratobranchial are removed. BA1-4 - branchial arches; CB5 - fifth ceratobranchial; HA - hyoid arch; HAB1-2 - hyohyoidei; HAD - hyohyoidei-adductores; IM - intermandibularis; Lig - ligaments; OV1-3 - obliqui ventrales; PH - protractor hyoidei; RC - rectus-communis; RV1-2 - recti ventrales; SH - sternohyoideus; TVa, TVm, TVp - transversi-ventrales; UH - urohyal.

thrum. The pharyngoclavicularis internus is a slender muscle that originates from the 5th ceratobranchial immediately medial to the externus element and tendinously inserts into the middle of the cleithrum.

Muscles of the occipital region (Fig. 3). - The levator pectoralis (LPC) originates from the posterolateral corner of the pterotic, epiotic, and intercalar. It is subdivided into two parts, which insert on the anterodorsal margin of the cleithrum (CL). The epaxialis inserts onto the epiotic, exoccipital, posttemporal, and posterior surface of the supraoccipital. No branches of the epaxialis extend to the temporal fossa. The obliquus superioris of hypaxialis (OBS) inserts on the ventrolateral surface of the exoccipital medial to the vagus foramen (Vf). The protractor pectoralis is obscure. The retractor dorsalis (RD) connects the posterior margin of the 3rd pharyngobranchial to the ventral surface of the 2nd - 4th centra (C). The levator posterior (LP) is a slender muscle that connects the posterodorsal surface of the 4th epibranchial to the intercalar of the cranium (CR).

Discussion

We compared the musculature of *Comephorus* with that of other cottoid and scorpaeniform fishes described by Yabe (1985), Shinohara (1994), Yabe and Uyeno (1996) and Imamura (1996), and identified certain peculiarities in the muscular morphology of *Comephorus*.

There are two unique aspects of the cheek muscles of *Comephorus*: the adductor mandibulae and the levator arcus palatini. As described by Mandrytza (1990) the adductor mandibulae section A1 of *Comephorus* is distinctly separated from section A2, and not connected with the ligamentum primordium except for its ventral insertion on the angular. The levator arcus palatini is subdivided into two distinct sections. This differs from its anatomy in the cottid fishes, in which the separation between sections A1 and A2 of the adductor mandibulae is incomplete, and section A1 is in close contact with the ligamentum primordium anteroventrally. The levator arcus palatini of cottoid fishes is present in the form of a single broad muscle mass (Yabe, 1985). Comparison of the Cottocomephorinae and

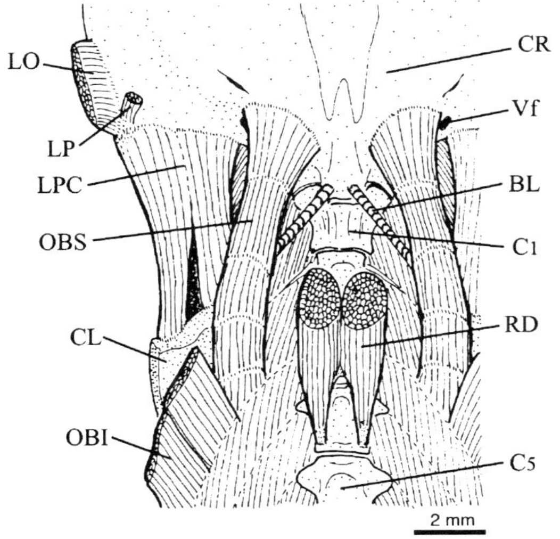

Fig. 3. Ventral view of the occipital region of *Comephorus dybowskii* (HUMZ 113882). BL - Baudelot's ligament; CL - cleithrum; CR - cranium; $C_{1,5}$ - first and fifth centra; LP - levator posterior; LPC - levator pectoralis; OBI - obliquus inferioris of hypaxialis; OBS - obliquus superioris of hypaxialis; RD - retractor dorsalis; Vf - vagus foramen.

Abyssocottidae specimens revealed almost the same anatomical features of the cheek muscles as in the Cottidae. Although the cheek muscles, especially the adductor mandibulae, are known to vary widely among fishes, no other scorpaeniform fishes have been observed to have the same anatomy as *Comephorus* (Winterbottom, 1974; Yabe, 1985). The individual muscles in the hyoid region do not form a developed muscular mass in *Comephorus*, and two peculiarities have been observed: the protractor hyoidei is not fused with its antimere element, and the first section of hyohyoidei abductores has a narrow area of fusion with the antimere element. Scorpaeniform fishes, including cottids have well-developed protractor hyoidei, muscle masses, which are broadly fused with the antimere in the ventral midline. The anatomy of the hyohyoidei abductores varies widely in teleostean fishes (Winterbottom, 1974). Simple fusion of the 1st element of the hyohyoidei abductores on the two sides is characteristic of the cottoid and hexagrammid Scorpaeniformes (Yabe, 1985; Shinohara, 1994). This anatomy is observed in *Comephorus*, but the fusion is extremely limited, whereas in cottoids and hexagrammids the muscle is broadly fused in the ventral midline. The muscles in the branchial region of *Comephorus* exhibit two unique anatomical characterisitcs. The second section of the recti ventrales connects the urohyal to the 3rd hypobranchial in *Comephorus*. The only Scorpaeniformes found to display this morphology are the cottoids (except the Ereuniidae and Rhamphocottidae) and the liparids (Yabe, 1985). A middle element of the transversi ventrales connecting the 4th to 5th ceratobranchials is observed in *Comephorus*, but has never been found in other scorpaeniform fishes. These peculiarities of the muscle morphology of the cheek, hyoid, and branchial regions of *Comephorus* are considered highly derived structures related to pelagic feeding behavior.

In *Comephorus* the obliquus superioris of the hypaxialis inserts on the ventrolateral surface of the exoccipital in the cranium. Similar anatomy has been observed in numerous fish groups, but not in cottoid fishes, with the exception of the most primitive cottid, *Jordania zonope*. In the cottoids, the obliquus superioris inserts on the posterodorsal margin of the cleithrum, and does not extend to the occipital region of the cranium. The cottoid fishes have an additional muscle element in the occipital region called the extrinsic muscle of the swimbladder that connects the occipital region to the cleithrum (Hallacher, 1974; Yabe, 1985). Among the endemic sculpins of Lake Baikal, Cottocomephorinae and Abyssocottidae fishes possess the extrinsic muscle of the swimbladder, but *Comephorus* does not. The presence of the obliquus superioris inserting on the cranium and the absence of the extrinsic muscle of swimbladder in the occipital region of *Comephorus* are regarded as plesiomorphic states in the Acanthomorpha.

The morphology of this region does not well support the monophyly of all endemic sculpins in Lake Baikal hypothesized by Taliev (1955). The myological peculiarities described here indicate that the origin or the possible sister groups of *Comephorus* cannot be limited to the endemic cottoids in Lake Baikal.

Acknowledgements
We wish to thank M. Grachev, K. Numachi, N. Miyazaki, K. Morino, and K. Mashiko for their aid through the program of the Baikal International Center for Ecological Research, and A. Goto and N. Nishida for their valuable discussion. This study was financed by grants-in-aid for the International Science Research Program (Nos. 04041035 and 07041130).

References

Berg L. S., 1940, Classification of fishes, both recent and fossil. Tr. Inst. Zool. Akad. Nauk. SSSR, 5:87-517.

Grachev M. A., S. Ya. Slogodyanyuk, N. G. Kholodilov, S. P. Fyodorov, S. I. Belikov, D. YU. Sherbakov, V. G. Sideleva, A. A. Zubin and V. V. Kharchenko, 1992, Comparative study of two protein-coding regions of mitochondrial DNA from three endemic sculpins (Cottoidei) of Lake Baikal. J. Mol. Evol., 34, 85-90.

Hallacher L. E., 1974, The comparative morphology of extrinsic gasbladder musculature in the scorpionfish genus Sebastes (Pisces: Scorpaenidae). Proc. Calif. Acad. Sci., ser. 4, 40, 59-86.

Imanura H., 1996, Phylogeny of the family Platycephalidae and related taxa (Pisces: Scorpaeniformes). Species Diversity, 1, 123-233.

Mandrytza S. A., 1990, The condition of the musculus adductor mandibulae of the scorpaeniform fishes. Proc. Zool. Inst., Leningrad, 222, 75-93.

Nelson J. S., 1994, Fishes of the world. 3rd ed. Wiley-Interscience, New York, xvii+600pp.

Shinohara G., 1994, Comparative morphology and phylogeny of the suborder Hexagrammoidei and related taxa (Pisces: Scorpaeniformes). Mem. Fac. Fish. Hokkaido Univ., 41, 1-97.

Sideleva V. G., 1982, Seismosensory systems and ecology of the Baikalian sculpins (Cottoidei). Izv. Nauka, Novosibirsk, 149 pp. (In Russian).

Sideleva V. G., 1994, Speciation of endemic Cottoidei in Lake Baikal. Arch. Hydrobiol. Beih. Ergebn. Limnol., 44, 441-450.

Sideleva V. G., 1999, A new species of the endemic Baikal genus *Batrachocottus* (Cottidae). Vopr. Ikhtiol., 39, 149-154 (in Russian).

Slogodyanyuk S. Ya., M. E. Pavlova and S. I. Belikov, 1994, Analysis of tenderm DNA repeats of cottoid fish in Lake Baikal by direct consensus sequencing. Mol. Mar. Bio. Biotech., 3, 301-306.

Slogodyanyuk S. Ya., S. V. Kirilchik, M. E. Pavlova, S. I. Belikov and A. L. Novitsky, 1995, The evolutionary relationships of two families of cottoid fishes of Lake Baikal (East Siberia) as suggested by analysis of mtDNA. J. Mol. Evol., 40: 392-399. Baikal by direct consensus sequencing. Mol. Mar. Bio. Biotech., 3, 301-306.
Taliev D. N., 1955, Sculpins of Baikal (Cottoidei), Akad. Nauk, SSSR, East Siberia Branch, Moscow, 603 pp.
Winterbottom R., 1974, A descriptive synonymy of the striated muscles of the Teleostei. Proc. Acad. Nat. Sci. Phila., 125, 225-317.
Yabe M., 1985, Comparative osteology and myology of the superfamily Cottoidea (Pisces: Scorpaeniformes), and its phylogenetic classification. Mem. Fac. Fish. Hokkaido Univ., 32, 1-130.
Yabe M. and T. Uyeno, 1996, Anatomical description of *Normanichthys crockeri* (Scorpaeniformes, incertae sedis: family Nomanichthyidae). Bull. Mar. Sci., 58(2), 494-510.

Lake Baikal
K. Minoura (editor)
2000 Elsevier Science B.V.

Morphometric comparison of skulls of seals of the subgenus *Pusa*

Amano, M.[1]*, Koyama, Y.[2], Petrov, E.A.[3], Hayano, A.[4], and Miyazaki, N.[1]
[1] Otsuchi Marine Research Center, Ocean Research Institute, the University of Tokyo, Otsuchi, Iwate 028-1102, Japan
Fax: 0193-42-3714, e-mail: amano@wakame.ori.u-tokyo.ac.jp
[2] Kagoshima Broadcasting Center, NHK, Kagoshima 890-0061, Japan
[3] Limnological Institute, Siberian Division of Russian Academy of Sciences, Irkutsk 664033, Russia
[4] Department of Zoology, Graduate School of Science, Kyoto University, Sakyo, Kyoto 606-8502, Japan
(*corresponding author)

Abstract

Skull measurements were compared among three species of seals of the subgenus *Pusa*. Baikal seals have larger relative growth coefficients for width of snout and nares. A larger zygomatic width, greater length of jugal and smaller orbital width are evidence of the larger orbit of this species. Analyses of covariance revealed that Caspian seals have narrower skulls, brain cases, and bullae. Ringed seals have a wider cranium, shorter snout, and larger bullae. Canonical discriminant analysis showed that inter-species differences are clearly greater than intra-species differences in ringed seals. Phenograms indicated slightly closer morphological affinity between Baikal and ringed seals than between Baikal and Caspian seals. This finding lends support to the hypothesis that the Baikal seal is a descendant of an Arctic ancestor rather than a Paratethys relict.

Introduction

Baikal seals, *Phoca sibirica*, Caspian seals, *P. caspica*, and ringed seals, *P. hispida*, share a small body size, cat-like face, delicate skull structure, and affinity for ice, and they are thought to be closely related (Scheffer, 1958). They are usually classified together in the subgenus *Pusa* (e.g. Burns and Fay, 1970; Frost and Lowry, 1981; Corbet and Hill, 1991), but some scientists have given them full genus status (e.g. Scheffer, 1958; Rice, 1999). The phylogenetic relationships among *Pusa* seals have not been elucidated, and their origins are still a matter of controversy. Although some studies have compared skull morphology, the results do not agree, and while most of them have suggested a closer relationship between Baikal and Caspian seals (Ognev, 1962; Pastukov, 1969;

Timoshenko, 1975), Chapskii (1955) concluded that Baikal seals more closely resemble ringed seals. The present study reevaluated the morphological relationships among the skulls of these three species by sophisticated statistical methods, including multivariate statistics.

Materials and Methods

The skull specimens of Baikal, Caspian, and ringed seals used in this study were collected during the Japan-Russia cooperative research expeditions in 1992 (Lake Baikal), 1993 (Caspian Sea), and 1995 (Dikson, Russian Arctic). All of the specimens were prepared and deposited in the National Science Museum, Tokyo. We also examined museum collections of ringed seals from other localities: the Museum of Comparative Zoology, National Science Museum, Tokyo, Swedish Museum of Natural History, University of Alaska Museum, and Zoological Museum of Helsinki University. Ringed seal specimens included the subspecies *P. h. hispida*

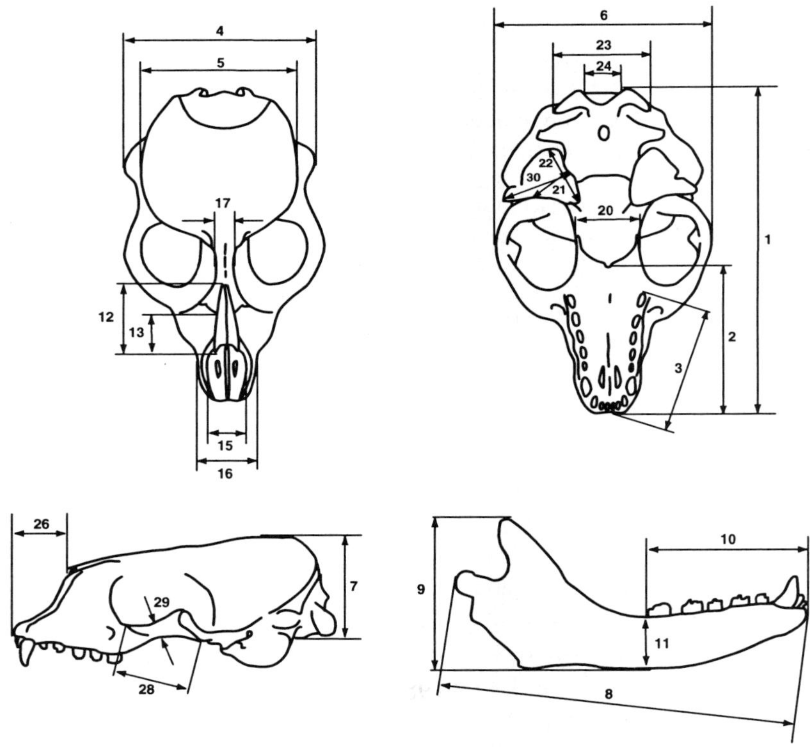

Fig. 1. Skull measurements of *Pusa* seals. Numbers refer to Table 1.

Table 1. Skull measurements of *Pusa* seals. Abbreviations are in parentheses.

1. Condylobasal length (CBL).
2. Palatal length (PL).
3. Length of upper tooth row (LUTR).
4. Greatest width at mastoids (GWM).
5. Greatest width of cranium (GWC).
6. Greatest zygomatic width (ZW).
7. Height of cranium (HC).
8. Length of mandible (LM).
9. Height of mandible at coronoid process (HMC).
10. Length of lower tooth row (LLTR).
11. Height of mandible (HM).
12. Overall length of nasals (LNAS).
13. Length of maxillo-frontal suture to anterior end of nasals (LMFN).
14. Width of nasals at maxillo-frantal suture (WNAS)*.
15. Maximal width of external nares (WEN).
16. Width of snout at canines (WSN).
17. Least interorbital width (IOW)*.
18. Greatest anterior-posterior length of second upper premolar (LPM)*.
19. Width of palate behind last molars (WPL)*.
20. Least width of palate at pterygoid hamuli (WPH).
21. Width of bulla notch anterior to auditory process - middle of carotid foramen (WB).
22. Greatest length of bulla (LB).
23. Greatest width at condyles (WCD).
24. Greatest width of foramen magnum (WFM).
25. Greatest height of foramen magnum (HFM)*.
26. Length of snout from anterior edge of nasals (LSN).
27. Distance from posterior end of vomerine septum to medial edge of palate (DVP)*.
28. Greatest length of jugal (LJ).
29. Heigh of jugal (HJ)
30. Width of bulla from tip of auditory process to anterior edge of carotid foramen (WB2).

* Not illustrated in Fig. 1.

(Alaska, Canadian Arctic, and Russian Arctic), *P. h. botnica, P. h. ladogensis, P. h. ochotensis,* and *P. h. saimensis*.

Thirty characters were measured with vernier calipers to the nearest 0.1 mm, as described by Burns *et al.* (1984) (Table 1, Fig. 1). Specimens with a condylobasal length (CBL) of 140 mm or greater were used so that the size ranges of the three species would be comparable. The ultimate sample size of each species was 26 Baikal seals, 37 Caspian seals, and 218 ringed seals.

We did not take sex into account, assuming that the inter-specifes differences would be much greater than sexual differences, although some differences have been reported in the skull measurements of Baikal and Caspian seals (Ognev, 1962; Timoshenko, 1975).

Differences in the relative measurements to CBL were examined by analysis of covariance using CBL as a covariate and by post hoc pair-wise comparisons by the Tukey method (Zar, 1996). Canonical discriminant analysis was used with species and localities of ringed seals as independent variables to demonstrate intra- and inter-species differences. Finally, phenograms of the three species were obtained by Neighbor-joining and UPGMA methods based on Maharanobis' distances using PHYLIP ver 3.5 (Felsenstein, 1993). To avoid bias of Maharanobis' distances by including multiple characters correlated with each other, measurements to calculate the distances were selected by stepwise discriminant analysis. SAS (SAS Inst. Inc., 1989) at the Kyoto University Data Processing Center was used to perform the statistical analyses.

Results and Discussion

Analyses of covariance revealed significant differences in the slopes of the regression lines of several measurements (Table 2) and in the adjusted means of most of the measurements (Table 3).

Table 2. Inter-specific differences among skull measurements of subgenus *Pusa*; significant differences ($P<0.05$) in relative growth coefficients (GC) detected by ANCOVAS with CBL as a covariate. Numbers and abbreviations for measurements refer to Table 1.

Measurement	*P. caspica*		*P. hispida*		*P. sibirica*		Difference
	N	GC	N	GC	N	GC	
3.LUTR	36	1.142	240	0.897	26	1.052	PC>PS>PH
9.HMC	37	1.744	221	1.454	25	1.729	PC>PS>PH
15.WEN	37	1.121	236	0.935	26	1.481	PS>PC>PH
16.WSN	37	0.991	239	1.068	26	1.397	PS>PH>PC
19.WPL	37	1.108	238	0.661	26	0.592	PC>PH>PS
20.WPH	37	0.771	231	0.518	24	0.857	PS>PC>PH
21.WB	37	0.169	237	0.512	24	0.553	PS>PH>PC
23.WCD	37	0.339	240	0.560	26	0.391	PH>PS>PC
24.WFM	37	-0.286	239	0.191	26	-0.337	PH>PC>PS

Baikal seals have larger slopes for width of snout and nares (WEN, WSN, WPH), which means that their nasal passage widens rapidly. Large nasal passages may allow seals to breathe faster and decrease surface time. In fact, this species has been reported to dive repeatedly for extended periods (Stewart *et al.*, 1996) and to have short surfacing intervals (Ponganis *et al.*, 1997). Larger zygomatic width, greater length of jugal, and smaller interorbital width indicate a larger orbit containing a large eye (Endo *et al.*, 1998). The large eyes have been thought to be an adaptation to the clear lake water environment, in which seals' sensation must be largely mediated by vision (Chapski, 1955).

Caspian seals have larger relative slopes for mouth measurements (LUTR, WPL) and smaller slopes for width of the bulla (WB). They also have smaller adjusted means for skull width (GWM, ZW), brain case (GWC, HC), and bulla (LB, WB2) measurements. Narrower zygomatic width and a shorter jugal indicate that this species has a smaller orbit than

Table 3. Inter-specific differences among skull measurements of subgenus *Pusa*; significant differences (P<0.05) in adjusted means (AM, log-transformed) detected by ANCOVAS with CBL as a covariate. Numbers and abbreviations for measurements refer to Table 1.

Measurement	*P. caspica*		*P. hispida*		*P. sibirica*		Difference
	N	AM	N	AM	N	AM	
2.PL	37	1.836	235	1.825	24	1.859	PS>PC>PH
4.GWM	37	1.928	241	1.994	25	1.953	PH>PS>PC
5.GWC	37	1.871	241	1.921	25	1.898	PH>PS>PC
6.ZW	37	1.932	240	1.982	24	1.996	PS>PH>PC
7.HC	37	1.735	237	1.757	25	1.755	PH, PS>PC
8.LM	37	2.002	217	2.000	25	2.020	PS>PC, PH
10.LLTR	36	1.644	215	1.636	26	1.657	PS>PC, PH
12.LNAS	35	1.566	234	1.546	24	1.603	PS>PC, PH
13.LMFN	37	1.407	237	1.359	25	1.441	PS>PC>PH
14.WNAS	37	0.831	239	0.767	25	0.736	PC>PH>PS
17.IOW	37	0.780	240	0.768	25	0.656	PC, PH>PS
18.LPM	35	0.757	225	0.784	26	0.863	PS>PH>PC
22.LB	37	1.446	237	1.537	24	1.475	PH>PS>PC
25.HFM	37	1.319	230	1.355	25	1.382	PS, PH>PC
26.LSN	37	1.377	238	1.322	25	1.382	PC, PS>PH
27.DVP	36	1.186	233	1.259	22	1.300	PS, PH>PC
28.LJ	37	1.619	233	1.640	26	1.673	PS>PH>PC
29.HJ	36	0.962	235	0.877	26	0.937	PC, PS>PH
30.WB2	37	1.431	240	1.546	24	1.490	PH>PS>PC

the others. Caspian seals have small anatomical features related to sensory functions.

Ringed seals have a wider cranium (GWM, GWC), shorter snout length (PL, LSN), and larger bulla (LB, WB2). The larger bullae suggest the importance of hearing ability in this species. The large portion of life of ringed seals are thought to be formed under considerable predatory pressure by polar bears and arctic foxes (Lydersen, 1995). Hearing ability may be important for mothers to detect predators' approach in the lairs where they nurse the pups.

A scatter plot between the first and second canonical variates showed that inter-species differences are clearly larger than inter-locality differences between ringed seals (Fig. 2). This suggests that Caspian and Baikal seals do not have affinity with particular subspecies or populations of ringed seals.

Both the Neighbor-joining and UPGMA phenograms indicated slightly closer morphological affinity between Baikal and ringed seals than between Baikal and Caspian seals (Fig. 3). This relationship did not change when we omitted the specimens with CBL smaller than 140 mm to diminish the ontogenetic variation, but distance between Baikal and ringed seals was shortened (Fig. 4). The branch lengths were not distinctly different, indicating that the three species are well differentiated from one anoth-

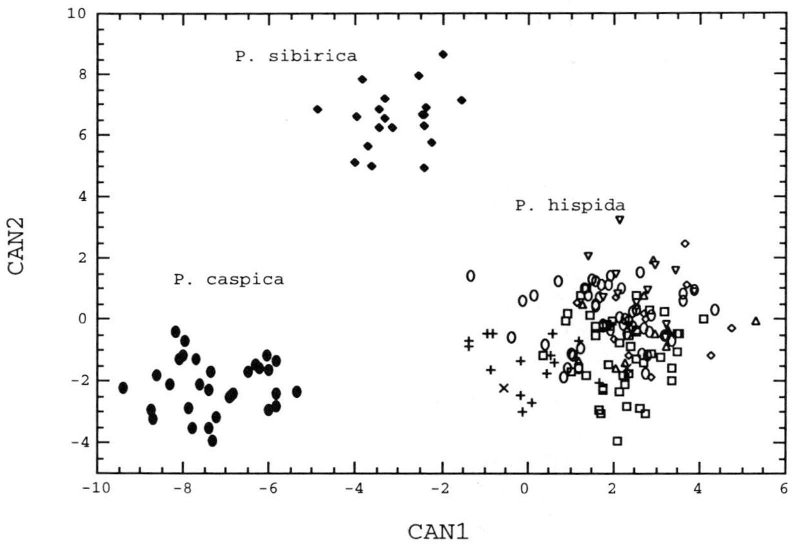

Fig. 2. Scatter plots of the first and second canonical discriminant scores. Symbols in P. hispida indicate different subspecies and localities.

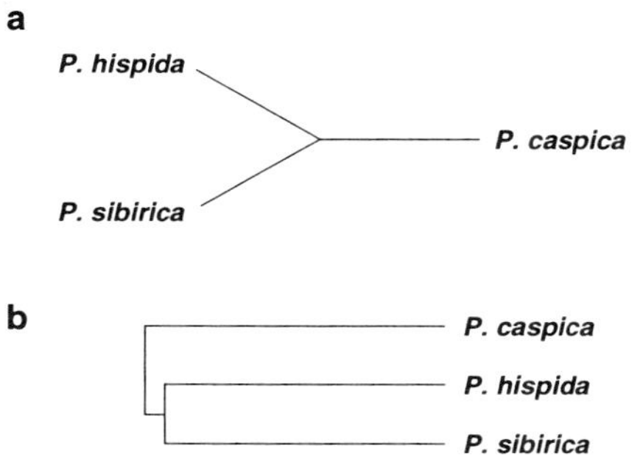

Fig. 3. Neighbor-joining (a) and UPGMA (b) phenograms based on Maharanobis' distances from 20 characters selected by stepwise discriminant analysis using specimens with a CBL of 140 mm or greater.

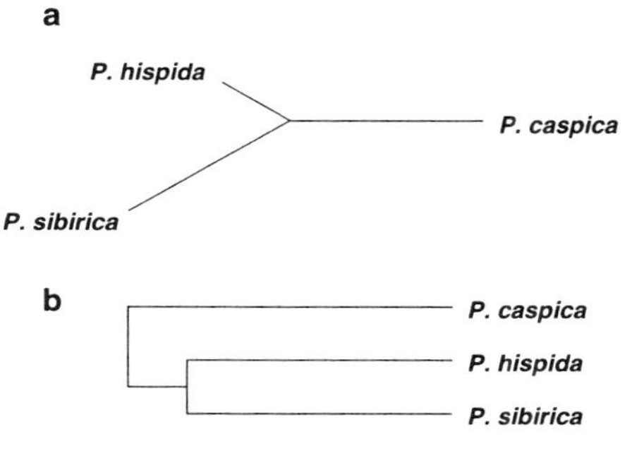

Fig. 4. Neighbor-joining (a) and UPGMA (b) phenograms based on Maharanobis' distances from 17 characters selected by stepwise discriminant analysis using specimens with a CBL of 160 mm or greater.

er morphologically. It is not surprising that the former studies using subjective or univariate comparisons yielded conflicting results for morphological similarity among *Pusa* seals.

The results of this study do not necessarily imply phylogenetic relationships among the species, because Baikal and Caspian seals have adapted to the distinct lake habitats, and several skull characters are considered to be the result of such adaptations, as discussed above. However, the preliminary results of a recent genetic study also indicate a greater similarity between ringed and Baikal seals (H. Sasaki and K. Numachi, per. comm.), and their overall morphological similarity may to some extent represent their phylogenetic relationships.

Two hypotheses based on the fossil record and geological information have been proposed for the origin of Baikal and Caspian seals. One states that recent *Pusa* seals share a common ancestor, *Phoca pontica*, which inhabited the Paratethys Sea during the Miocene and Pliocene epochs, and that the recent Baikal and Caspian seals evolved directly from them (Chapskii, 1955). The other states that the common ancestor of the these seals once moved into Arctic from the Paratethys and then migrated southward along the Ob and Yenisey River system during glaciation (Davies, 1958; Repenning *et al.*, 1979). Both the results of this study and genetic findings showing closer affinity between Baikal and ringed seals seem to support the latter hypothesis.

Acknowledgements

The authors thank M. Grachev, A. Timonin, M. Ivanov, and other colleagues at the Limnological Institute, Siberian Division of the Russian Academy of Sciences, and S. Khuraskin of the Caspian Scientific Research Institute. S. Tanabe, H. Sasaki, H. Nakata, M. Ichikawa, S. Belikov, and A. Boltunov assisted us in the fieldwork. We also thank K. Numachi and T. Kawai for their support and arrangements for the research. We examined museum collections under the care of Joe Cook and Gordon Jarrel of the University of Alaska Museum, Ann Forstèn of the Zoological Museum, University of Helsinki, Maria Rutzmoser of the Museum of Comparative Zoology, Adam Stanczak of the Swedish Museum of Natural History, and Tadasu Yamada of the National Science Museum, Tokyo. The present study was financially supported by a grant-in-aid from the International Scientific Research (Project nos. 04041035, 07041130 and 09041149) and Scientific Research (B)(2) (09460086) of the Ministry of Education, Science, Sports and Culture of Japan.

References

Burns J. J. and F. H. Fay, 1970, Comparative morphology of the skull of the ribbon seal, Hsitriophoca fasciata, with remarks on systematics of Phocidae. Journal of Zoology, London, 161, 363-394.

Burns J. J., F. H. Fay and G. A. Fedoseev, 1984, Craniological analysis of harbor and spotted seals of the North Pacific region, In: Soviet-American cooperative research on marine mammals, eds. Fay F. H. and G. A. Fedoseev, NOAA Tech. Rep. NMFS 12.

Chapskii K. K., 1955, Contribution to the problem of the history of development of Caspian and Baikal seals. Trudy Zoologicheskogo Instituta Akademii Nauk SSSR, 17, 200-216. (In Russian).

Corbet G. B. and J. E. Hill, 1991, A world list of mammalian species. Oxford University Press, Oxford, 243pp.

Davies J. L., 1958, Pleistocene geography and the distribution of northern pinnipeds. Ecology, 39, 97-113.

Endo H., H. Sasaki, Y. Hayashi, E. A. Petrov, M. Amano and N. Miyazaki, 1998, Macroscopic observations of the facial muscles in the Baikal seal (*Phoca sibirica*). Marine Mammal Science, 14, 778-788.

Felsenstein J., 1993, PHYLIP ver 3.5.

Frost K. J. and L. F. Lowry, 1981, Ringed, Baikal and Caspian seals *Phoca hispida* Schreber, 1775; *Phoca sibirica* Gmelin, 1788 and *Phoca caspica* Gmelin, 1788, In: Handbook of marine mammals volume 2: seals, S. H.

Ridgway and R. J. Harrison, eds., Academic Press, London, 359pp.
Lydersen C., 1995, Energetics of pregnancy, lactation and neonatal development in ringed seals (*Phoca hispida*), In: Whales, seals, fish and man, eds. A. S. Blix, L. Walløe and Ø. Ulltang, Elsevier Science B.V., Amsterdam, 720pp.
Ognev S. I., 1962, Mammals of U.S.S.R. and adjacent countries, volume III Carnivora. Israel Program for Scientific Translation, Jerusalem. 640pp.
Pastukhov V. D., 1969, Craniometric characteristics of *Pusa sibirica* (Pinnipedia, Mammalia). Zoologicheski Jhurnal, 48, 722-733. (In Russian).
Ponganis P. J., G. L. Kooyman, E. A. Baranov, P. H. Thorson and B. S. Stewart, 1997, The aerobic submersion limit of Baikal seals, *Phoca sibirica*. Canadian Journal of Zoology, 75, 1323-1327.
Reppening C. A., C. E. Ray and D. Grigorescu, 1979, Pinniped biogeography. In: Historical biogeography, plate tectonics and the changing environment, eds. J. Gray and A. J. Boucot, Oregon State University Press, Corvallis, 512pp.
Rice D. W., 1999, Marine mammals of the world, systematics and distribution. Special Publication No. 4, the Society of Marine Mammalogy, Lawrence, KS, 231pp.
SAS Inst. Inc., 1989, SAS/STAT User's Guide, Version 6, 4th edition. Vol.2. SAS Institute Inc., Cary, NC, 846pp.
Scheffer V. B., 1958, Seals, sealions, and walruses. Stanford University Press, Stanford, 179pp.
Stewart B. S., E. A. Petrov, E. A. Baranov, A. Timonin and M. Ivanov (1996) Seasonal movements and dive patterns of juvenile Baikal seals, *Phoca sibirica*. Marine Mammal Science, 12, 528-542.
Timoshenko U. K., 1975, Craniometric features of seals of the genus Pusa. Rapport et Rapport et Procès-verbaux des Réunions Conseil International Pour L'Exploration de la Mer, 169, 161-164.
Zar J.H., 1996, Biostatistical Analysis, Third Edition. Prentice-Hall, Inc., New Jersey, 662pp.

Lake Baikal
K. Minoura (editor)
2000 Elsevier Science B.V.

The importance of habitat stability for the prevalence of sexual reproduction

Martens, K.* and Schön. I.
Royal Belgian Institute of Natural Sciences, Freshwater Biology, Vautierstraat 29, 1000 Brussels, Belgium
Fax: +32-2-64 64 433. Email: martens@kbinirsnb.be

* to whom the correspondence should be addressed.

Introduction

Why sex has become the dominant reproductive mode in both plant and animal kingdom, is still a largely unsolved enigma (Maynard Smith, 1998). Through the presence of recombination, sexual reproduction has a more flexible genetic variability, but the presence of males makes that up to 50% of the population does not actively contribute to the production of offspring (the so-called two-fold cost of sex; Kondrashov, 1993). Furthermore, recombination and crossing-over during meiosis break up favourable allele combinations as much as new ones are created (Hudson and Peck, 1996). Asexual lineages do not have these problems and have the additional advantage that they are potentially better dispersers, because in sexual species, two individuals of opposite gender must colonise the same habitat and must find each other in a spatially and temporally very diluted environment.

Nevertheless, more than eight different theoretical considerations explain why sex is the most advantageous reproductive mode and predict that asexual lineages are invariably doomed to early extinction (Butlin et al., 1998 - Table 1). Most of these theories and hypotheses deal with intrinsic biological features of organisms themselves and can be considered as 'variation and selection' models. Only few take habitat-related aspects into account: the sib-competition, and especially the tangled bank (lottery) models refer to the advantage of sexual reproduction in habitats with large complexity, i.e. with a high number of microhabitats and niches. We argue that ecological and geological habitat stability are important factors in the prevalence of sexual reproduction and illustrate this by using ancient lake ostracods.

Non-marine ostracods (small, bivalved crustaceans) are an ideal model group to investigate the paradox of sex (Martens, 1998a), as they display a broad variety of reproductive modes, ranging from ancient asexuals (which have managed to escape extinction for millions of years) over species with mixed reproduction (with two types of females and geographical parthenogenesis) to exclusively sexual species. Moreover, ostracods are not only one of the few extant groups with an extensive fossil record, but at the same time constitute one of the rare fossil groups with such a large extant diversity. The group thus reflects the evolutionary history of reproductive modes over hundreds of millions of years.

Ancient lakes are, in their turn, ideal sites to study such questions as they are, most likely, the cradle in which their endemic flocks originated and allow only limited habitat tracking. When conditions change, organisms will generally migrate and seek places with their preferred environment (track their habitat) rather than change and adapt. In ancient lake basins, extant faunas will have the choice between emigration or extinction on the one hand, or adaptation to conditions on the other. We will thus have a high degree of certainty that extant faunas originated in that basin and that they have a history of constant adaptation to changing environmental conditions.

The hypothesis

We hypothesise that sexual reproduction has an advantage over asexuality in long-lived habitats (geological stability) which are unstable on ecological time scales (Martens, 1998a). There are two main (theoretical) reasons for this. Firstly, processes favouring sexuality over asexuality work relatively slowly, so that short-lived habitats (for example Holocene lakes) simply do not allow enough time for sexuality to prevail over asexuality. Secondly, there is mounting evidence that standing genetic diversity is similar in both sexuals and asexuals (except for ancient asexual darwinulids, which most likely combine low genetic variability with a general purpose genotype), so that only the plasticity, i.e. the tempo of change, of variability differs between reproductive modes. In ecologically stable conditions, asexuality will mostly prevail.

The evidence

There are six main lines of evidence. (1) Endemic species of ancient lakes nearly invariably occur in bisexual populations, (although the sex ratio can show significant deviations) and ostracods are no exception (Martens, 1994). (2) Although ancient lake ostracods themselves reproduce sexually, their closest, extra-lacustrine relatives can be exclusively parthenogenetic. (3) This pattern is also compatible with higher, taxonomic levels. The Darwinuloidea, an ancient asexual lineage which persisted for at least 100 million years without sex, hardly occur in ancient lakes, and certainly have no endemic species flocks there. The family Cyprididae, which has mixed reproduction, but with a dominance of asexual lineages, are nearly completely absent in spite of the fact that they constitute up to 80% of the specific diversity of non-marine ostracods in short-lived habitats (Martens *et al.*, 1998). The most extensive ostracod radiations in ancient lakes are represented by Cytheroidea and Candonidae, two groups in which sex is the most dominant reproductive mode (Martens, 1994). (4) *Cytherissa lacustris* constitutes a recent branch of the extensive (50+ species) Baikalian *Cytherissa*-flock (Schön *et al.*, in press). Outside of the Baikalian province, throughout the Holarctic, it reproduces exclusively parthenogenetically, whereas in Baikal only sexual populations can be found. This means that the ability to reproduce asexually has survived genetically throughout the 3 million year evolution of Baikalian *Cytherissa*, but that it can only be expressed outside of that lake, namely in lakes of Holocene age. (5) Holocene water bodies are geologically young (c 10,000-15,000 yrs) and as the Holocene has been a relatively stable interglacial period (Horne and Martens, 1999), a high incidence of asexuality in non-marine ostracods (> 80%) occurs. (6) Intuitively, asexual reproduction is strongly linked to life in temporary environments. Short-lived pools indeed have high numbers of asexual species, but geologically old temporary pools, such as those in southwestern Africa, have high incidence of (endemic) sexual ostracod and phyllopod species (Martens, 1998b).

Discussion

Clearly, sexual reproduction is essential for the persistence of species flocks in ancient lakes. Theoretical considerations explain these patterns by referring to the more flexible gene pool, which will allow sexuals to adapt more rapidly to alterations in the environment. However, ancient lakes are generally considered to be very stable habitats. It is important to distinguish between geological and ecological stability: the former term deals with the continued persistence of a water body over long periods of time (> 1 million years), whereas the latter refers to environmental fluctuations over shorter time frames, either predictable (cyclic) or unpredictable (catastrophic). There is mounting evidence for long-lived lakes such as Tanganyika and Baikal that both types of ecological fluctuations can be very common in ancient lakes (Martens, 1997). Climatically induced cyclic fluctuations in limnological conditions (temperature, oxygen, salinity) have been demonstrated from the analysis of long cores in Baikal (Williams *et al.*, 1997) and short cores in Tanganyika (Lezzar *et al.*,1996). Circumstantial evidence indicates recurrent catastrophic changes in both lakes: sharp lake-level fluctuations, with salinity crises, in Tanganyika and changes in temperature regime, accompanied by changes in oxygen availability, in Baikal. A combination of such geological stability and (two-fold) ecological instability thus occurs in both ancient lakes. The comparison of the situation in ancient lakes with that in other extant aquatic habitats indeed confirms the relevance of habitat stability for the incidence of asexuality.

There are some further considerations. Firstly, the above demonstrates that sexuality is necessary for the persistence of lineages in ancient lakes: ecological instability gives sexuals an edge over asexuals due to their higher genetic flexibility, while geological stability allows sexuals to obtain selective advantage over longer time frames. However, Table 1 also shows

Table 1: summary of the most commonly cited hypotheses explaining the prevalence of sexual reproduction and the early extinction of asexual lineages (adapted after Butlin et al. 1998). The presently discussed hypothesis is also added.

Variation and Selection models
1. Müller's ratchet
Random loss of mutation-free genotype.
2. Mutational load
Accumulation of deleterious mutations.
3. Fisher-Müller Accelerated Evolution
Recombination combines new (advantageous) mutations rapidly, which in asexuals would have to occur sequentially.
4. Red Queen (arms races)
Continuous adaptive evolution is required to survive in constantly changing biotic interactions.
5. Fluctuating Selection
Abiotic environments, and thus the optimum phenotype, constantly fluctuate.

Immediate benefit
6. DNA repair
Repair can occur from homologue chromosomes during meiosis.

Ecological models
7. Sib-competition (lottery model)
Sexuals with more diverse offspring will have a better chance to survive in a patch which can support limited numbers of adults.
8. Tangled-bank
Diversification is a better strategy in a complex (saturated) habitat.
9. Habitat stability
Geological stability combined with ecological fluctuations will allow sexuals to outcompete asexuals.

that several theories refer to the higher evolutionary potential of sexuals (eg. Fisher-Müller accelerated evolution, sexual selection leading to speciation). Because only sexuals can survive in ancient lakes, such biota are also more likely to speciate faster than aquatic biota in short-lived habitats, in which a larger fraction reproduces asexually. As a consequence, the high rate of speciation can be seen as coincidental, resulting from a combination of necessity of sexuality to survive in ancient lakes, together with reduced habitat tracking. Speciation in ancient lakes, and the resulting high standing endemic diversity, may thus be regarded as accidental side effects.

Secondly, it must be noted that asexuality does occur in ancient lakes, however always in conditions which are atypical for the lake as such. Ancient lakes are generally large and not uniform. Asexual darwinulid ostracods are found in places of the lakes which are either polluted (river mouths), or relatively recently inundated (Martens, 1994). Asexual hybrids of endemics snails are found in zones of Baikal which have experienced relatively recent tectonic disturbance (Zubakov *et al.*, 1997). Such sites fail to meet the criterion of geological stability.

Thirdly, it could be argued that the tangled bank hypothesis (Table 1) is sufficient to explain the prevalence of sexuality in ancient lakes. Certainly, this could at least partly be true. However, there are almost 200 species of ostracods in Lake Baikal, and following recent estimates, more than one thousand species of amphipods (Vaïnölä and Kamaltynov, 1999). The tangled bank would require that all of these species are adapted to one species-specific and very specialised niche in order to allow both their sympatric persistence and their origin. Although especially the amphipods have known extensive adaptive radiation, it is hard to imagine that there are enough sufficiently different niches in this lake to account for such a high number of (sexual) species. For example, in Baikalian and Tanganyikan ostracods there is almost no trophic specialisation, and only limited bathymetric and sedimentological specialisation (Martens, 1994). Other factors must therefore also be of importance and we argue that the two types of habitat stability are involved.

Acknowledgements
IS acknowledges Marie Curie Fellowship no. BIO4-98-5086. KM is grateful to the organisers of the meeting for inviting him and acknowledges the grant from the Japanese government which allowed his participation in the symposium. A more extended version of the paper will be published elsewhere.

References

Butlin, R. K., Schön, I. and Griffiths, H.I. (1998) Introduction to reproductive modes. In: Martens, K. (ed.) Sex and parthenogenesis ‑ evolutionary ecology of reproductive modes in non-marine ostracods, 1-24. Backhuys Publ., Leiden.

Horne, D. J. and Martens, K. (1999) Geographical parthenogenesis in European non-marine ostracods: post-glacial invasion or Holocene stability? Hydrobiologia, 391,1-7.

Hudson, L.D. and Peck, J.R. (1996) Recent advances in understanding of the evolution and maintenance of sex. Trends Ecol. Evol. 11, 46-52.

Kondrashov, A.S. (1993) Classification of hypotheses on the advantage of amphimixis. J. Hered. 84, 372-387.

Lezzar, K. E., Tiercelin, J. -J., De Batist, M., Cohen, A. S., Bandora, T., Van Rensbergen, P., Le Turdu, C., Mifundu, W. and Klerkx, J. (1996) New seismic stratigraphy and Late Tertiary history of the North Tanganyikan Basin, East African Rift system, deduced from multichannel and high-resolution reflection seismic data and piston core evidence. Basin Res. 8, 1-28.

Maynard Smith, J. 1998. Evolutionary genetics, 2nd edition. Oxford Univ. Press.

Martens, K. (1994) Ostracod speciation in ancient lakes: a review. In: Martens, K., Goddeeris, B. & Coulter, G. (eds.) Speciation in ancient lakes, Adv. Limnol. 44, 203-222.

Martens, K. (1997) Speciation in ancient lakes. Trends Ecol. Evol. 12, 177-182.

Martens, K. (1998a) Sex and ostracods: a new synthesis. In: Martens, K. (ed.) Sex and parthenogenesis - evolutionary ecology of reproductive modes in non-marine ostracods, pp. 295-322. Backhuys Publ., Leiden.

Martens, K. (1998b) Diversity and endemicity of recent non-marine ostracods (Crustacea, Ostracoda) from Africa and South America: a faunal comparison. Verh. int. Ver. Limnol., 26(4), 2093-2097.

Martens, K., Horne, D. J. and Griffiths, H. I. (1998) Age and diversity of non-marine ostracods. In: Martens, K. (ed.) Sex and parthenogenesis - evolutionary ecology of reproductive modes in non-marine ostracods, 37-55. Bakhuys Publ., Leiden.

Schön, I., Verheyen, E. and Martens, K. Speciation in ancient lake ostracods: comparative analysis of Baikalian *Cytherissa* and Tanganyikan *Cyprideis*. Verh. int. Ver. Limnol. (In press)

Vaïnölä, R. and Kamaltynov, R. M. (1999). Species diversity and speciation in the endemic amphipods of Lake Baikal: Molecular evidence. Crustaceana. (In press)

Williams, D. F., Peck, J., Karabonov, E. B., Prokopenko, A. A., Kravchinsky, V., King, J. and Kuzmin, M. I. (1997) Lake Baikal record of

continental climate response to orbital insolation during the past 5 million years. Science, 278, 1114-1117.
Zubakov D. Yu., Sherbakov D. Yu. and Sitnikova T. Ya. (1997) Analysis of phylogenetic relationships of Baikalian endemic molluscs fam Baicaliidae, Klessin 1878 (Gastropoda, Pectinibranchia) based upon partial nucleotide sequences of mitochondrial gene CO I. Molekul. Biol., 31, 1097-1102

Index of Authors